D1426747

# Geography of India

# Geography of India

**SECOND EDITION**

**RANJIT TIRTHA**
*Eastern Michigan University, USA*

*and*
*Regional Chapters by*

**Gopal Krishan**
*Panjab University, India*

**RAWAT PUBLICATIONS**
Jaipur • New Delhi • Bangalore • Hyderabad • Guwahati • Kolkata

ISBN 978-81-7033-600-6 (HB)
ISBN 978-81-7033-601-3 (PB)

**Second Edition, 2002**

**Reprinted, 2017**



First edition published by
Conpub, 3595 Foxhunt Dr., Ann Arbor, MI, 48105, USA

*Reprint Permissions:*
Figure 3.1: Courtesy of the Superintendent of Documents, U.S. Government Printing Office.
Figures 6.3 and 6.4: Courtesy of the Center for South and Southeast Asian Studies, the University of Michigan, from Hemlata C. Dandekar, *Men to Bombay, Women at Home: Urban Influence on Sugao Village, Deccan Maharashtra, India,* 1942-1982 (Ann Arbor: CSSEAS Publications), 1986.

*Published by*
Prem Rawat for **Rawat Publications**
Satyam Apts., Sector 3, Jawahar Nagar, Jaipur - 302 004 (India)
Phone: 0141 265 1748 / 7006 Fax: 0141 265 1748
E-mail : info@rawatbooks.com
Website: rawatbooks.com

*New Delhi Office*
4858/24, Ansari Road, Daryaganj, New Delhi 110 002
Phone: 011-23263290

Also at *Bangalore, Hyderabad, Guwahati* and *Kolkata*

Typeset by Rawat Computers, Jaipur and
Printed at Nice Printing Press, New Delhi

ॐ श्री साईं समर्पणम् ।।

# Contents

## III

## RESOURCES AND ECONOMIC DEVELOPMENT



## IV

## REGIONS OF INDIA

# Preface

The present revised edition of *Geography of India*, first published in 1992, is coming out keeping in view the strong popular demand. Although it retains the basic structure, substance and style of its predecessor, it has been transformed to a degree, that in its reincarnated form, it is virtually a new book.

History, geography and culture have conspired to make India a major force in South Asia. Since Independence in 1947, it has been trying to become an increasingly important player in international affairs, especially in the developing world. It includes, within its borders, one-sixth of humanity. Its struggle toward economic development is of special interest to the developing countries which also experienced colonial past and now face problems of development. The Indian experiment in democracy and struggle toward unity, stability and progress presents a case study for a critical examination.

Events have moved rapidly since the publication of the previous edition. More than ever, India is at crossroads. Its economy is passing through a difficult phase of transition. Political and social unrest is posing new challenge to nation's unity. This new edition examines the emerging shape of India's socio-economic geography. Discussion is focused primarily on nation's socio-cultural groups, politico-administrative structures and regional-economic disparities.

Completely revised and updated, it includes the results of the 1991 census and other recent surveys. Several new computer-assisted maps have been added. Hopefully, the inadequacies and errors of the old edition have been removed. The updating and inclusion of new material may have increased its size a little but an overall pruning through necessary revision has resulted in more muscular if not a leaner text.

While this edition carries my initial gratitude to several authorities who reviewed the draft and offered useful suggestions for the improvement of the earlier edition, Professor Joseph E. Schwartzberg may be singled out for the incisive comments he made in the printed copy of the previous text. This edition has gained immensely from his suggestions to the previous edition and I am enormously indebted to him for his meticulous scrutiny. I also thank Professor Andrew Nazzaro, who introduced me to computer-assisted mapping, David Allen who executed several of the computer-generated maps, Joanne Davies who carried out most of the cartographic work, Professor K.G. Janardan for scrutinizing the final draft, Dr. Rajiva Tirtha for his moral and material support, and Kevin and Kate Hazelton for their unstincted cooperation in transforming a disoriented manuscript into a publishable form. The new text has also been influenced directly or indirectly by several other scholars who evinced a special interest in the book. I am grateful to all of them. For all the deficiencies and errors that remain, however, I alone am responsible. Mr. Pranit Rawat—the publisher—deserves a special vote of thanks for his steadfast devotion, constant ecouragement and skillful editorial support for publishing a large manuscript into a good-looking textbook.

Written essentially for students at the undergraduate level, the book will be useful for general courses in South Asian Studies and for lay persons interested in India's current events. It is intended to be not just a geography text but a general reader too. It is hoped that it will not only be read but enjoyed as well.

Ann Arbor **Ranjit Tirtha**

## A WORD TO THE READERS

Maps are kept to a minimum. These are carefully designed to function as necessary adjuncts to the text as well as to whet readers' appetite to use a standard atlas like the *School Atlas* by the Government of India or *Goode's World Atlas*. For an ambitious student, *A Historical Atlas of*

*South Asia,* edited by Joseph E. Schwartzberg and *A Social and Economic Atlas of India* edited by S. Muthiah are particularly recommended; the former for its wealth of scholarship and extensive cartographic record of India's socio-economic and political history, and the latter for mapping India's current economic conditions.

Spellings usually conform to the *School Atlas* mentioned above or the publications of Census of India. Unless otherwise indicated choropleth maps represent data by districts. The use of footnotes has been eschewed.

# List of Figures

# I

# Introduction

# 1

# Perspective

## *Size, Status, and Location*

India is a large country and the home of a long and rich civilization. It has often amazed and intrigued the rest of the world by the infinite variety of castes, and creeds, a tradition of religious toleration, a capacity for survival and by the maintenance of its timeless traditions. Today it is the world's second largest nation in population and seventh largest in area (Table 1.1). In industrial production it ranks among the top ten nations, although 70 per cent of its population lives off the land. On the average, an Indian earns US $ 370 per year—a figure that places it among several poor nations. By comparison an American earns $ 27,100 a year.

Ranking high among the Asian countries in size, India is a colossus among South Asian nations. It is four times larger in area, and nearly eight times in population than those of Pakistan which ranks next to it in South Asia (Table 1.2). It is also one of the most crowded countries in the world; surpassed only by nations having small territory like Japan, Bangladesh and Sri Lanka which contain higher population densities. Table 1.1 lists population densities of the seven countries with the largest territories in the world, and Table 1.2 gives the density figures for seven South Asian nations. Japan's population density of 867 persons per square mile (in 1997) is higher than that of India's figure of 845. But, in 1991, India's population was growing at a rate nearly seven times that of Japan (2 per cent a year for India as compared to 0.3 per cent for Japan).

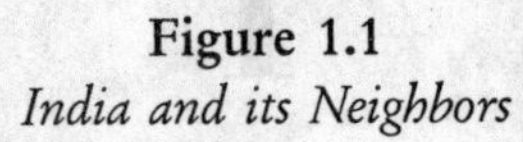

**Figure 1.1**

*India and its Neighbors*

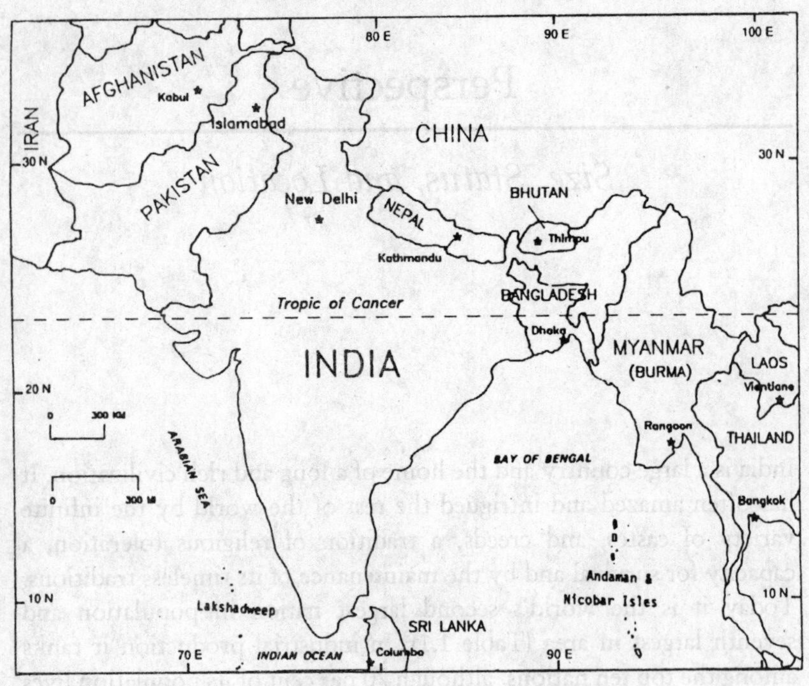


**Table 1.1**

*World's Largest Countries Compared in Area, Population and Income Levels*

| *Rank in area* | *Name* | *Area in miles (million)* | *Area in square kms (million)* | *Population in mid-1997 (million)* | *Population density (people per sq. mile)* | *Per capita GNP in US $ (1997)* |
|---|---|---|---|---|---|---|
| 1 | Russia | 6.5 | 16.8 | 147.3 | 23 | 2240 |
| 2 | Canada | 3.8 | 9.9 | 31.1 | 8 | 19380 |
| 3 | China | 3.7 | 9.6 | 1237.0 | 343 | 620 |
| 4 | U.S.A. | 3.5 | 9.3 | 268.0 | 76 | 26980 |
| 5 | Brazil | 3.3 | 8.4 | 160.3 | 49 | 3640 |
| 6 | Australia | 2.9 | 7.7 | 18.5 | 6 | 18720 |
| 7 | India | 1.3 | 3.3 | 970.0 | 845 | 360 |

*Source:* *World Population Data Sheet, 1997*, Washington, D.C., 1997.

Table 1.2

*South Asian Countries Compared in Size and Population*

| *Rank in area* | *Name* | *Area in square miles* | *Area in square kilometers* | *Population in mid-1997 (estimate in millions)* | *Population density (people per sq. mile)* | *Per capita GNP in US $ (1997)* |
|---|---|---|---|---|---|---|
| 1 | India | 1,269,340 | 3,287,540 | 970 | 845 | 370 |
| 2 | Pakistan | 310,400 | 803,936 | 137 | 463 | 460 |
| 3 | Bangladesh | 55,600 | 144,004 | 122 | 2432 | 240 |
| 4 | Nepal | 52,920 | 140,792 | 23 | 429 | 200 |
| 5 | Sri Lanka | 25,330 | 65,604 | 18.7 | 748 | 700 |
| 6 | Bhutan | 18,000 | 47,913 | 0.8 | 46 | 420 |
| 7 | Maldives | 120 | 311 | 0.3 | 2342 | 990 |

*Source:* *World Population Data Sheet, 1997*, Washington, D.C., 1997.

Population is, therefore, a major concern, not only in terms of its size, or densities, but also in its growth rate that has averaged 2 per cent per year since the 1960s; though modest in comparison with several developing nations, it continues to add more than 15 million people a year to the nation's population. Covering only 2.5 per cent of the world's land, India contains 16 per cent of its population. Based on the current rate of population increase, it is likely to contain about 1 billion people in less than ten years from now, registering an increase of 160 million. By 2020, it is likely to surpass China to become the world's most populous nation.

How does India compare economically with other developing nations? Table 1.3 puts into perspective the status of the basic economic features of selected developing countries. Like China, agriculture is the dominant sector of India's economy, accounting for more than one-third of her gross domestic product (GDP) and two-thirds of employment. Its gross domestic savings is the smallest among the listed nations and its per capita income is the lowest (Table 1.3). It is the least trading nation, too. Its outstanding external debt in 1996 stood at about 94 billion dollars, though quite high, is lower than that of Brazil and Mexico—two large developing nations with rapidly growing populations.

**Table 1.3**

*Selected Development Indicators, 1995-96 for India and Some Developing Countries*

| | India | China | Brazil | Mexico | Rep. of Korea | Pakistan |
|---|---|---|---|---|---|---|
| Population (millions) (1996) | 945 | 1215 | 161 | 93 | 96 | 134 |
| Geographical Area (thousand sq km) | 3288 | 9561 | 8512 | 1973 | 98 | 804 |
| Gross Domestic Product (GDP) (billions of US$) (1996) | 356 | 815 | 749 | 335 | 485 | 65 |
| Gross National Product (GNP) Per Capita US$ (1996) | 270 | 310 | 1640 | 2080 | 2150 | 380 |
| Share in GDP (per cent) (1996) | | | | | | |
| Agriculture | 28 | 21 | 14 | 5 | 6 | 26 |
| Industry | 29 | 48 | 38 | 26 | 43 | 25 |
| Services | 43 | 31 | 50 | 69 | 51 | 49 |
| Life Expectancy at Birth (years) (1996) | 59 | 69 | 65 | 67 | 69 | 61 |
| Urban Population (per cent of total population in 1996) | 27 | 31 | 76 | 79 | 82 | 35 |
| Population Growth Rate (annual per cent) (1995-96) | 2.0 | 1.3 | 1.8 | 1.5 | 1.1 | 2.5 |
| Labor Force in Agriculture (per cent)(1996) | 61 | 71 | 26 | 24 | 12 | 48 |
| Labor Force in Industry (per cent) (1996) | 9.1 | 13.5 | 13.4 | 16.0 | 12.0 | 11.0 |
| Merchandise Exports (billions of US$) (1996) | 29.0 | 14.9 | 46.5 | 98.0 | 129.7 | 6.4 |
| Total Outstanding External Debt (1996) (billions of US$) | 93.8 | 94.6 | 96.6 | 94.2 | 24.0 | 23.7 |
| Capita Growth Rate (1995-96) (in per cent) | 6.9 | 9.7 | 3.0 | 5.1 | 5.5 | 6.0 |
| Inflation Rate (1995-96) | 9.0 | 10.1 | 23.0 | 52.0 | 4.5 | 13.0 |

*Sources:* 1. World Bank, *World Development*. New York: Oxford University Press, 1998.
2. *World Economic Outlook Indicators.* International Monetary Fund, 1997.

Despite relentless demographic pressures, and stunted economic growth, India today is a leader among developing nations, dominating the South Asian scene with nuclear capability, and armed forces of estimated 1.5 million personnel.

## Location and Boundaries

The triangular peninsula projecting into the Indian Ocean with its apex pointing southward from the continent of Asia, delineated in the north by the steeply rising Himalayan mountains, and in the east and west by its flanking offshoots, has been historically known as the Indian subcontinent. Within the limits thus marked off by these physical features are included the countries of Pakistan, Bangladesh, Nepal, Bhutan, Sri Lanka and India. The cultural and political fortunes of these countries have been so intertwined as to define the area in terms of an Indian culture realm.

India is the largest of these political constituents and the most centrally located and is apparently destined to play a major role on the subcontinent (Figure 1.1). The Indian mainland (excepting the island groups of Andaman-Nicobar and Lakshadweep, in the Bay of Bengal and the Arabian Sea) lies between 8° and 37° north, and 68° and 97° east, the Tropic of Cancer roughly dividing the landmass into two unequal parts. To its north the Himalayan kingdoms of Nepal and Bhutan are sandwiched between the Tibetan region of the People's Republic of China and the plains of north India. The small Himalayan principality of Sikkim which lies between Nepal and Bhutan was incorporated into India in 1975. In the northwest, India shares a sensitive and occasionally volatile boundary with Pakistan, created across the Indus basin as a consequence of the partitioning of the subcontinent in 1947. The former principality of Jammu & Kashmir has been disputed between India and Pakistan since their births. It has been *de facto* partitioned between the two since 1950. In the northeast, beyond the eastern offshoots of the Himalayas, lies Burma. Bangladesh, which broke off from Pakistan in 1971 as an independent nation, is almost an enclave within India, landlocked between Indian territory except to the south where it is accessible through the Bay of Bengal. Over 51,800 sq km of disputed territory lies along India's 9,425 mile land-border shared with China in the north and northeast. This mountainous and inhospitable territory has been the scene of

Sino-Indian hostilities. Sri Lanka lies just off India's southern tip in the Indian Ocean.

The coastline of over 3,500 miles along the Arabian Sea in the west and the Bay of Bengal in the east, the two arms of the Indian Ocean, has only sporadically remained active in trade or communications. Despite vigorous trans-oceanic connections established by Indian kings in Southeast Asia and the implanting of Indian art, architecture, religions, and social traditions in these colonies, Indian coasts effectively insulated the Indian culture "realm" from the outside world. Since the Indian Ocean has remained a natural defensive base for the subcontinent, any power-play in it, such as the consolidation of the U.S. naval base at Diego Garcia or the Soviet movements off the coast of East Africa, is looked upon with apprehensive disfavor by the Indian administration.

## India: A Historical-Geographical Expression

Virtually isolated from the two mainstreams of Asian history, the civilization of China and the Islamic world of the Middle East, the Indian culture essentially represents a third stream of Asian history. Largely emanating with the Aryanization of the Indus-Ganga basin as early as 1500-2000 B.C. but not excluding the Harappan civilization, and based on the subsequent development of Hindu thought, early Indian civilization was broadened by the incorporation of religious, artistic, political, and social mores of invaders from the west, and gradually evolved throughout ancient, medieval and modern times into a distinctive Indian culture.

Ancient Greeks associated the term "Indoi" for the Indian people and territories with the Indos (Indus) river, adopted apparently from the Sindhu of the ancient Sanskrit texts. The land of the Indoi and others represented the Indian subcontinent, all the area lying east of the Sulaiman range. The later Muslim version became "Hindu" for the people, and "Hindustan" for the area, more specifically for the territory of Muslim consolidation in north India. The southern peninsular part was *Dakshinapatha* (literally, the south) or Deccan in ancient Hindu writings. The entire subcontinent was styled as *Bharata Varsha* by ancient Hindu writers, the land of the legendary King Bharata, stretching from Kashmir in the north to Kanya Kumari, the southernmost tip, and Afghanistan in the west to Assam in the east.

The concept of *Bharata Varsha*, or a pan-India, has remained ever since an ideal among India's rulers.

"India" as a geographic expression has its base in the territorial layout of the subcontinent, and the distribution of physical features within it. Effectively contained within the physical boundaries, the subcontinent has been insulated from the historic forces emanating from China and the Middle East. Isolation from the north has been more complete. Except for localized trade between India and Tibet, and the diffusion of Mahayana Buddhism, the India-China exchange was minimal over the high, inaccessible, snow-clad passes and difficult terrain. Even Buddhism traveled to China centuries after its origin in India. The two civilizations matured independently and exclusively, the Himalayan barriers or oceans separating the two. In fact, the Indians and the Chinese both remained oriented largely inward, exclusive, with little interest in commerce. Contracts over the rain-swept forests of the Assam hills in the northeast segment of the Himalayas, and its eastern offshoots separating the subcontinent from Burma were also slight. Buddhism diffused to Burma and southeast Asia by sea.

Isolation has been less restrictive on the western and northwestern side. Lying to the east of the plateaus of Afghanistan and Iran, several of the mountains, the Sulaiman, the Hindu Kush and the Kirthar, although rocky, bare and harsh in appearance, are breached by several accessible passes. A succession of invasions flowed through these passes (the Khyber and the Bolan being more notable), starting from the distant Aryans c. 1500 B.C. to the Mughals in the sixteenth century. In due course of time, the invaders became Indianized, absorbed much of the social systems, and transmitted some of theirs to the existing Indian mores.

Trans-oceanic contacts between India and other places have been, however, notable, especially with the Mediterranean World and southeast Asia. As early as the first century A.D. flourishing trade in Indian spices, ivory, silks, and precious stones existed between South Indian ports of Malabar and the Roman ports. Even prior to that, commercial contacts between the Indus valley civilization and the Sumerians as early as 2500 B.C. had been established. In the eleventh and twelfth centuries, Hindu kings, traders, priests and Buddhist missionaries went to Indonesia, Burma, Cambodia, Thailand and

Malaya. Large Hindu empires were set up. Indian social and political institutions, legal systems, art forms and architectural monuments still persist there in varying forms. The temple town of Angkor Wat in Cambodia and Borobudor in Java are eloquent testimonies to the times of Indianization of southeast Asia.

During the sixteenth and seventeenth centuries, European powers encroached upon India through the ocean routes. The culmination of European colonial expansion was the consolidation of a vast British empire over the subcontinent in the eighteenth and nineteenth centuries. The geographical and cultural expression "India" was indeed perpetuated by the establishment of a long history of a single administration, political systems and economic linkage under the British.

The lack of an integrated and continuing pan-Indian administration during the long and chequered history of the subcontinent is reflected in its regional diversities of language, religion and race. Despite the obvious contradictions and diversities, an undercurrent of Indian civilization kept social and political life as a functioning identity for all historic times. Even after the emergence of the three major countries of Pakistan, Bangladesh and India in the aftermath of the post-colonial period, the historical and geographical expression "India" for the subcontinent clearly defined as it is by physical and human forces, remains largely valid. The three major countries of south Asia have shared a common historical, artistic and linguistic heritage and face common problems in the future. Common boundaries, cultural associations and economic links will continue to intertwine their futures.

### *References*

Government of India, *Economic Survey, 1989-90*. New Delhi, 1990.

Government of India, *A Pocket Book of Economic Information, 1973-74*. New Delhi, 1975.

Government of India, *India, 1997: A Reference Annual*. New Delhi, 1990.

Muthiah, S. *et al.* (eds.), *A Social and Economic Atlas of India*. Delhi, 1987.

Schwartzberg, J.E. (ed.), *A Historical Atlas of South Asia*. Chicago, 1978.

*Statistical Outline of India, 1997-98*. Mumbai, 1997.
Wolpert, S., *A New History of India*. New York, 1986.
World Bank, *World Development Report*. New York, 1997.
*World Population Data Sheet, 1997*, Population Reference Bureau. Washington, D.C., 1991.

# 2

# Physical Environment

The arrangement of relief features of the subcontinent is relatively simple, if grand, in design (Figure 2.1). The southern peninsular part, occupied mostly by the Deccan plateau, is an old, stable block of subdued relief sloping generally gently toward the east coast. It is flanked on both the eastern and western sides by mountain ranges. In the northern parameters of the subcontinent lie the long, sweeping mountain chain of the Himalayas with its western and eastern off shoots, and containing several ranges of young folded, deeply dissected and unstable structures attaining the world's highest elevations. Between the two lie the vast alluvial plains of the Indo-Ganga river basins, which, according to geologists, resulted from the in-filling of the structural trough (a long, deep depression) between the peninsular block of the Deccan to the south, and the northern mountain ranges of the Himalayas. Within this rather simple tripartite division exist diversities in structure, drainage, soil conditions, topography and geologic history.

According to the widely accepted geological theory, the Deccan tableland was once a part of an ancient supercontinent, Gondwanaland, which included the present-day continents of Africa, Australia, Antarctica and parts of Brazil. Its subsequent fracturing and movement, and the formation of mountains and other relief features are attributed to continental drift or movements of the plates which composed the ancient supercontinent. Such forces are even now at

**Figure 2.1**
*Relief Features*

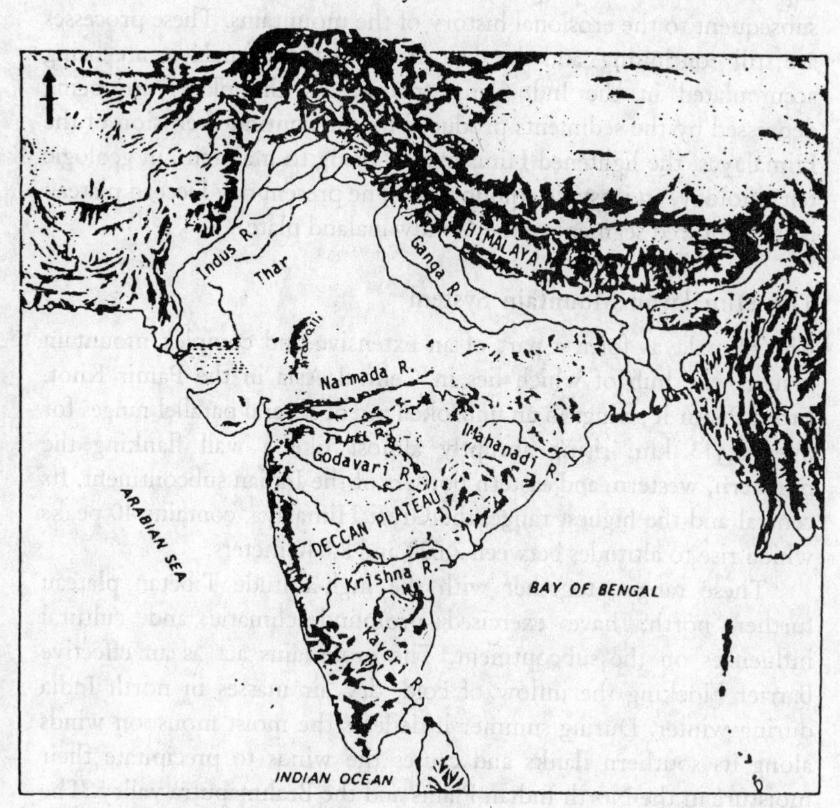


work. To the north of Gondwanaland was another vast plate of ancient Angaraland. Between these two ancient supercontinents existed a huge geosynclinal depression, into which were poured sediments by rivers of the two flanking supercontinents. The colossal infilling of the sediment in the depression disturbed the gravity equilibrium of the crust or the plates of the continents, creating forces of mountain building. The northern plate advanced toward Gondwanaland, and the sediments of the intervening trough or "geosyncline" were buckled up to form the present-day Himalayas. As the Himalayas was rising under the impact of the advancing plate from the north, erosional processes were removing sediments for deposition in the marine gulfs separating the northern plate from the southern one. The present-day Indo-Ganga plains resulted from the deposition and infilling by the sediments in the geosyncline subsequent to the erosional history of the mountains. These processes are still continuing, and the deposits from the Himalayas are being accumulated in the Indo-Ganga plains. As the plains are being depressed by the sediments produced by the continued erosion of the Himalayas, the lightened Himalayas is rising to maintain, in geologic terminology, an isostatic equilibrium. The present-day Deccan plateau is the ruptured segment of the Gondwanaland plate.

## The Himalayan Mountain System

The Himalayas form a part of an extensive and complex mountain system, the hub of which lies in Central Asia in the Pamir Knot. From Pamir it swings in an unbroken arc of several parallel ranges for over 2,413 km, rising abruptly almost like a wall flanking the northern, western, and eastern borders of the Indian subcontinent. Its central and the highest range, the Great Himalaya, contains 40 peaks which rise to altitudes between 7,620 and 8,840 meters.

These ranges, together with the high-altitude Tibetan plateau further north, have exercised profound climatic and cultural influences on the subcontinent. The mountains act as an effective barrier blocking the inflow of cold, dry air masses in north India during winter. During summer it deflects the moist monsoon winds along its southern flanks and causes the winds to precipitate their moisture in the North Indian Plains and the Brahmaputra valley. The Himalayas also cause the jet stream circulation of the upper atmosphere to assume a main course south of it in winter, and helps the passage of a series of cyclonic storms. The passage of jet streams in summer, the monsoonal precipitation in summer, coupled with the high altitudes of the Himalayan ranges—all these meteorological factors contribute to the accumulation of large amounts of snow and the formation of permanent ice-fields and glaciers in the Great Himalayan ranges. The Himalayas thus become the feeding ground of major perennial rivers which are extensively utilized for year-round irrigation in the subcontinent.

In structure and relief, the Himalayas offer a sharp contrast to the Deccan plateau. Its topography is youthful, rugged and dissected, containing V-shaped valleys in highly folded and faulted structures resulting from their uplifts and compressions and the youthful erosional processes prevailing on their surfaces whereas relief of the Deccan plateau is slight and the rock structures stable and old. The unusually high elevations of the Himalayas have resulted from the mountain building forces during Tertiary times in geologic history (about 60-65 million years ago). The intense folds and compression of rocks suggest a buckling up and uplifting of enormous sedimentary strata once these were deposited in a geo-synclinal depression, the ancient sea or Tethys. The material was compressed in the form of rockwaves when Angaraland, the plate of the northern supercontinent moved toward Gondwanaland. Marine fossils embedded in the Himalayas at 4,260 to 5,181 meters altitudes indicate the existence of such an ancient Tethys sea, separating the two ancient supercontinents, at the location of the present-day Himalayas. The movement of Angaraland, and subsequent compression and buckling of the sediments previously deposited by rivers from the supercontinents into the Tethys, as indicated earlier, resulted in the Himalayas. The trendlines as reflected in rock structures and surface features of the Himalayas run parallel to the margins of the North Indian Plains, and suggest that the kinetic forces producing the Himalayas were transverse in direction, i.e., came from the north. Subsequent geologic history has been explained by the gravity,

The Himalayan mountain system is arranged in parallel ranges. The ranges of the Main or Great Himalaya lie immediately to the south of the Tibetan plateau and contain the highest elevations. The Inner or Lesser Himalaya lies to its south is about 60 miles in width and contains a series of overthrust folds and jagged relief with several interconnected ridges, rising to 3,048 to 4,572 meters. The Outer Himalaya consists of low-lying ranges of altitudes ranging between 609 and 1,067 meters at the foothills of the previous two. In the western segment of the Outer Himalaya such hills are known as Siwalik hills, which essentially form a zone of low basins and depressions or "duns". Their origin has been ascribed to the recent warping and faulting of the unconsolidated gravelly materials of the southern parts of the Himalayas. At various places within the Himalayan mountain system erosional and structural valleys are found. Good examples of these valleys are: in Kashmir, the valley of Kashmir, and in the Central Himalaya, the Kathmandu valley. The valley of Kashmir is a synclinal depression covered by a broad floor of young deposits, the center of which is a floodplain of Jhelum river. To the south the valley is enclosed by a mountain range Pir Panjal, blocking the valley's accessibility to India. To the north of the valley lies the Great Himalayan chains of Zaskar, Ladakh and Karakoram, often attaining elevations of over 6,096 meters. Like most Himalayan structures, these are youthful, and are deeply dissected and folded.

The Himalayas also contain a few passes which connect the subcontinent with the Tibetan plateau, most lying at high altitudes between 4,877 and 5,639 meters, passable only during a brief period of summer, and are difficult to traverse. Their climatic and cultural influence is to seal off the subcontinent from the north.

The western flanks of the northern mountain wall are in Pakistan. The two major are systems are the Sulaiman and Kirthar ranges. Their northeast-southwest trendline, with convexity eastward suggests that a compressional force acted from the west when these mountains were being formed. Although similar in structure to the Himalayas, these are lower in relief and elevation, rising at their highest to 3,352 meters. Historically, the several easily traversable passes in these bare and rocky ranges have frequently been used by invaders from the west. The eastern flanks of the northern mountain wall lie close to the India-Burma border, and often form a succession of tangled, arcuate northsouth ranges, such as the Patkai and Naga mountains. Rarely rising over 2,134 meters in altitude, the highest peak being 3,553 meters, the mountains are clothed with dense forests, coupled with difficult terrain, have acted as an effective physical and cultural barrier between India and Burma.

### *The Indo-Ganga Plains*

The vast plains of northern India, Pakistan and Bangladesh form a major topographic and physiographic division of the subcontinent's landscapes, the other two being the Himalayan mountain system and the Deccan plateau. These lowlands include the basins of the two major river systems of northern India, the Ganga and the Indus, together with their numerous tributaries. Occupying only one-fifth of the area, but supporting half the population of southern Asia, the plains have historically been the core region of the subcontinent's political, cultural and economic activities.

In scale, the plains are immense, covering a crescent-shaped area of over 1,036,000 sq km, 1,610 km long and 161-232 km wide, stretching-from the Indus delta in the west through the plains of Pakistan and Ganga plains in northern India to the Ganga-Brahmaputra delta in Bangladesh. Though smaller than the structural plains of North America, the Indo-Ganga plains form the largest stretch of alluvial plain in the world.

Geologically, the plains belong to the Tertiary and Quaternary periods, the youngest element in the structure of the subcontinent. The vast plains were built by deposits in a depression, the Tethys or geosynclinal sea, lying between the Himalayas in the north and the peninsular block of Deccan plateau in the south. Gradually, the Tethys was filled up by the sediments brought by streams from the northern and southern blocks on either side of the geosynclinal depression, burying the hard rocks of the peninsular Deccan block except in a few places, such as Aravalli hills in Rajasthan state and its remnant visible near Delhi as Delhi ridge, and the Shillong plateau in Maghalaya state. While the process of sediment in-filling was going on, the peninsular block to the south was forced downward by tectonic pressures exerted by the northern block. The sediments deposited are among the deepest in the alluvial plains of the world, reaching as much as 2.286 meters. The depths vary, from 457 to 610 meters in the basin of the Indus river to over 1,524 meters in the Ganga delta. The alluvium (fine sediments) is so deep that for miles one may not find a small stone on the surface in the lower Ganga basin. The Ganga delta is still sinking under the continuing sedimentation.

Topographically, these vast plains are flat for hundreds of miles, forming a long, nearly straight horizon. The flat, even monotony of the plains is relieved only by minor local modifications in relief, introduced by river bluffs, hollows, and at. places, by badlands formed by gully erosion. Notable examples of badland relief formed by gully erosion are in the lower Chambal basin. Relief gradients are so gentle that the waterparting between the basins of the Ganga and Indus rivers, lying about 161 km northwest of Delhi, is only 274 meters in elevation above sea level, although it is located 1,448 km from the Arabian Sea and over 1,600 km from the Bay of Bengal. From this water-divide, the Ganga plain slopes gently to the southeast to the sea with a gradient of inches per mile, and the gradient of the Indus plain which slopes southwestward from the water-divide, is only a little greater.

The Indo-Ganga topography consists, in general, of a succession of floodplains and the slightly higher ground of 45-61 meters of local relief in the interfluves (known as *doabs*) lying away from the rivers. These bluffs are not subjected to periodic flooding and are composed of older alluvium (*bhangar*) and they may contain gravels, sand and coarser materials. The newer alluvium (*khadar*) lies in the immediate vicinity of the river courses, natural levees and the adjoining territory. Gentle gradients help promote river meanders, the occasional shifting of river courses, and the creation of meander scars and ox-bow marshes. As the rivers gently meander through their journey toward the deltas, they have become braided into several channels forming large aggradational plains. The Indus and its tributaries are aligned approximately northsouthward. The Ganga and its tributaries initially run parallel to the water-divide between the two river basins, but swing to the east toward the Bay of Bengal. The southern tributaries of the Ganga river run southwest to northeast. In general, the rivers have cut into their right banks due to the rotation of the earth, leaving these higher in elevation than the left banks. It is thus easier to build irrigation channels from the left banks.

The Ganga receives its greatest volume from its northern snow-fed Himalayan tributaries. Its volume progressively increases as it proceeds toward the delta in area of increasing rainfall. In contrast, the Indus river traverses an area of decreasing rainfall, receiving little flow from the west. Its Himalayan tributaries lie mostly north of latitude 28° N. In its lower journey it passes through semi-arid territory.

At the foot of the Himalayas in the Ganga basin lies an irregular zone of low, piedmont alluvial fans, an area of marshy underground seepage, known as the *Tarai*. This marshland has long been a neglected area. The work of draining and mosquito control is still in its infancy in this potentially fertile and cultivable area.

East of the lower Indus basin lies Thar desert, cutting across the Indo-Pakistani border. Much of its surface is covered by a veneer of loose soil, interrupted by rocky projections and wind-blown sandy ridges. Underlying this veneer, hard, crystalline rocks similar to those of the Deccan plateau are found. The Aravalli mountain range belongs to this category. A prehistoric river channel, the Ghaggar, passes through the desert. It can be traced for 600 miles from the Punjab in India to Sind in Pakistan. The Ghaggar basin once drained into the Rann of Kutch, and probably formed a part of the Indus basin. The Rann is now a salty marshland inundated seasonally during the summer monsoon season.

The partitioning of the subcontinent in 1947 created physical and human problems, both in the Ganga-Brahmaputra delta and in the middle Indus basin. The main problem hinged around the division of the waters between the new countries. In Punjab, or the land of five tributaries of the Indus river, political partitioning created problems of water irrigation. The partition divided the Indus irrigation system, cutting many canals in Pakistan from their headworks in India. Pakistan was, thus, left at the mercy of India for the flow of waters needed critically for her agriculture. The Indus Waters Agreement in 1960 resolved the problem by specifically allocating the waters of the eastern rivers to India and the western ones to Pakistan and by the provision of link canals to keep water supply adequate to both sides.

### *Deccan Plateau*

The peninsular part of the subcontinent is mostly a raised tableland of old, stable structure, known generally as the Deccan plateau. The western and northwestern parts of the plateau extend up to Kutch in Gujarat and the Aravalli range in Rajasthan. Its northern limits run parallel to and 50 miles south of the course of Ganga-Yamuna rivers. Several outliers of the plateau rise above the sediments of the Indo-Ganga plains, e.g., Delhi ridge, Kirana hills (in Pakistan) and Shillong plateau. Almost the entire peninsula south of the line formed by the Vindhya-Satpura-Mahadeo hills is termed the Deccan plateau.

The tableland is, in fact, made up of several plateaus of undulating relief, with elevations ranging between 305 and 915 meters above sea level, flanked on the west and east sides by the coastal ranges. The low, gentle gradients are produced by prolonged weathering and erosion. In the south, where the two coastal ranges appear to merge, are two high granitic massifs, the Nilgiri hills and the Caradamon hills, rising to elevations of about 8,500 feet. The western flanking ranges, the Western Ghats or the Sahyadri mountains rise as a bold escarpment of 1,220 to 1,525 meters above the western coast, leaving a very narrow coastal plain. Toward the interior, the Western Ghats assume a hill-like appearance. The ranges flanking the eastern side of the Deccan plateau are the Eastern Ghats and are not as rugged, or high, or as continuous as the Western Ghats. These are discontinuous hills with elevations between 609 to 762 meters. In their northern section, the Eastern Ghats and the adjoining interior plateau area contain several minerals, notably iron ore, manganese and mica.

Geologically, the Deccan plateau is the oldest part of the subcontinent, a "shield" of old, stable rocks. Gneisses and schists dating from pre-Cambrian times (500-2,000 million years ago), cover half of its area. The structures are similar to the Laurentian uplands of Canada, and plateaus of South Africa and western Australia. Other common rock formations of the Deccan plateau are as old as 200-500 million years.

Surface configuration consists of undulating hills with rounded, broad, low summits of peneplanes and residual blocks. The general slope, influenced in part by fracturing and tilting, is toward the east. The major rivers, Godavari, Krishna and Kaveri start within 80 km of the Arabian Sea and flow eastward for distances of 482-964 km across the tableland to the Bay of Bengal. The rivers flow in broad, shallow valleys, their grading disturbed by gaps and enclosed escarpments. Entirely dependent on rainfall, these are nearly dry in the hot weather, and are of limited value for irrigation. The north central section of the Deccan plateau is tilted northward, resulting perhaps from the tectonic pressures of the ancient northern supercontinent, and the drainage is toward the Ganga basin (rivers Son and Chambal) of the North Indian Plains.

The northwestern part of the Deccan plateau, about 647,500 sq km, is covered by thick horizontal beds of lavas, the Deccan trap. The thickness may be as great as 1524 meters. It is thought that in times of Tertiary crustal instability associated with the formation of the Himalayas, a succession of lava flows spread over the area, solidifying into the trap. The lava formations have been eroded and give the impression of a mesalike topography. In the northern limits of the Deccan plateau, the two mountain ranges, the Vindhya and the Satpura, rise in sharp escarpments. The Satpura is a structurally uplifted area, a horst, whereas the river Tapti flows in a structural trough or graben; both are parallel to the Narmada river, and drain westward to the Arabian Sea, unlike most rivers of the Deccan plateau. The Narmada waterway has been a historic routeway from the Bay of Cambay on the western coast to Varanasi in Ganga plains. The Aravalli mountains bordering the Rajasthan desert approximately define the northwestern limits of the Deccan plateau. The Aravalli is a worn-out stump of the ancient mountains and rises to 1,706 meters at is highest altitude.

The peninsula has almost continuous coastal plains on its margins, developed from the alluvial fans deposited by the rivers of the plateau. The eastern plains are broader, especially where the deltas have been built by the east flowing rivers, and progression of the deltaic plains is still continuing. The west coast is narrower, except around the Gulf of Cambay. The water-divide between the Arabian Sea and the Bay of Bengal drainage is close to the Arabian Sea, with few, shorter west-coast rivers, leaving little scope for their depositional action.

Most of India's mineral wealth is found in the Deccan plateau. The richest area is in its northeastern section, the Chota Nagpur region. Several coalfields lie in the downfaulted sedimentary rocks of Chota Nagpur region in south Bihar. Other mineral-rich areas of the Deccan plateau are in Mahanadi valley and north Andhra Pradesh. In addition, mica, iron ore, manganese, and gold are obtained in several parts of Deccan in Orissa, Madhya Pradesh, Andhra Pradesh, Karnataka, and eastern Maharashtra. Most mineral deposits are associated with the metamorphosed Cambrian rocks.

## Climatic Characteristics

A distinguishing characteristic of South Asia's climate is the prevalence of monsoonal (from the Arabic word *mausim* meaning "season"), or seasonal wind system, the alternating dry cool continental winds in winter and water-laden, tropical, oceanic winds in summer. This term is applicable to most, but not all, parts of the subcontinent. This idealized perception of the monsoon climate best applies to the areas of northern and western India, Pakistan, and Bangladesh. In the extreme south, in Kerala, tropical rain forest climate prevails, whereas along the southern parts of the eastern coast, the retreating monsoon brings rainfall in winter after picking up moisture over the Bay of Bengal. In the northwestern parts of the subcontinent, in Pakistan and the Punjab plains, western disturbances brings winter rainfall. These exceptions apart, the monsoonal winds have almost a universal effect, and are of great human significance. Most summer weather crops (*kharif*) depend on rainfall associated with the monsoon winds. Any major deviation from the seasonal rhythm, a late arrival or an early retreat of the monsoons, resulting in a deficiency of rain, can well spell foodgrain shortages or even famine.

In general, the monsoon may be described as similar to, but on a larger continental scale, the land-and-sea breeze mechanism. It consists of annual reversal of seasonal wind systems, between cool, dry land-winds in winter and warm, rain-bearing winds in summer. Traditional hypotheses ascribe their origin to the development of a distinct thermal low pressure system over the Indus plains during summer, effectively separated from the Central Asian counterpart by the Himalayas and the Tibetan plateau. This results in a gigantic monsoonal indraught of the moist winds of the Arabian Sea and the Bay of Bengal during summer. This Indus low pressure system induces the indraught of the moisture-laden summer winds which bring rainfall to the western coast and the Western Ghats. The Deccan remains comparatively dry as it lies in the rain-shadow of the Western Ghats. Another monsoon current curves up the Ganga plains, progressively losing moisture during its travel toward the Indus plains. The greatest impact of the Bay of Bengal "current" is in Shillong plateau, Brahmaputra valley and the Ganga-Brahmaputra delta. The Himalayas act as a barrier and the current is diverted left-ward to the Ganga plains. In the Himalayas, this tropical, maritime air is present up to elevations of 4572 meters. Beyond that, most of the moisture has already fallen. The Tibetan plateau is dry as it lies in the Himalayan rain-shadow area. In winter, the subcontinent presents a reversed situation. The high pressure system of the subcontinent induces the flow of cool, dry air from the northwest across the Indo-Ganga plains. A notable meteorological factor during the winter is the blocking off of cold, continental air originating from the Siberian high pressure cell by the Himalayan mountain wall, thus helping north India to escape the bitter cold winters that characterize Central Asia.

This simplistic concept of the monsoon winds as related to the development of "low" and "high" pressure cells south of the Himalayas during the summer and winter seasons, giving rise to the

**Figure 2.2**
*Jet Stream Movement*

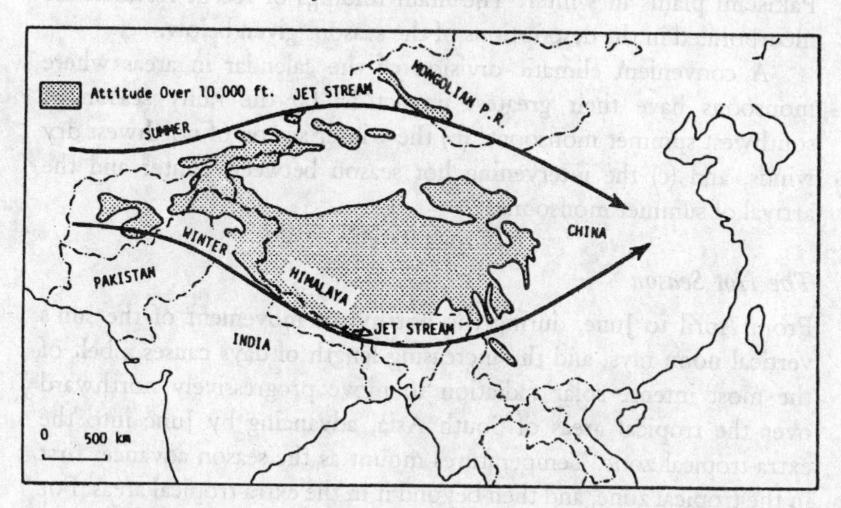


seasonal wind systems and associated climatic features, fails to explain several of the characteristics of the Indian monsoons. The traditional model which emphasizes purely thermal factors responsible for the development of surface pressure conditions, does not fully explain the characteristic vagaries connected with the timing and incidence of summer rainfall, its "bursts", "lulls", and "breaks". Furthermore, the traditional models are inadequate in explaining the distribution and behavior of the depressions which bring wide-ranging rains to the plains of Pakistan and north India. Recent advances in the study of the dynamics of the upper atmosphere and its relationship with surface air circulation have yielded fruitful explanations of the behavior of the Indian monsoons. Tropical regional meteorology of the subcontinent has now been analyzed within the global framework, and the role of the Himalayas and its association with air waves and currents at various altitudes has been better understood. While a review of the recent theories concerning the mechanism of Indian monsoons is outside the scope of our treatment, it may be pointed out that the Himalayas have effectively controlled the passage routes of the upper jet stream circulation, and has divided the jet stream into two seasonal offshoots; a northern route is established in summer, and one south of Himalayas in winter (Figure 2.2). The latter affects principally the flow of easterly surface depressions across the north Indian and Pakistani plains in winter. The main findings of recent research are incorporated in the descriptions of the seasons given below.

A convenient climatic division of the calendar in areas where monsoons have their greatest impact is: (a) the rainy season of southwest summer monsoons, (b) the winter season of northwest dry winds, and (c) the intervening hot season between winter and the arrival of summer monsoons.

### *The Hot Season*

From April to June, during the northward movement of the sun's vertical noon rays, and the increasing length of days causes a belt of the most intense solar radiation to move progressively northward over the tropical areas of South Asia, advancing by June into the extra-tropical zone. Temperatures mount as the season advances first in the tropical zone, and then beyond it in the extra-tropical areas. For example, in March the highest day-time temperatures of over 40°C are recorded in central Deccan, Maharashtra and central Andhra Pradesh. In April, highest daytime temperatures 43.5°C are experienced in Madhya Pradesh, and in June these are 48°C in Rajasthan and Sind (Pakistan), and about 45° to 47°C in most of the Indo-Ganga plains. In north India, and away from the coast, the dirunal ranges also increase. The highest temperatures are recorded in the Thar desert, on the Indus plains, and occasionally in the middle Ganga plains. In June, Deccan also records high temperatures (37°C daily maximum) although somewhat lower than the Indo-Ganga plains (41°C or over as daily maximum). The only areas with daily maximum temperatures below 37°C are the coastal plains and the higher altitudes of the mountains. During these summer months afternoons are unbearably hot, and almost all human activity is suspended until the sunset. Men and beast take shelter indoors. This description applies to most parts of India except the high mountains and coastal areas.

Meteorologically, this is the season of continental dry, stable hot winds from the northwest, which permit the intense heating of the land before the summer solstice (June 22). The sun gradually moves northward toward the Tropic of Cancer (until June 22) causing a reduction in the dynamic power of the cold polar airmass which held sway during winter. Until May a dynamic anti-cyclone centered in

**Figure 2.3**
*Summer*

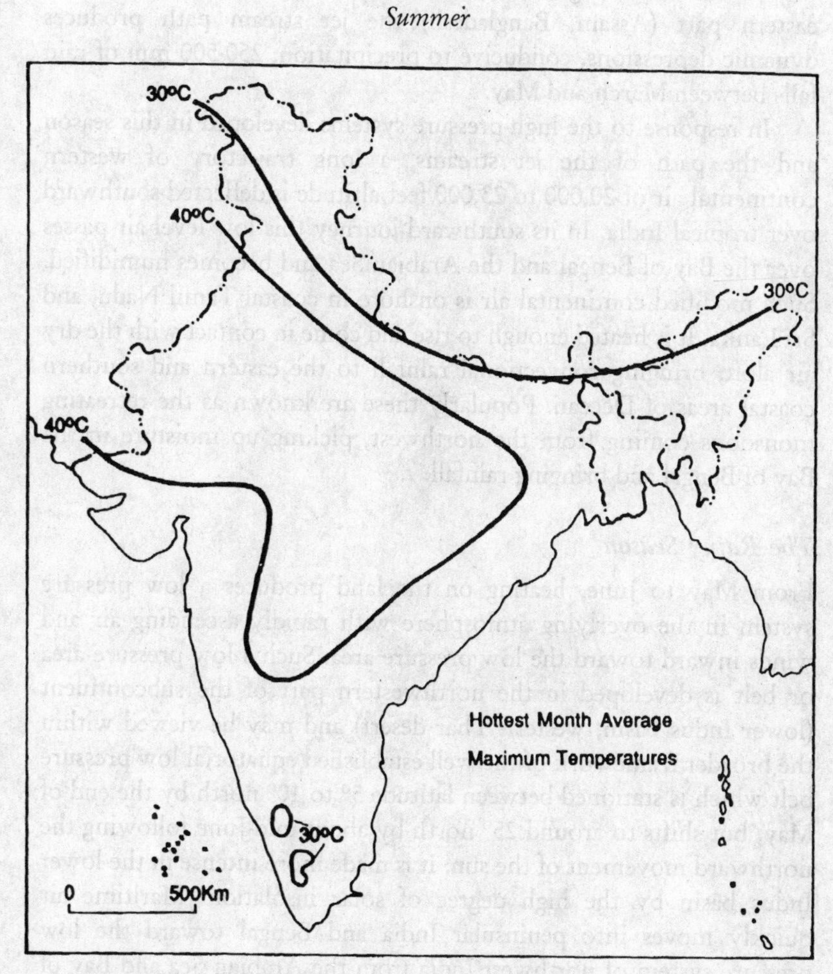


Afghanistan-Iran continues to affect the climate of the northwestern part of the subcontinent. Summer (hottest month) average maximum temperatures are shown in Figure 2.3.

Throughout these months the Himalayas act as a topographic barrier to the low-altitude air and the jet stream which has moved north of it. As the season advances, a thermal low is being gradually established at the surface over the northwestern part of the subcontinent, while the jet stream swings south of the Himalayas.

In the remaining sections of the subcontinent, especially the eastern part (Assam, Bangladesh) the jet stream path produces dynamic depressions, conducive to precipitation; 250-500 mm of rain falls between March and May.

In response to the high-pressure systems developed in this season and the path of the jet streams, a long trajectory of western continental air of 20,000 to 23,000 feet altitude is deflected southward over tropical India. In its southward journey this low level air passes over the Bay of Bengal and the Arabian Sea and becomes humidified. Such modified continental air is onshore in coastal Tamil Nadu, and Sri Lanka. It is heated enough to rise and come in contact with the dry air aloft, bringing convectional rainfall to the eastern and southern coastal areas of Deccan. Popularly these are known as the retreating monsoons coming from the northwest, picking up moisture in the Bay of Bengal and bringing rainfall.

### *The Rainy Season*

From May to June, heating on the land produces a low pressure system in the overlying atmosphere with rapidly ascending air and winds inward toward the low pressure area. Such a low pressure area or belt is developed in the northwestern part of the subcontinent (lower Indus basin, western Thar desert) and may be viewed within the broader framework of the well-established equatorial low pressure belt which is stationed between latitude 5° to 10° north by the end of May, but shifts to around 25° north by about mid-June following the northward movement of the sun. It is made more intense in the lower Indus basin by the high degree of solar insolation. Maritime air quickly moves into peninsular India and Bengal toward the low pressure system of northwest India from the Arabian Sea and Bay of Bengal. This change in the development of pressure systems and consequently the direction of winds is further facilitated by the northward movement of the intertropical front. Winds are southward over the Arabian Sea and Bay of Bengal. The southeast trade winds of the southern hemisphere cross the equator responding to the northward migration of the intertropical front and change their direction to southwest as monsoon winds.

This traditional explanation of the thermal causation of the summer monsoons fails to account for the "pulsation" of the monsoon, its "breaks" and "bursts". Pierre Pedelaborde has recently proposed a more reasoned explanation of the pulsation phenomena which is based on our increased knowledge of upper air movements. While not discarding thermal factors, the new theory emphasizes the significance of the role of the jet stream and of the Himalayan-Tibet highland as the topographic barrier in explaining the erratic behavior of the timing and incidence of Indian monsoons. According to this "upper air and perturbation hypotheses", the Himalayas, the Tibetan plateau, and several high altitude mountains like the Kunlun, the Pamir Knot, the Hindu Kush, are able to alter the flow of jet stream which normally travels between latitudes of 20°-40° north at altitudes of 7,620 to 10,668 meters. The jet stream is bifurcated; the summer branch lies north of the Tibetan highlands, connected as it is with the northward movement of the surface winds but traveling northward to Central Asia, whereas the southern branch stays south of the Himalayas during the winter.

By the second week of June, the jet stream moves entirely north of the subcontinent into Central Asia. This causes the indraught of free equatorial air to the north and northwest of the continent leading to the formation of dynamic depressions which had already been established by thermal factors. The situation, therefore, aggravates the triggering of the "burst" of the monsoon, allowing as it does the in-flow of equatorial air deep into India. The "pulsations" of the monsoons are attributed to the dynamic waves originating from interaction of the upper air jet stream and the surface air which develop in the inter-tropical front, rather than a juxtaposition of a frontal surface between two contrasting surface air masses.

The summer monsoon winds arrive in two major southwest streams; one southwesterly stream from the Arabian Sea brings large amounts of precipitation to the windward plains along the western coast and Sri Lanka; and a second more southernly stream from the Bay of Bengal brings heavy rainfall to Bengal, Bangladesh and the Brahmaputra valley and its adjoining mountains. The moist air stream that strikes the western coast is forced to rise abruptly by the Western Ghats and produces heavy orographic rainfall. At elevations of 610 to 915 meters in the Western Ghats and Shillong plateau and Meghalaya hills rainfall is over 3,000 mm during these months. Cherrapunji (elevation 4,300 feet), surrounded on three sides by high mountains, receives the greatest impact of the summer monsoons, experiencing a world record rainfall of about 7,600 mm during these five months out of a total annual precipitation of 10,400 mm. Although much precipitation results from orographic downpours, the warm, flooded landscape also provides a base for moisture intake by air through evaporation.

Most of the Deccan plateau lies in the "rain-shadow" of the southwest air stream of the Arabian Sea. Most moisture is shed until the crest of the Western Ghats is reached. To the east of the crest in the interior air is subsiding and dry. East of the Western Ghats, peninsular India remains, in general, dry, excepting the extreme southern tip and parts of the eastern coast.

The Arabian Sea current does not bring rainfall in the Thar desert. The only topographic barrier, the Aravalli mountains are aligned parallel to the monsoons. Only a portion of high altitudes and its southern tip capture some precipitation.

A part of the Bay of Bengal monsoons stream strikes the Himalayas after crossing the Ganga delta, is uplifted and sheds copious precipitation. High altitudes of the Himalayas also divert the moist stream westward into the Ganga plains of north India. This diversion, or channeling effect carries rainfall farther west. As the monsoon stream moves westward into the Ganga plains, it continues to shed moisture, becoming increasingly dry.

Most places in the middle Ganga plains receive between 650-1,300 mm of rainfall during the rainy season. By the time the stream reaches the Indus basin it has lost most of its moisture, although some precipitation occurs in the Kashmir mountains. Thar desert in Rajasthan and Pakistan remains practically dry as it is too far westward in the journey of Bay of Bengal stream of the monsoons. Rainfall, in general, is less than 375 mm between June to October. Despite the establishment of an intense surface "low" over the Thar desert and the lower Indus basin, the absence of relief barriers to the inflow of the southwestern monsoon contributes to scanty rainfall.

The amount of summer rainfall, therefore, varies in the different parts of the subcontinent, depending as it does on their locations with reference to distance from the coast or from the Himalayas, their altitudes, the direction of the monsoon winds and whether the locations lie on the windward or leeward side of mountains. Figure 2.4

**Figure 2.4**
*Average Annual Rainfall*

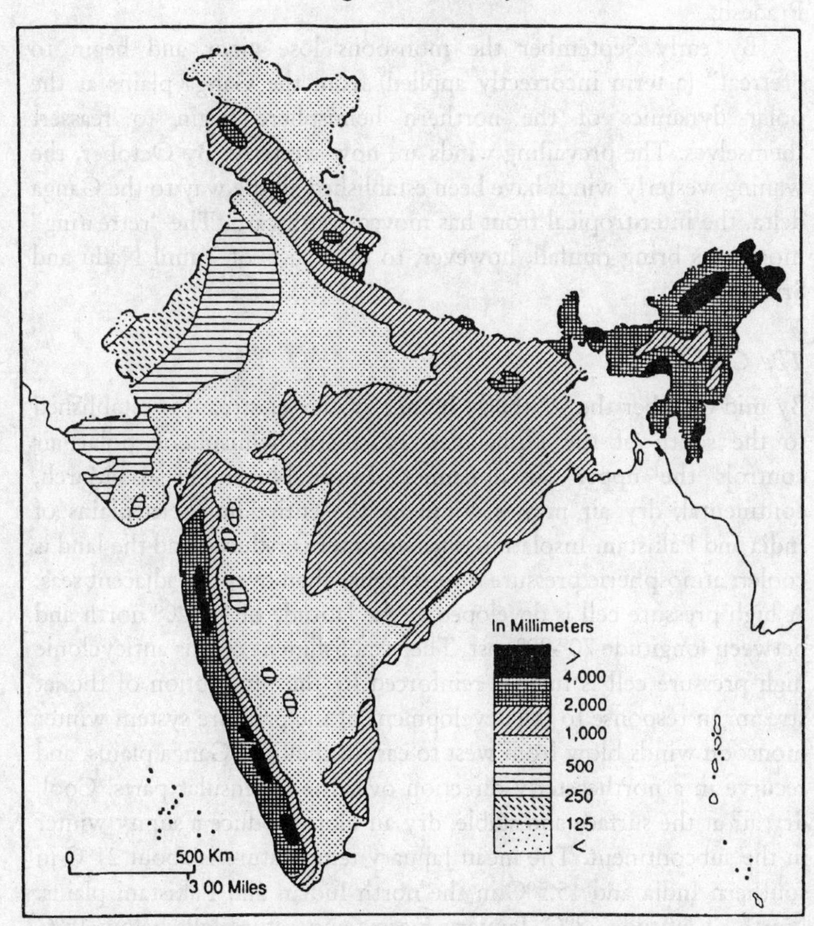


shows the distribution of annual average rainfall in the country.

Compounding this irregular distribution of the incidence of summer monsoons is their proverbial irregular timing, mention of which has been made earlier, and the prevalence of tropical cyclones of the Bay of Bengal which bring, on the average, about a dozen tropical storms annually to Ganga-Brahmaputra delta region. These storms have repeatedly caused large scale destruction. A storm in November 1979, carrying winds of over 240 km per hour and 6 meters waves, killed over 200,000 persons in Bangladesh in addition to completely annihilating settlements in a large area. A few tropical storms also affect the coastal areas in Tamil Nadu and Andhra Pradesh.

By early September the monsoons lose vigor and begin to "retreat" (a term incorrectly applied) from the Ganga plains as the polar dynamics of the northern hemisphere begin to reassert themselves. The prevailing winds are now westerly. By October, the waning westerly winds have been established all the way to the Ganga delta. the inter-tropical front has moved southward. The "retreating" monsoons bring rainfall, however, to the coasts of Tamil Nadu and Sri Lanka.

### *The Cool Season*

By mid-October the southerly branch of the jet stream is established to the south of the Himalayas. Northern hemispheric polar air controls the upper air dynamics. From November to March, continental, dry air masses dominate the entire northern plains of India and Pakistan. Insolation during winter is reduced and the land is cooler; atmospheric pressure is thus higher than over the adjacent seas. A high pressure cell is developed at the latitude of 15°-20° north and between longitude 70°-80° east. The establishment of this anticyclonic high pressure cell is further reinforced by the resumption of the jet stream. In response to the development of the pressure system winter monsoon winds blow from west to east in the Indo-Ganga plains, and recurve in a northeasterly direction over the peninsular parts. Cool, dry air at the surface and stable, dry air aloft produce a sunny winter in the subcontinent. The mean January temperature is about 21°C in southern India and 15.5°C in the north Indian and Pakistani plains. North of latitude 28°C, January mean temperature falls below 15°C and light frosts occur in the plains. In the Himalayas, temperatures decrease sharply with altitude. Figure 2.5 presents the winter (January) coolest month average temperatures.

The Tamil Nadu region exhibits a different rainfall region than anywhere else in India. The pattern consists of a dry season between January and June, followed by a summer of light monsoon rains, and succeeded by a major rainy season between October and December.

As noted earlier, the northwestern part of the subcontinent, the Punjab plains and the upper Indus basin receive winter rainfall caused

Figure 2.5
*Winter*

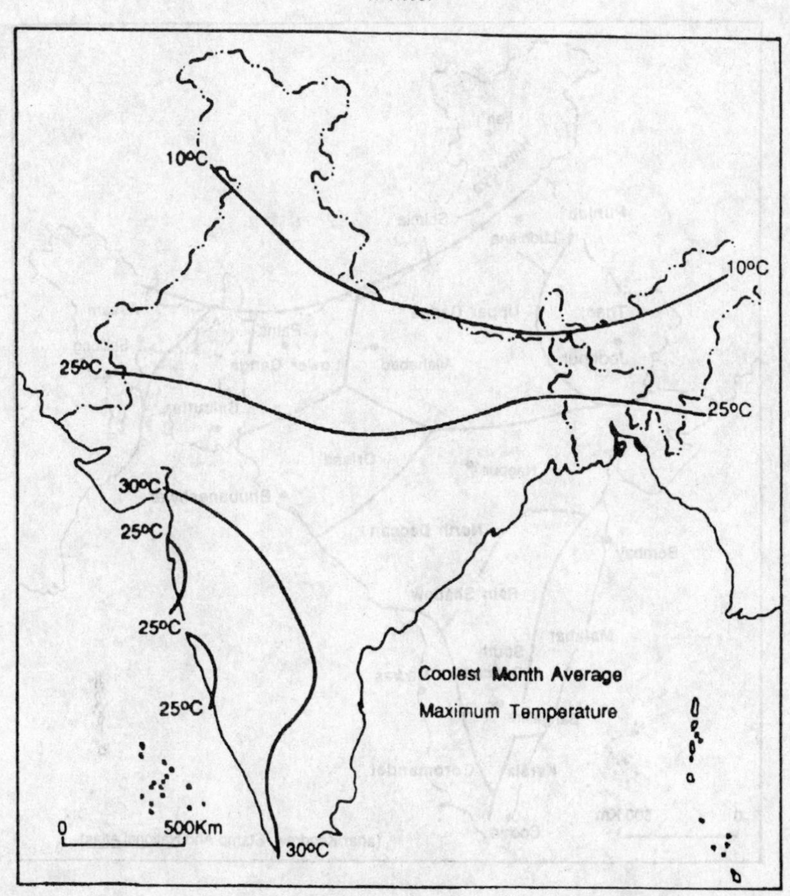


by the cyclonic depressions which follow the path of the jet stream. These average 4 to 5 a month between December and March, and travel from the Mediterranean Sea through Turkey, Iran, Afghanistan, Pakistan, western Kashmir, and the Himalayan flanks of northwestern India. These bring only light precipitation which is of great value to the winter crops such as wheat, cotton and mustard.

## Climatic Regions

The foregoing review of factors controlling the climates suggests the existence of a wide variety of regional distributions of rainfall regimes,

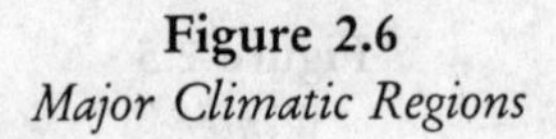

**Figure 2.6**

*Major Climatic Regions*

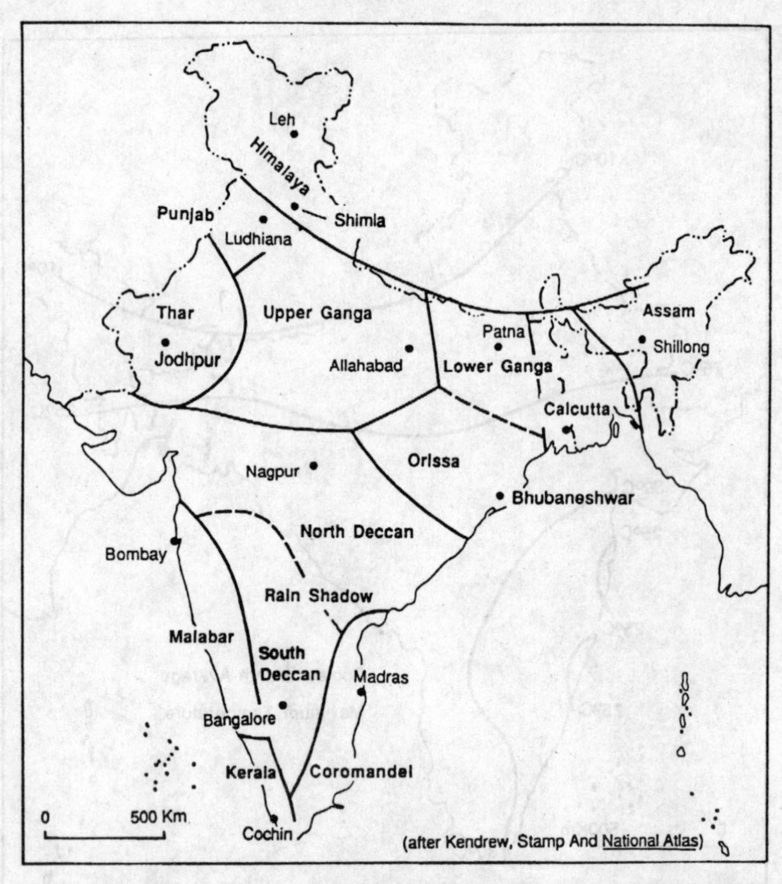


temperature ranges and pressure wind systems over the subcontinent. It is, therefore, most convenient to study the climate diversities by dividing the subcontinent into major climatic regions. Among the best known climatic classifications are those of Koppen, Thornthwaite and Kendrew. The following functional classification, a modification of the Koppen and Kendrew-Stamp classifications, clearly shows the distinction between the more "continental" north and the peninsula. This distinction follows in general the 25°C isotherm for maximum average monthly temperature for winter (Figure 2.6). In the north, the continental part, winters are cooler, fit for the growth of such temperate corps as wheat and barley, although wheat growing also extends farther south in the western Deccan. The continental-peninsular thermal division of special significance in human geography, especially, in its relation to the dietary and health conditions of the people in the two macro-divisions.

The climatic data-tables (Table 2.1) list several stations exemplifying the climatic regions mentioned below, and the figure of major climatic regions (Figure 2.6) gives their locations.

1. *The Himalaya type* follows the Himalayan ranges from Kashmir to Arunachal Pradesh. Rainfall is orographically induced and occurs with seasonal variations throughout the year. Rainfall incidence depends on altitude, the alignment of mountain ranges in respect to the direction of the monsoon winds, and the relative distance from the sea. Summers are mild, winters are cold. A typical station is Shimla.
2. *Punjab type* is a continental climate. Ranges of temperatures are high. It has some winter rainfall, induced by western depressions, although the rainiest season is in summer from the Bay of Bengal stream of the southwest monsoon. Rainfall is between 500-750 mm annually. The mean January temperature at Ludhiana is below 15°C. This region lies in the upper Indus basin, extending eastward to include parts of adjoining Punjab state in India.
3. *Upper Ganga type* receives moderate rainfall (between 700 and 1,000 mm annually), most of which falls during the summer months. Winters are cool, with mean January temperature between 15° and 25°C. The summers are very hot. Daily maximum temperature rises to over 42°C during late May or June. Average daily mean maximum for the same period is 40° to 42°C. Allahabad is a good example.
4. *Lower Ganga type* is transitional, with milder winters (16°-20°C January mean), heavier annual rainfall and a smaller seasonal range of temperatures than in the upper Ganga type. In the southern part (Orissa, eastern Madhya Pradesh) winters are milder (20°-23° in January). There is little or no winter rainfall.
5. *Calcutta type* lies in Ganga-Brahmaputra delta and Central East India. It is an area of high rainfall, over 1,500 mm annually. The temperature conditions are similar to the previous two types. In March and April violent cyclonic "nor-westers" frequently invade

this area, as also parts of Meghalaya (Assam type noted later) bringing heavy downpours. Although these cyclones are very useful to the growing rice and jute crops, cause large scale destruction of human and animal life, and settlements. Calcutta is an example.

6. *Coromandel type* prevails along the southern segment of the eastern coast. Winters are warmer than in all previously noted regions, with mean January temperatures over 24°C. Summers are hot, but do not attain the extreme temperature of upper Ganga type. Seasonal ranges also are smaller. Months of greatest rainfall are in winter (November and December) received from the northeasterly monsoons which pick up moisture over the Bay of Bengal. Chennai is a typical station of this region.
7. *Kerala type* lies in the southern segment of the western coast. A region of very heavy rainfall and tropical high temperature with little seasonal variation. January temperatures average above 24°C. Most precipitation (over 3,000 mm annually) is received from the vigorous southwest onshore summer monsoon winds during the months of May through July, although there is no dry month. Cochin is a good example.
8. *Malabar-Konkan type* lies in the Malabar-Konkan coast, or the northern segment of the western coast and also resembles the Kerala type except that it has a distinct dry season. Mumbai is a good example of this region. Most of the precipitation (over 1,875 mm annually) is received in the summer months between June and September from the summer monsoons. Winters are as warm as in the Kerala type.
9. *Deccan type* prevails in the interior of the peninsula. The area immediately to the east of the Western Ghats is a rain-shadow belt and includes southern Maharashtra, Karnataka and western Andhra Pradesh. It is a dry region with annual rainfall less than 650 mm, received mostly during the summer. Bellary is a good example. Winters are cooler than in the eastern and western coastal regions (Coromandel, Kerala and Malabar-Konkan types), whereas summers are warmer, due to the interior locations. This segment of the Deccan type may be called the "rain-shadow sub-type" region. To the north and northeast of this sub-type is an area of moderate rainfall (ranging between 650 and 1,300 mm

annually). Nagpur is a good example. Gujarat, north Maharashtra, the southwestern parts of Madhya Pradesh belong to this sub-type, which may be called the "north Deccan sub-type".

10. *Assam type* occurs in Assam, Manipur, Arunachal Pradesh, Nagaland, Meghalaya and Mizoram. It is an area of very heavy rainfall (over 2,000 mm annually), most of which falls in summer, between May and August, from the northeast monsoons. The "nor-westers" also bring rainfall between March and May. Shillong is an example.

**Table 2.1**

*Climatic Data for Selected Stations*

*(Temperature in degrees celsius and rainfall in mm)*

| *Type Station* | *Jan* | *Feb* | *Mar* | *Apr* | *May* | *June* | *July* | *Aug* | *Sept* | *Oct* | *Nov* | *Dec* | *Annual* |
|---|---|---|---|---|---|---|---|---|---|---|---|---|---|
| | | | | | | *Himalaya* | | | | | | | |
| *Shimla* | | | | | | | | | | | | | |
| Av. daily max. | 8 | 9 | 14 | 18 | 22 | 23 | 21 | 19 | 19 | 17 | 14 | 11 | |
| Av. temp. | 5 | 6 | 11 | 15 | 18 | 20 | 19 | 17 | 17 | 14 | 11 | 8 | |
| Av. rainfall | 61 | 69 | 61 | 53 | 66 | 175 | 424 | 434 | 160 | 33 | 13 | 28 | 1574 |
| | | | | | | *Punjab* | | | | | | | |
| *Ludhiana* | | | | | | | | | | | | | |
| Av. temp. | 13 | 16 | 21 | 27 | 33 | 34 | 31 | 30 | 30 | 26 | 20 | 15 | |
| Av. rainfall | 35 | 35 | 29 | 11 | 9 | 54 | 191 | 173 | 136 | 35 | 3 | 14 | 725 |
| | | | | | | *Upper Ganga* | | | | | | | |
| *Allahabad* | | | | | | | | | | | | | |
| Av. daily max. | 24 | 26 | 33 | 40 | 42 | 40 | 33 | 32 | 33 | 32 | 28 | 24 | |
| Av. temp. | 16 | 19 | 25 | 31 | 35 | 34 | 30 | 29 | 29 | 26 | 20 | 16 | |
| Av. rainfall | 23 | 15 | 15 | 5 | 15 | 127 | 320 | 254 | 213 | 58 | 8 | 8 | 1032 |
| | | | | | | *Calcutta* | | | | | | | |
| *Calcutta* | | | | | | | | | | | | | |
| Av. daily max. | 27 | 29 | 34 | 36 | 36 | 33 | 32 | 32 | 32 | 32 | 29 | 26 | |
| Av. temp. | 20 | 22 | 28 | 30 | 30 | 30 | 29 | 29 | 29 | 28 | 24 | 20 | |
| Av. rainfall | 10 | 31 | 36 | 43 | 140 | 297 | 325 | 328 | 252 | 5 | 20 | 114 | 1582 |

Contd...

Contd...

| | | | | | | | | | | | | | |
|---|---|---|---|---|---|---|---|---|---|---|---|---|---|
| | | | | | | *Coromandel* | | | | | | | |
| *Chennai* | | | | | | | | | | | | | |
| Av. daily max. | 29 | 31 | 33 | 35 | 38 | 38 | 36 | 35 | 34 | 32 | 29 | 29 | |
| Av. temp. | 24 | 25 | 28 | 31 | 33 | 33 | 31 | 31 | 30 | 28 | 26 | 25 | |
| Av. rainfall | 36 | 10 | 8 | 15 | 25 | 48 | 91 | 117 | 119 | 305 | 356 | 140 | 1233 |
| | | | | | | *Kerala* | | | | | | | |
| *Cochin* | | | | | | | | | | | | | |
| Av. daily max. | 32 | 32 | 33 | 33 | 32 | 29 | 29 | 29 | 29 | 31 | 31 | 32 | |
| Av. temp. | 27 | 28 | 29 | 30 | 29 | 27 | 26 | 27 | 27 | 28 | 28 | 28 | |
| Av. rainfall | 23 | 20 | 51 | 125 | 297 | 724 | 592 | 353 | 196 | 340 | 170 | 41 | 3106 |
| | | | | | | *Malabar-Konkan* | | | | | | | |
| *Mumbai* | | | | | | | | | | | | | |
| Av. daily max. | 28 | 28 | 30 | 32 | 33 | 32 | 29 | 29 | 29 | 32 | 32 | 31 | |
| Av. temp. | 24 | 24 | 26 | 28 | 30 | 29 | 27 | 27 | 27 | 28 | 28 | 26 | |
| Av. rainfall | 10 | 18 | 15 | 15 | 20 | 224 | 371 | 290 | 203 | 56 | 20 | 13 | 2078 |
| | | | | | | *Deccan Rainshadow* | | | | | | | |
| *Nagpur* | | | | | | | | | | | | | |
| Av. daily max. | 28 | 32 | 37 | 41 | 43 | 37 | 31 | 31 | 32 | 32 | 29 | 27 | |
| Av. temp. | 21 | 24 | 29 | 33 | 36 | 32 | 28 | 28 | 28 | 26 | 23 | 20 | |
| Av. rainfall | 10 | 18 | 15 | 15 | 20 | 224 | 371 | 290 | 203 | 56 | 20 | 13 | 1251 |
| | | | | | | *Assam* | | | | | | | |
| *Shillong* | | | | | | | | | | | | | |
| Av. temp. | 9 | 12 | 16 | 19 | 20 | 20 | 21 | 24 | 20 | 17 | 13 | 10 | |
| Av. rainfall | 15 | 28 | 59 | 136 | 325 | 545 | 395 | 335 | 315 | 220 | 35 | 6 | 2416 |
| | | | | | | *Thar* | | | | | | | |
| *Jodhpur* | | | | | | | | | | | | | |
| Av. daily max. | 21 | 27 | 32 | 37 | 41 | 40 | 36 | 34 | 34 | 35 | 31 | 26 | |
| Av. temp. | 17 | 19 | 24 | 29 | 34 | 34 | 32 | 30 | 29 | 27 | 22 | 18 | |
| Av. rainfall | 3 | 5 | 3 | 3 | 10 | 36 | 102 | 122 | 61 | 8 | 3 | 3 | 364 |

## Major Vegetation Types

Historical evidence indicates that an almost continuous belt of forest-cover was spread over north India at the time of Alexander's invasion in 326 B.C. Even during the tenth and fourteenth centuries, the Ganga plain was covered with vast patches of forest. However, only a few areas at high altitudes or the coastal swamps now remain where plant cover reaches its climatic potentialities. Elsewhere, man's modification of and interference with the "original" plant cover has resulted in its removal or degradation creating serious problems of soil erosion and the growth of stunted jungly vegetation (thickets, bushes and dwarf trees). At present, forests occupy only about 18 per cent of the total land of the country.

In general, natural vegetation types follow the distribution of rainfall, higher altitudes providing the major exceptional areas. Based largely on annual rainfall distribution, four broad categories of vegetation are commonly identified. Evergreen broad-leaved forests occupy areas receiving over 2,000 mm annual rainfall. Deciduous broad-leaved trees are associated with areas of 1,000 to 2,000 mm rainfall. Dry deciduous broad-leaved forests and open thorny scrubland occupy areas with 500 to 1,000 mm annual rainfall. Thorny scrubs and bushes are found in the semideserts and deserts receiving less than 500 mm annual precipitation. Within these four broad categories of vegetation, 16 major vegetation types my be recognized, shown in Figure 2.7 which is modified from the *National Atlas of India*.

*Tropical wet evergreen* forests are typical rain forests with their best stands growing in areas receiving over 2,500 mm annual rainfall. In their "natural" form their main area of distribution is an elongated strip along the Western Ghats south of Mumbai at altitudes between 457 and 1,371 meters. Parts of Assam, the Annamalai hills, Coorg, and the Andaman-Nicobar Islands in the Indian Ocean are some other areas of concentration. On their drier side, these are bordered by *tropical semi-evergreen* forest along the western coast from Bombay to Kerala. Elsewhere this type is found in Assam and Meghalaya. Both these types contain dense forest cover of tall trees rising to about 100 feet. The semi-evergreen type contain mixed stands of deciduous trees, especially toward their drier edges. Most of these forests are administered by the forestry department of the state governments. Woods for plywood and rattan are the main products. Adjoining these two types in Assam, the Western Ghats, Annamalai hills, and Coorg lies the *tropical moist deciduous* type characterized by widely spaced

**Figure 2.7**
*Vegetation Types*

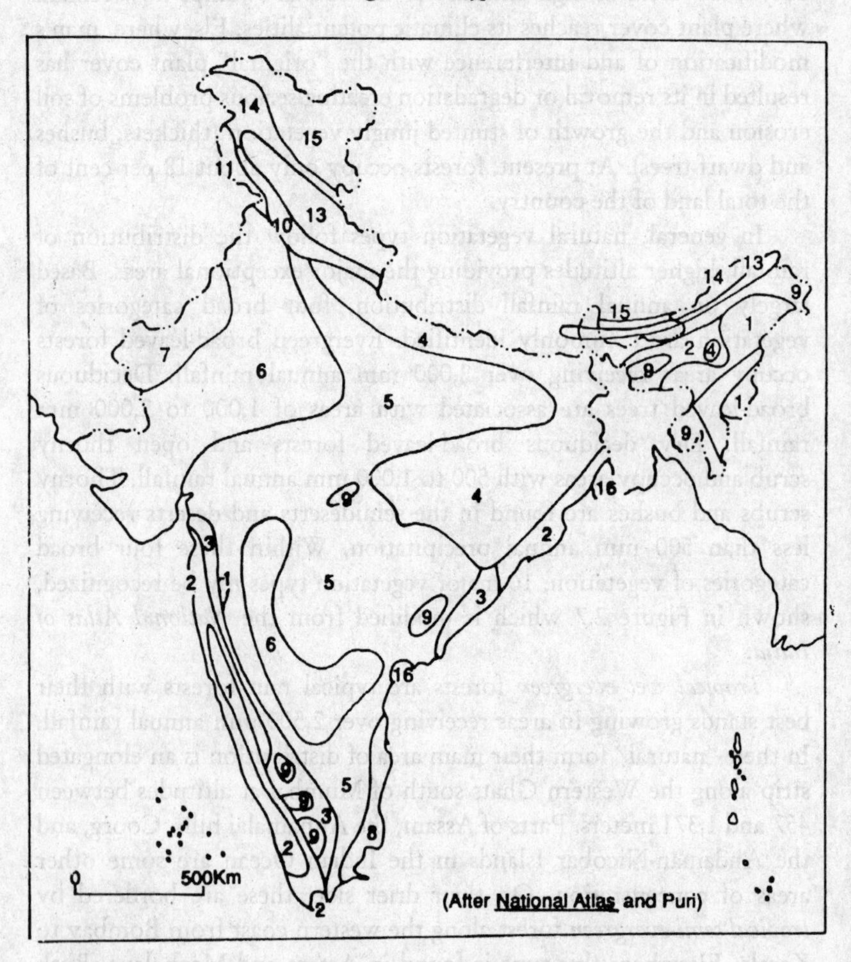


(1) Tropical Evergreen, (2) Tropical Semi Evergreen, (3) Tropical Wet Deciduous, (4) Tropical Moist Deciduous, (5) Tropical Dry, (6) Tropical Thorn, (7) Tropical Desert, (8) Tropical Dry Evergreen, (9) Sub Tropical Wet, (10) Sub Tropical Pine, (11) Sub Tropical Dry, (12) Wet Temperate, (13) Moist Temperate, (14) Dry Temperate, (15) Alpine, and (16) Tidal.

(Type 12 is not represented by country)

and broad-leaved trees, like laurel, and ebony, in an area receiving an adequate precipitation of 1,000 to 2,000 mm annually but has seasonal rhythm. A variation of the moist deciduous vegetation type covers a large part of Orissa, Chota Nagpur plateau, south Bihar, eastern Madhya Pradesh and the submontane Terai areas of Siwaliks along the southern Himalayan fringes in Himachal Pradesh and Uttar Pradesh. *Sal* is a typical tree of these areas.

Toward the drier parts of the moist deciduous type lies the *tropical dry deciduous type* which covers large portions of the subcontinent from the Ganga plains and the states of Madhya Pradesh, Maharashtra in central India to Tamil Nadu in extreme south. Although *sal* is still an important specie of forests of this type, especially toward the moister margins, several other trees gain regional importance. Examples of these trees are: *sandal*, *shisham*, satinwood and teak. Mysore state is a major producer of teak and sandal trees. Most of these trees are now utilized for commercial uses, the manufacture of furniture and handicraft artifacts.

*Tropical thorn* and *tropical desert* types are associated with semi-deserts and deserts receiving rainfall less than 500 mm annually. Main areas lie along the Pakistan border and western Madhya Pradesh, and Gujarat states. The Great Thar desert is a good example of the extremely dry section of these types. In the *thorn* type, broad-leaved, deciduous thorny bushes, generally widely-spaced and with heights ranging from 3 to 5 feet, make up most of the vegetation. In true desert conditions their size shrinks and the stunted trees and bushes give way to thorny and very widely-spaced vegetation.

*Tropical dry evergreen* is associated with the coastal region of Tamil Nadu and southern parts of Andhra Pradesh. Rainfall is less than 2,000 mm annually, mostly ranging between 1,000 and 1,500 mm, but characterized by a distinct but, short dry period summer. Most of the land has been cleared for agriculture during centuries of human occupancy. Areas of the *sub-tropical wet* type are spotty, clinging to high altitudes (between 3,000-5,000 feet altitude) in the Shillong plateau, Nilgiri hills, and in the Himalayas. Oak, beech and ash are the main trees. Toward the drier margins but in the mainly higher altitudes, lies the *sub-tropical pine* type. Oak tree is an example of this category. Most trees are exploited for the production of furniture, railroad sleepers, resin, and paper. *Sub-tropical dry* type occurs in a few spotty areas in the west Himalayan foothills at elevations between 457 to 1,524 meters. Acacia and scrubby trees form most of the vegetation in this type.

*Wet temperate* type occurs in the Nilgiri and Annamalai hills, at altitudes between 914 and 1,676 meters in areas experiencing rainfall over 1,250 mm annually. Magnolias, laurels and rhododendrons are the major trees.

*Moist temperate* variety is largely found in the Inner Himalayas, Kashmir, Sikkim, generally on the leeward side. Conifers like deodars and junipers predominate. The *Alpine* type is a very high altitude variety, lying beyond 3,505 meters altitude in the Himalayas. The vegetation consists of a shrubby forest of firs, junipers, birches and rhododendrons. The *Tidal* forest type consists mainly of mangroves and swamps in the Ganga-Brahmaputra delta. Main trees are hardwood palms used mostly for fuel and boat-making.

## Vegetation Ecology

Systematic deforestation of the woodlands, shifting cultivation by man, and overgrazing by animals for thousands of years over most of the subcontinent has stripped it of its natural vegetation and turned it into a vast scrubland. Woodland is confined to the riverine strips, the Tarai, village groves (of *tamarind*, *banyan*, mangoes), and the coastal areas. The exceptions are the areas in the Himalayas, the Western Ghats, and other hills (as in Assam and the Nilgiris). The Deccan plateau is dotted with short grasses and scattered trees. Most of north India has been cleared for crop cultivation, drily a few tree groves cling to the village settlements.

Estimated to affect 150 million acres, soil erosion has become a major problem. It is especially acute in the Punjab Siwaliks, Chambal valley, and parts of Assam and Burma borderlands. Shifting cultivation and *jhuming* (slash and burn) activities have been and are widely practiced in the tribal areas of Assam, Arunachal and central India. Sheet erosion in most of these areas has produced lowering of the water-table and has reduced these areas to a stubby wasteland.

Ancient scriptures have extolled the importance of "natural" vegetation and trees. Medieval literature also abounds in the beauty of groves, trees and forests. Most deforestation probably occurred after the medieval period. During the British rule a few studies and reports pointed out these problems but little effort was devoted to remedy the situation. Since Independence, the government has lent moral support to the "greening" of the country by rhetorical exhortations. Official efforts have largely remained halting, half-hearted and ineffectual. What is needed is a concerted, coordinated, comprehensive, national plan integrated into the national Five Year Plans.

## Major Soil Patterns

During the British rule in India, a vast body of fascinating accounts of soils, rich in local names and folklore, and often replete with shrewd observations, had emerged in the district gazetteers and official reports. These accounts, often based on centuries-old farming experiences, were generally directed toward the assessment of differential soil fertility and land revenue collection, but did not attempt classification and distribution of soil types in the country.

Since its establishment in 1956 the Soil Survey of India has been engaged in the production of scientific analysis of the soil distribution of a few selected areas of the country. Its final goal is to prepare detailed and accurate soil maps, based on the latest soil classification systems for the entire country. Given the dimensions of the problem and its modest budget, the Survey has been able to produce few such maps.

Recent attempts to divide the country into broad categories of soil types are contained in the *National Atlas of India* (1957), *The Irrigation Atlas of India* (1972) and Spate's *India, Pakistan and Ceylon* (1976). The last two have utilized the 7th approximation soil classification developed by the U.S. Department of Agriculture (USDA). Virtually ignoring the role of man in the formation of soil types, traditional soil grouping (zonal, azonal and intrazonal) relied essentially on the genesis and development of soil profile. The 7th approximation defines soil classes strictly in terms of their morphology and composition as produced by a set of natural and human forces the classification is determined by quantifiable criteria.

Detailed mapping of the country on the basis of the 7th approximation scheme will surely take several years, if not decades, and given its heavily terse, and tongue-twisting terminology, may find it difficult for wide acceptance. Through centuries of farming experience, the Indian farmer has been able to discover the

**Figure 2.8**

*Soil Regions (Simplified)*

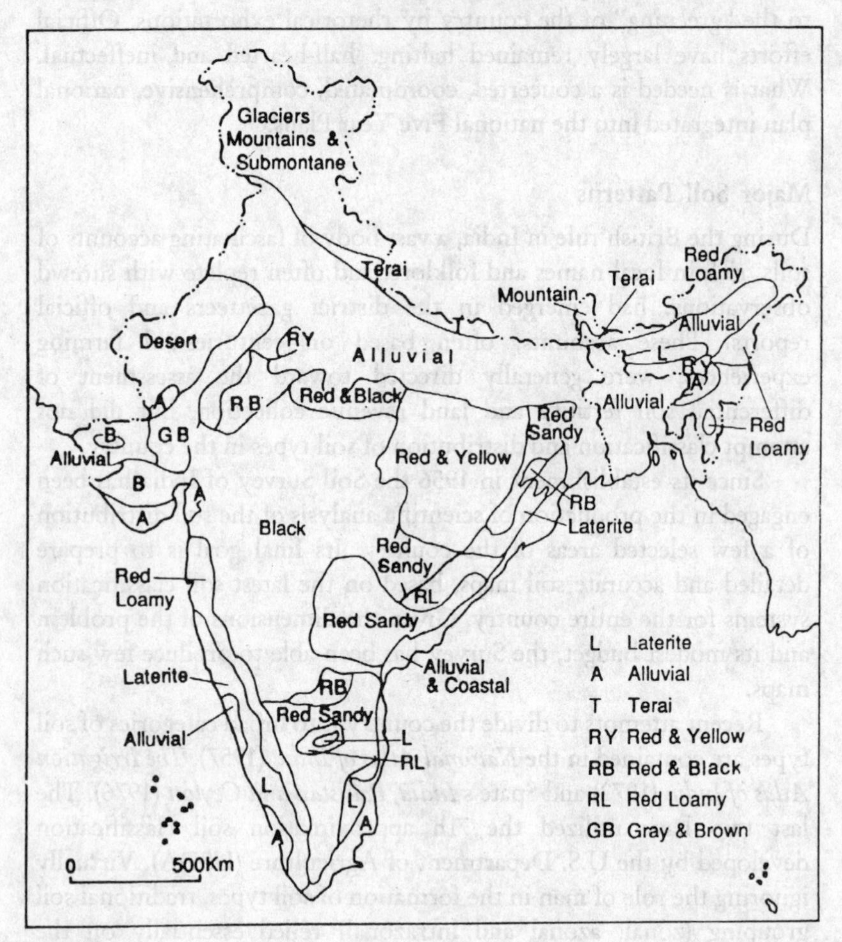


relationship between the fertility and texture of different soils. His main concern, therefore, is how soils of different textures be put to use. An attempt to divide the country into broad soil regions based primarily on the physical and chemical properties (texture, structure, color reaction and plant food), and suitability for farming is given in Figure 2.8.

*Alluvial* soils occupy most of the Indo-Ganga plains (from Punjab to West Bengal) and the Brahmaputra valley in Assam. Produced by water and wind-transported materials, these soils consist of alluvial gravels, sands, silts or clays and show little profile development. Containing an immature horizon, developed as these are under water-sufficient-or water-surplus-conditions, soils are best formed on these plains. An ideal example is that of the Ganga delta region where continual sedimentation is at work. Here, numerous river tributaries deposit alluvium of sands, and silts during the floods. Upstream, in the middle and upper Ganga plains, the volume of the riverborne materials is much smaller, and humidity lower, sedimentation is still at work although at a lower pace. The soils are very productive, composed of "newer" alluvium (*khadar*), and have been intensively cultivated for centuries. The "old" alluvium (*bhangar*) occupy the interfluves (*doabs*) or inter-stream portions where local relief reaches up to 61 meters above the rivers in the otherwise monotonously flat topography. *Khadar* soils are more sandy than the *bhangar* ones. Both contain calcareous concretions (*kankars*), which are a useful source of building lime and used in whitewashing houses in the Indian village. Both are fertile, but lower water-table in *bhangar* and its higher elevation makes its higher elevation makes it difficult to irrigate. A variety of dry-season crops such as wheat, cotton, groundnuts, pulses and sugarcane are extensively cultivated in the middle and upper Ganga plains under irrigation. In the more humid Ganga delta and the lower Ganga plains irrigation is not essential and wet crops such as rice and jute are grown without irrigation. The *alluvial* soils also cover smaller deltas of the eastern coast and few other scattered areas along the Malabar and Gujarat coast.

The *black soils* or *regur* cover a large territory (about 518,000 sq km) of the northwestern Deccan plateau in Maharashtra, western Madhya Pradesh, north Karnataka, and most of Gujarat. The parent matter is derived from the Deccan lavas (or alluvium derived from it) and contain a high content of black-colored clay minerals which shrink greatly when dried out and swell when saturated. The soils, therefore, form wide cracks during the dry season. Remarkably "self-ploughed" by the loosened particles fallen from the ground into the cracks, the soil "swallows" itself and retains soil moisture. The iron, lime, and aluminum content is high and the soils are quite productive. The dark color of the soils is attributed generally to the parent matter, although recent research ascribes a more potent role to the organic matter in the development of the soil-color. The upper horizon is three to six feet thick. A combination of unusual physical and chemical properties enumerated above makes the soil particularly useful for the cultivation of such crops as cotton, millets, pulses and groundnuts.

Peninsular India, with the exception of the *black soil* region, is covered largely by red soils and their variants. Figure 2.8 identifies three distinct red soil regions: (a) *red sandy* soils cover a large part of Tamil Nadu, Karnataka and Andhra Pradesh, parts of Orissa, Madhya Pradesh and Bihar; (b) *red loamy* soils occupy a long stretch of area in the foothills along the Western Ghats, and some pockets in eastern Karnataka, and Orissa; and (c) *red and yellow* soils are spread over parts of Orissa and Madhya Pradesh. Outside the peninsular India, *red and yellow* soils are distributed in southern Bihar and southeastern Uttar Pradesh. Mixed *black* and *red* soils are found in the Krishna-Tungabhadra valley and some scattered areas in Tamil Nadu. These soils are produced under conditions of high temperatures. Leaching and oxidation robs the soil of humus and soluble plant food particularly in areas of abundant rainfall, and forest cover rendering soils much of their fertility. The residual surface is usually a hard crust of porous rocks varying in color from reddish brown to yellow which is rich in iron and aluminum. Such soils are known as the *laterite soils* although these are more of a geological formation than soils in the true sense of the term. The process of producing such laterite surfaces is known as *laterization*. *Laterites* cover several areas along the coast of Kerala, Karnataka, parts of southern Maharashtra, Tamil Nadu, Orissa and West Bengal.

*Desert soils* cover nearly 259,000 sq km in western Rajasthan. These are, in general, sandy and alkaline. Under careful irrigation these are quite suitable for agriculture, although constant irrigation can contribute to soil salinity. To the east of the *desert soil* region is a zone of *gray-brown* soils, reflecting the increasing humid conditions and changes in surface geology.

The soils of the Himalaya mountain are complex. Vertical zonation introduces several soil types on the slopes. Areas of very high altitude, glacier-bound sections and snow fields are mostly denuded of soils or contain immature, *skeletal soils*. Valley floors of the river basins contain recent alluvium deposited by the streams. The soils of the agriculturally productive Jhelum valley in Kashmir belong to this category. An elongated zone of swampy, immature soils known as *tarai* soils lies at the foothills of the Himalayas mountain.

In terms of surface area, the *alluvial* soils occupy nearly 42 per cent of the country, and the black soils about 14 per cent. Thus, more than half of the country is overlain by fertile soils which are potentially capable of yielding much higher returns than at present under the stimulant of scientific technology. Red soils cover nearly 10 per cent, *laterites* and *red* and *yellow* soils another 15 per cent. The latter, though poor in nutrients, respond favorably to irrigation in areas where these are deep enough for cultivation. *Mountain soils*, *tarai soils* and *gray* and *brown* soils are unproductive and occupy only 9 per cent of the country's land surface, whereas the remaining 10 per cent is covered with *desert soils*.

## Environmental Degradation

Pollution of the physical environment resulting from an unbridled use of modern technology has become a major contemporary concern. The accelerating pace and power of technological change have tended to exacerbate many environmental problems in recent years. Although the impacts are both global and regional transcending political systems and stages of development, the developing nations that adopted modern technology (undoubtedly initially to improve their quality of life) without proper safeguards have been left with a large cleanup job which they can financially ill-afford to handle. The liquefied gas fire in Mexico City in November 1984, and the deadly gas leak at Bhopal (India), in December 1984 are two, among the several instances, of an increasing number of environmental hazards in the developing countries.

Recent investigations of the air quality of North Indian Plains suggest that air smog severity, acidification severity (popularly known as "acid rain") and corrosion severity have been on the increase (Darmstadter, 1988: 10). Although "acid rain" activity has been present through the past century in the Delhi area and Calcutta region, a trend in increasing acidity is related to the rapidly growing industry. In decades to come, sulphur and nitrogen oxides produced from motor vehicles will also become an important factor in Indian air quality of larger urban centers. In New Delhi alone, 2,200 tons of such particles are released into its air daily, putting it very close too the list of most polluted places. An estimated one-third of its inhabitants suffer from chronic respiratory diseases. According to a January 1988 article in the *New York Times*, the number of vehicles is increasing by about 10,000 every month. Coal combustion is another major contributor to the deteriorating air quality, although biomass burning, petroleum combustion, and unwise agricultural practices are other contributory factors.

During the last forty years the potentially hazardous chemical and fertilizer industries have expanded in the densely populated urban centers e.g. near Delhi, Kanpur, Bhopal, Chennai, Nagpur, Calcutta, Vadodara, Hyderabad and Mumbai. Plants in Madras produce caustic soda, alkali and fertilizers. More than forty companies manufacture or deal with toxic chemicals and pesticides without proper safeguards, and without proper provision of their waste disposal. Technological mishaps are common place in Calcutta-Damodar-Subarnrekha industrial region (Karan, 1986: 196-200).

In all, there are about 7,000 potentially hazardous plants in the country that produce or deal with explosive, toxic or flammable substances. These industries include metallurgical plants, oil refineries and petrochemical factories; most located near or in major cities. The number of accidents at these plants has been increasing. Government regulations regarding limits imposed on factories producing or dealing with toxic substances are very lax. According to *Wall Street Journal* of 19 December, 1984, the concentration of toxic dusts and gases in an average Indian factory air contain several times the limits imposed in the United States (four to seven times).

While the technological hazards have brought catastrophic consequences to the environment, whether by malfunctioning or by unbridled use of technology, nearly most environmental problems stem from human activities. The result has been an ever-increasing and widespread pressure on the Indian ecosystem. Population pressure, intensive and unwise use of land, overgrazing, indiscriminate deforestation, excessive fuel-wood collection, or soil erosion have contributed to environmental degradation. On conservative estimates one-half to one per cent of the country's area is turning into wasteland every year. The various processes of industrial activity—mining, combustion, the use of chemicals, unwise deforestation (and resulting soil-erosion) are the major contributors to forms of environmental pollution. Humans have been evolving with and modifying the Indian landscape for so long that they take it pretty much for granted, which has, so far, resulted in a lack of public awareness and government priority.

That the poorer nations like India do not have the financial resources to clean up the environmental pollution, was forcefully argued at the United Nations Council for Environment and Development (UNCED) meeting in June, 1992 at Rio de Janeiro (popularly called the "Earth Summit") attended by 180 nations. Poorer countries maintained that the richer, industrialized nations, have a greater responsibility to pay for the environmental clean-up job, since these countries contribute far more to the environmental damage that modern technology and industries have caused in the first place. On the issue of depletion of ozone layer in the atmosphere and global warming, the poorer nations asserted, that the industrialized nations must reduce their emissions of carbon dioxide instead of placing future limitations for poorer nations, since constraints on the developing countries would adversely affect their economic development programmes.

## *References*


Chatterji, S.P. (ed.), *National Atlas of India*. Calcutta and Dehradun, 1957.

Darmstadter, J., "Atmospheric Stress and Economic Development in India's Gangetic Plain," *Resources*, Vol. 91, Spring 1988, Washington, D.C., pp 8-11.

Geddes, A., "The Alluvial Morphology of Indo-Gangetic Plain", *Transactions*, I.B.G., Vol. 28,1960, pp. 262-63.

Government of India, *Irrigation Atlas of India*. Dehra Dun, 1972.

Karan, P.P., *et al.*, "Technological Hazards in the Third World", *Geographical Review*. Vol. 76(2), April 1986, pp. 195-208.

Kendrew, W.G., *Climates of the Continents*. Oxford, 1961.

Learmonth, A.T.A., *The Vegetation of the Indian Sub-Continent*. Canberra, 1964.

McKnight, T.M.. *Physical Geography*. 2nd ed., Englewood Cliffs, N.J., 1987.

Muthiah, S. *et al.* (eds.), *A Social and Economic Atlas of India*. New Delhi, 1987.

Pedelaborde, P., *Les Moussons*. Paris, 1958.

Puri, G.S., *Indian Forest Ecology*. New Delhi, 1961.

Shrinivasan, V., "South West Monsoon Rainfall in the Gangetic West Bengal", *Indian Journal of Meteorology and Geophysics*, Vol. 11, 1969, pp. 5-18.

Singh, R.L., (ed.), *India: A Regional Geography*. Varanasi, 1971.

Spate, O.H.K., and A.T.A. Learmonth, *India and Pakistan and Ceylon*. 3rd ed., London, 1967.

Schwartzberg, J.E. (ed.), *A Historical Atlas of India*. Chicago,1978.

Trewartha, G.T., *Earth's Problem Climates*. Madison, 1961.

Wadia, D.N., *Geology of India*. 3rd ed., London, 1961.

# 3

# Historical Setting

The Indian civilization, like those of the Mesopotamian, Assyrian, Egyptian and Chinese civilizations, is ancient, has had a continuous, although not entirely recorded, history dating back to several centuries before the Christian era. The Indian subcontinent which includes present-day Pakistan and Bangladesh has been inhabited since at least 5000-6000 B.C. Archaeological excavations during the last thirty years over a large area from Baluchistan in Pakistan to Bengal and Sri Lanka have produced evidence of food production and sedentarization as early as c. 7500 B.C.

## Beginnings of Urban Civilization

Developed first in the Indus delta area during the third millennium B.C., the Indus valley civilization, representing the earliest known civilization in the subcontinent, appears to have spread north and then east during the succeeding centuries. This civilization persisted for 500 to 700 years and covered an area of over a half million square miles from Baluchistan to the Ganga basin, northeast of Delhi (Figure 3.1). Remarkably uniform in several cultural traits (agricultural products, inscriptions on seals, figurines and artifacts of presumably religious significance), the Indus valley was an urban and commercial civilization. Internal communication links were developed along the river. Overland and by sea trading connections with the outside world were established through several ports along the Arabian Sea from the

**Figure 3.1**
*Indus Valley Civilization*

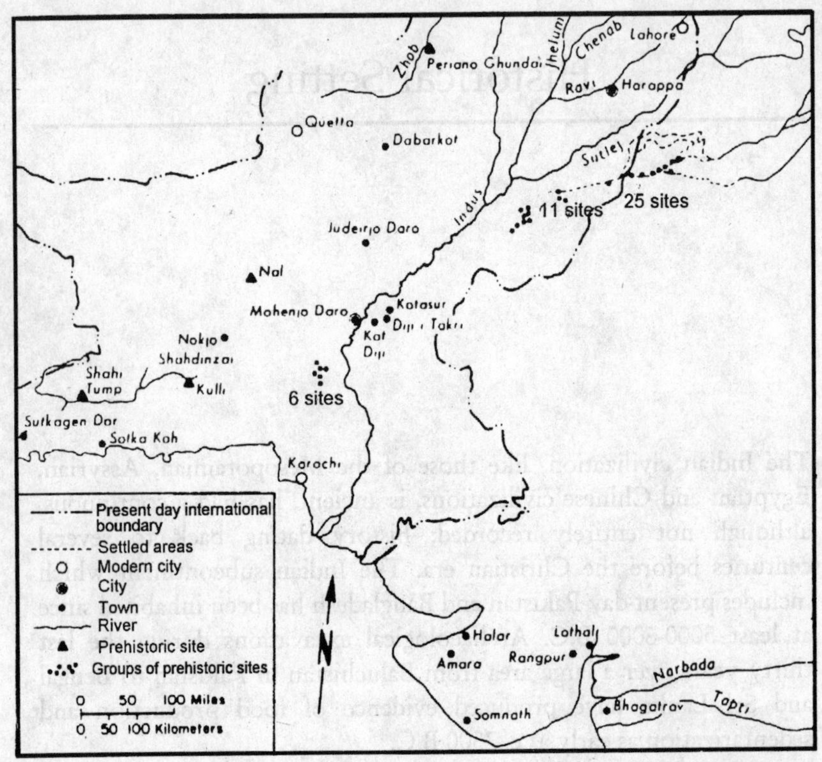


Reprinted with permission.

Gulf of Cambay to Sumeria.

The planning and design of the Indus valley cities are characteristic: their gridiron layout, northsouth axis, dimensions, size of about one and a half mile square area and the existence of a large fortified "citadel mound" to the west with the larger city toward the east. Common building material consisted of burnt and unburnt brick, with an occasional timber lining. Stone was rarely used. A wide range of house sizes, from single-room dwelling unit to a large house, containing private wells, cooking and bathing areas and a large compound have been unearthed. A well-organized sewage system of drains disposing wastes from houses into the main street drains served most of the city. A wide variety of stores and craft shops (potters' kilns, dyers' materials, and metal works) formed an important section of the city. The "citadel" contained several public buildings which were used for administrative purposes as well as for storing foodgrains. In Mohenjodaro, a large centrally located "great bath", which was probably used for ritual group bathing, has also been found. Estimates, based on the count of dwelling units, suggest that about 30,000 to 40,000 persons lived in each of the two large cities of Harappa and Mohenjodaro.

The cities were supported by a vast agricultural base of the fertile river basins. Wheat, rice, dates, melons and leguminous plants were the principal crops grown. Cattle, sheep and donkeys were the chief domesticated animals. The Indus seals depicting tigers, bulls and elephants have also been discovered. Copper, silver and bronze were used to produce implements, seals and artifacts. Pottery was mass produced. Indications are that the use of woven fabric of cotton was common. The larger settlements were interconnected by river transport and land routes, facilitating inland commerce which involved acquiring raw materials in exchange for finished products and was supplemented by a much wider external trade in lead, gold and precious stones with the Mesopotamian cities. A regular system of weights and measures was developed for commercial purposes.

Several of the unearthed seals and figurines exhibit the prevalence of a fairly developed level of artistic and social activity. Scenes of mythological and religious significance depicted on a number of seals suggest the existence of some forms of religious practices which have been carried over later into Hinduism. Relationships with the later Hindu god Shiva and the goddess Durga have been perceived by some authorities. The existence of several burial sites indicated Harappan belief in man's after-life. Attempts to decipher the Harappan script have so far failed, but there is a probable link to the Dravidian languages of south India.

Between c. 2500 and 1700 B.C., the agricultural settlements and the associated urban economic structure of this civilization declined quite suddenly and by c. 1500 B.C. geological and ecological factors, such as changes in the river courses, salinization and waterlogging of soils and expansion of the desert and climatic changes making the area physically inhospitable, have been advanced for the disappearance or total transformation of the Indus valley civilization.

## Early Indian Civilization

The advent of the Indo-Aryans marks the beginnings of Indian history. Although little is known about the Indo-Aryans, linguistic and literary evidence point to their unmistakable links with the Indo-Europeans as well as to the fact that they came from the northwest, the direction of the Iranian plateau, in successive waves between 1500 and 600 B.C. Early Vedic literature, as well as the contemporaneous Iranian literature of the Avesta, attest to such migrations, although precise source regions and timing of these movements are still obscure. Most scholars favor the idea that these migrations originated in Central Asia; literary and linguistic evidence suggests that they were probably a part of great and complex folk movements which brought ancient Persians into Iran and the Greeks into Greece. Such common customs among these widely spread ancient people, such as cremation of the dead, and their dress forms, as well as their institutions as king, council and assembly, indicate their common origins.

The Vedic literature (1400-700 B.C.) is our primary source of information regarding the behavior of these migrations: the direction of migratory currents, the gradual diffusion of migrants over the north Indian plains, and the settlement patterns and socio-economic structure of the society. Additional knowledge is obtained from the two Hindu epics—the *Ramayana* and the *Mahabharata* (400-300 B.C.) As the Aryan invaders gradually settled over the North Indian Plains, they pushed the indigenous people, possibly the early Dravidians, to the east and the south and assimilated them in part. Gradually, they established a new socio-economic order and introduced new languages, literatures, philosophies and ideas of social behavior. The growth of Sanskrit, a language of the Indo-European linguistic family, parallels the growth of the Vedic literatures and philosophy during this early period of the Aryanization of India's history.

The early Aryans were nomadic pastoralists. As they moved into the North Indian Plains and finally penetrated into southern India beyond the Vindhyan hills, they gradually cleared the forests and started establishing agrarian communities. To some scholars the story of the *Ramayana* illustrates the historic Aryan penetration of south India. While Aryans were settling in the northwestern parts of the subcontinent, through a process of their contact and interaction with the relatively underdeveloped aboriginals, society became divided into stratified classes, first along apparently occupational lines, but eventually organized along strict caste-based rules.

Meanwhile, several philosophical systems regarding nature, matter and spirit were being evolved and refined. A vast body of rituals, worship and caste regulations was written in the *Upanishads*, treatises which form the basis and, some would say the culmination of Hindu thought. Doctrines of *maya* (illusion), *karma* (moral consequences) and transmigration were expounded. Several deities, such as Vishnu and Shiva, also emerged, transcending the importance of the early Aryan deities (e.g., Usha, Indra and Varuna) which represented the various forms of nature.

During the sixth century B.C., two sects—Buddhism and Jainism—stood up against the ritualistic and polytheistic forms of Hinduism; each attempted to offer a satisfactory solution to the mystery of life and death. Each also struck against the entrenched caste system of the society. In this last effort they ultimately failed, for the caste system remained a bedrock of society.

**The Mauryan Empire**

The first major attempt to consolidate the tribal and political units of north India into a single political entity was realized by the Mauryan kings (321-181 B.C.). A detailed account of the administration and the prevailing socio-economic structure of society can be obtained from the contemporary accounts of Kautilya (who wrote a famous treatise on the administrative role and techniques of the rulers) and Magasthenes, a Greek ambassador from the Seleucids to the Mauryas who stayed in India during the Mauryan times and recorded extensively on the nature of society. The Mauryan government was a centralized bureaucracy pivoted around the king. The imperial system was sustained financially by the income derived from the land taxes, internal commerce and foreign trade. A number of cities flourished in the empire, and were connected by a well-developed system of highways. Artisans formed a major section of the urban populations. Several urban guilds of artisans, bankers and merchants existed and exercised a powerful influence on the city administration. The state maintained a large standing army. Under Ashoka, the greatest of the Mauryan rulers, the empire was subdivided into a least four large provinces, each administered by a royal prince. The provinces were sub-divided into districts. The provincial and district administrations, which became the basis of Mughal and British administrative systems later, were reminiscent of the Mauryans.

Ashoka (c. 274-236 B.C.) was one of the most illustrious kings in India's history. During his rule, the Mauryan empire was extended to cover almost the entire subcontinent. He became a convert to Buddhism and had the central principles of the Buddhist religion inscribed on rock pillar edicts at prominent places within his empire. The designs of these pillars and the insignia used on rock edicts are now used as the symbols of Indian tradition in government documents. By preaching the idea of *Dhamma*, an attitude of social piety and conscience, he tried to weld the nation into a united force.

In 326 B.C., Alexander of Greece invaded northwestern India through the passes from Afghanistan. He could not advance beyond the river Beas in present-day Punjab and stayed in the country for only a short while. His quick and fleeting stay and sudden departure, left only a few marks on Indian society. After his invasion Greek influence started to percolate slowly into the country. The Gandhara sculpture patronized by the Kushans, who ruled northwest India a few centuries later, bears unmistakable Hellenic influence. The Gandhara sculpture tradition blended the Greek facial features and dress style with the traditional native motifs in the statues of Buddha.

By 150 B.C. the vast Mauryan empire had started to crumble. A number of small states had emerged. The disintegration of the subcontinent into several political units along regional lines became a recurrent pattern in Indian history.

An important development during Mauryan and post-Mauryan times was the development of a commercial economy and the establishment of trade routes both within the country and with the outside world. A flourishing civilization had independently developed earlier in south India which maintained commercial and cultural contacts with Aryan north India during and after Mauryan times. Several land routes joined the Ganga plains with Taxila (now in Pakistan) and Kabul (in present-day Afghanistan) and, across Iran, with the Black Sea ports. The maritime trade routes from the Indian ports to the Persian Gulf and further westward to the eastern Mediterranean Sea were also firmly established. The Roman Emperor

Augustus appears to have received two trade missions from India between 25 and 21 B.C.

After a lapse of several centuries, large sections of the country, at least in north and central India, were united by the Gupta kings, who attempted to revive the imperial idea of a *Chakravartin* (supreme ruler) of a pan-Indian empire. In addition to its military glories, the Gupta Age (320-540 A.D.) achieved notable successes in artistic and humanistic spheres and is often acclaimed as the Classical or Golden Age of India. Indian art, literature, architecture and philosophy reached a high level of perfection. Treatises on metallurgy, astronomy and mathematics were written. Several systems of painting, sculpture, music, dance, drama, legal theory and philosophy were perfected. It was a time of intense intellectual and cultural renaissance. The famous Buddhist University of Nalanda was founded. A detailed and lively account of the society and economy during the Gupta Age has been handed down to us by a Chinese Buddhist pilgrim, Fa-hsien, who traveled India from 405 to 411 A.D.

After the collapse of the Gupta empire around 540 A.D., the country once again returned to a state of disorganization and political fragmentation. Invasions from Central Asia by the Huns and other groups had already started. Several tribes including the Huns, the Shakas and the Scythians migrated to India in large numbers and set up small principalities in northwest India. Gradually, they adopted Indian ways and were assimilated into society. The rise of the Rajput families, who later became the ruling dynasties in Rajasthan, has been attributed to the social changes resulting from the movements and assimilation of the groups into Hindu society during post-Gupta times.

The southern regions of Peninsular India had, through the centuries, developed their own linguistic, philosophic and cultural traditions. Very rarely had the military arm of the north Indian emperors interfered in their political and administrative affairs. Powerful kingdoms like the Pallavas, Chalukyas and Pandyas, later the Cholas and Rashtrakutas were the south Indian empires established between the fourth and tenth century A.D. Within these kingdoms, a distinctive Dravidian culture took root. Tamil and associated Dravidian languages had reached a high degree of development.

By the twentieth century northern India was split into several small regional kingdoms associated with several Rajput families and the descendants of the Huns, the Shakas and the Scythians. Noteworthy among those were the Solankis, Paramars, Chauhans and Tomars. Regional feelings were also expressed in the development of local languages such as Gujarati, Bengali, Rajasthani, Punjabi and Hindi which began to be used in preference to Sanskrit.

## Early Muslim Period

By the 8th century, Islam had been firmly rooted in Arabia and Iran and had started to encroach upon the lower Indus basin. Until the 10th century the Islamic invasions were sporadic and did not take root in the subcontinent beyond its northwestern fringe. Increasingly, these became more regular and intense, and during the succeeding centuries the petty, warring Hindu principalities could not long withstand their onslaughts. Islam came to India mainly from Central Asia, rather than from Arabia. By the tenth century the Turks had begun penetration from the northwest, from present-day Afghanistan and Pakistan, through the several passes in the area. After repeated attempts, they were successful in consolidating their power in northwest India and eventually in occupying the throne at Delhi. Several Turkish dynasties ruled over the Delhi sultanate and expanded their authority over a large portion of the Indus-Ganga plains. Their rule was generally marked by the rigid enforcement of the Islamic law, increase in the power of the nobility and succession disputes. Invasions from Central Asia continued, often crippling the Delhi sultanate. Notable among such invasions was the one by Timur, a ruler of a vast empire in Central Asia who struck Delhi in 1398 A.D. with a devastating blow in the wake of which the Delhi empire crumbled and broke into several regional Muslim and Rajput states.

While northern India was being preyed upon by Islamic invasions, South India led a separate life. Islam came to south India eventually, however, centuries later. The Khaljis and Tughlaks penetrated much of the south and during the fourteenth and fifteenth centuries, the establishment of the Muslim kingdom of Bahmani (c. 1347-1527 A.D.) represented a major political thrust of Islam into southern India. As a challenge to the Bahmani empire, the vast empire of Vijayanagar (c. 1336-1646 A.D.) became the chief bastion of Hinduism against this Islamic thrust from north India. The Vijayanagar empire was notable in many respects. The present-day linguistic regions of the Dravidian speech and culture experienced renewed development, and its rulers vigorously pursued the development of fine arts and cultural traditions.

An undercurrent of Hindu revival was reasserting itself during the period 1350-1450 A.D., a century of political turmoil for northern India, beneath the rise and fall of dynastic rules. Basically, a religious movement aimed at self-preservation, Hindu resurgence manifested itself in various aspects of life, literature, art, philosophy and architecture. Rajasthan and south India were the bastions of Hindu resistance against the Islamic foreign encroachments. Its final outcome was the evolution of an Indian civilization in which Muslims became a major element in Indian national life, although the Hindus remained a dominant constituent in a plural society.

Although this period was not, by any reckoning, the most glorious of Islamic chapters in India's artistic, philosophic and linguistic traditions, it clearly laid down the foundations for the development of a Hindu-Muslim cultural synthesis. The spread of the Persian language, the evolution of Urdu as a court language, the growth of the *Sufi* religious belief through the Hindu-Muslim interaction, and the establishment of new architectural forms developed during the period substantially enriched the evolving Indian civilization.

**The Mughal Empire**

The establishment of the Mughal rule in north India in 1526 A.D. opened a new and long chapter in Indian history. Originally natives of Central Asia, the Mughals established a vast empire encompassing not only the northern part of India, but during its greatest extension almost all of southern India as well. They adopted, in general, less restrictive religious policies than the the Turks. Akbar (c. 1556-1605 A.D.), perhaps the greatest of the Mughal rulers, vigorously pursued tolerant policies conducive to greater harmony between the various communities. His policies were responsible for achieving greater political unity, social progress and economic development in the country. India, once again, became one of the richest and most powerful countries in the world with which Persia, Turkey and several European nations sought friendly relations and commercial links.

Akbar expanded the empire by military annexations as well as by matrimonial alliances with the princely households of the Rajputs. The empire was sub-divided into provinces (*suba*), several districts (*sarkar*), and sub-districts (*pargana* or *mahal*). The *subas* were run by efficient, centrally-appointed governors (*subadars*). The army was well-trained and organized. The Mughal imperial system and bureaucracy of courts, revenue department and police system became the forerunners for the British administrative machinery. The present-day structure of the Indian and Pakistani administrations is basically patterned after the Mughal arrangement.

The Mughal rulers were patrons of art, literature and architecture. They introduced a new court language, Urdu, which arose out of an interaction of Hindi (the existing language) and Persian brought by the Mughals and Turks into India. The graceful "Mughal" style of architecture flourished under their active patronage. They constructed hundreds of beautiful monuments, forts, tombs, mosques, gardens and public buildings reflecting this form of architecture. The Shalimar Gardens of Srinagar and Lahore, the famous Taj Mahal mausoleum and the mosques at Agra, and the abandoned city of Fatehpur Sikri are eloquent examples of Mughal patronage of art.

Aruangzeb (1657-1707 A.D.), the great grandson of Akbar, was able to extend the Mughal empire to its greatest extent, covering most of the subcontinent. His fanatically intolerant religious policies toward non-Muslims, however, sowed the seeds of rebellion in several parts of the country; among the Rajputs in Rajasthan, the Jats near Delhi, the Sikhs in the Punjab and the Marathas in western Deccan. Soon regional chiefs began asserting themselves in anti-Mughal revolts. Even during his lifetime his vast empire began to crumble. A large Maratha kingdom, which was established in the northwest Deccan, became a source of great irritation to Aurangzeb. This Maratha empire (later confederacy) persisted for over 150 years and finally succumbed to the onslaughts of the Muslim states of Deccan (chiefly Hyderabad) and the penetration of the British East India Company. By 1817 A.D., after the Third Maratha War, the British had extingusilied the Maratha Confederacy.

## Establishment of the British Rule

The Europeans arrived in South Asia in search of profitable trade in spices and textiles. They possessed superior maritime techniques which were helpful to them in this quest. Their commercial activity started in the 16th century. The Dutch, the Portuguese, the French and the English competed for power and economic gains in the subcontinent. After, passing through stages of intrigue and rivalry with the other European companies, the British East India Company had emerged victorious as the single most important force in the country by 1750. The Company successfully interceded in the internal affairs of the weak rulers of Bengal and swiftly gained a stronghold in the lower Ganga valley as well as the strategic ports of Bombay (now Mumbai) and Madras (now Chennai) along the coast. The extension of British rule in north India was facilitated by the political vacuum and civil strife in the area following the collapse of the Mughal empire.

Between 1750 and 1803, the British consolidated power in the Indo-Gangetic plains and the coastal areas, and was eventually extended to the entire subcontinent by 1849. Although the country had fallen to British control, within the British-occupied territory lay 650 units which were internally administered by the native princes. By the treaties of subsidiary alliance their rulers owed allegiance to the British. The British not only controlled the external affairs of these states, but also their internal security. Furthermore, they maintained a tight surveillance over the states through residents representing the Crown, who often meddled in the states' succession of disputes and internal affairs. By the establishments of military stations known as the cantonments and by the institution of grants and favors to the rulers, the Crown virtually controlled the princely states and in time of grave political disorder could even take over their internal administration.

By 1914, the British control had achieved its maximum extent in space as it had also crystallized firmly in policy and practice. Historians are often puzzled at the question of the imposition of a completely alien control over a large area containing a rich and unique civilization. Prior to the British control, the conquering rulers had become Indianized and merged into the native society. But the new masters remained socially aloof. In fact, they were able to take advantage of the divisiveness of the society, rent by religious, class and caste distinctions. Often one prince was pitched against the other and preferred an unwelcome outsider to defeat by an unwelcome neighbor. There were other factors for the relatively swift British acquisition of supremacy over India. Their technical superiority, military organization, naval supremacy and discipline contrasted with the religious cultural background of the natives with their firm belief in *kismat* ("fate").

The British did not particularly change the prevailing administrative or revenue systems established during the Mughal period, but gave these a new direction and set a more paternalistic tone. The administration at the sub-district and village level went on as before. At the district level, the main administrative functions were controlled by Englishmen; at the lower level these continued to be in the hands of the Indians. The military and police departments were, however, tightly controlled by the British.

Among the notable socio-economic changes introduced by the East India Company were the new laws of taxing, buying, selling and inheriting property, particularly in respect to agricultural lands. This eventually widened significantly the existing class distinctions in the Indian society. The commercial classes were the prime beneficiaries. A large class of land proprietors grew up who were initially intermediaries between the landed peasantry and the government and acted as revenue collectors. New laws enabled them to confiscate lands from delinquent payers.

A second and perhaps a more far-reaching change in the economic sphere was the introduction of new mechanized industries. A direct result of this was the decline of ancient crafts, especially of cotton goods. The British also established a new legal system together with a vast network of courts which were in part based on the Mughal and traditional systems, but proved more efficient. The new civil service and the organization of the army were patterned somewhat on the European model. On the social front, only a few attempts were made to intrude into the existing mores of the society. Notable among the new ordinances were the abolition of the practice of *sati*, the burning of Hindu widows on the funeral pyres of their husbands, and the ritual human sacrifice at certain religious ceremonies.

Even more far-reaching than the welfare programs was the introduction of English as a medium of transaction of government affairs at the national and provincial levels. This spurred the growth of English-speaking schools. Indirectly it helped diffuse western ideas regarding arts, sciences and government through books written in English. The English-medium schools churned out a vast army of clerks, accountants, revenue officers and lawyers for the ponderous bureaucracy of the administration. The cumulative effects of these changes were the growth of landed aristocracy, English education and the growth of military and civil social clubs. All these created even deeper rifts among the already caste-ridden society. Slowly but surely, there emerged a small, but powerful, western-educated elite among the Indians. Trained in western education, the children of this elite group, however, slowly began clamoring for the greater participation of Indians in the administration of the country.

The diffusion of the western education system and political ideology was also achieved by Christian missionary organizations through their vast network of church-affiliated schools. The Christian missionary activity was primarily aimed at conversions among the native population.

## Consolidation of the British Empire

The first jolt to British rule in India was experienced in 1857 in the form of a military revolt which became a turning point in the administrative functioning of the period. Soon, the military revolt turned into a more popular rebellion. The dispossessed princes of the Ganga plain and adjoining areas rallied at the back of the mutineering soldiers. The landowning class in northern India were already seething with hostility as a protest against the land revenue settlement. The western innovations had militated against the socio-religious customs of the society. Missionary activity was deeply and widely resented. After the revolting armies were subdued, the British Parliament transferred power from the East India Company to the Crown and initiated a new social and constitutional policy aimed in particular at a more efficient government management of executive, legislative, and judicial affairs. Gradually, following the passing of the administration directly into the hands of the British government, the central administration became an increasingly centralized bureaucracy, run by a tight, efficient but complex hierarchy of the Indian Civil Service, which was manned exclusively at first and later predominantly by the English.

These changes led to radical constitutional readjustments. Strict and uniform laws had to be enforced in order to rule effectively a large colony. The existing village, local and regional economic structure based on handicrafts and subsistence economy slowly crumbled as new mechanized industries were introduced, which in turn gave rise to new market centers. The development of these market centers and the effective control of the country required a large and efficient professional and clerical class.

By 1914, the British imperial power, the *Raj*, had become the largest stretch of British possessions in the world, and India a source of large revenues, derived mainly from the vast agricultural production, expanding trade and industrial development. About half of the enormous land revenues were spent on supporting a huge standing army. Taxes on opium and salt also yielded large revenues. Personal income tax, introduced around 1880, became a contributor to the official revenues.

The country was divided into provinces (presidencies in the case of Bombay, Bengal and Madras) which were administered by governors appointed by the Viceroy. The governors were political appointees and were directly responsible to the Secretary of State in England.

The crowning glory of British rule was the construction of a vast network of railroads across the subcontinent. The railroads accelerated the pace of raw material extraction from the rich agricultural countryside, and speeded up the transition from subsistence foodcrop production to a commercial agricultural economy, thus contributing substantially to the economic development of the country. This efficient system of transportation dealt a death-blow to indigenous handicraft industries. Trains brought vast quantities of cheap manufactured goods from England to the inland towns for distribution to the villages. Millions of craftsmen were thus forced to return to the soil for a living. Village after village, known for handicrafts, lost its traditional markets. Railroads also provided the military with swift means of movement in case of revolt or emergency.

There were beneficial results too. The government acquired greater access to the countryside to transport food-grains for relief in case of drought or famine. Furthermore, rich coalfields of Bihar and the Bengal were developed during the early twentieth century. The iron and steel company in Bihar and textile industries in Mumbai (then Bombay) and Calcutta were also established. Highways, steamboats, banks, insurance agencies, printing presses, modern stock exchanges and telegraph connections all accelerated the process of transformation of the economy and modernization of the production and exchange system.

### The Independence Movement

Meanwhile, an elite class of English-educated Indians was gradually emerging which increasingly competed with Englishmen for employment in the Indian Civil Service, legal services and education. The universities of Bombay, Calcutta and Madras were established in 1857. Many Indians, including some who broke away from the Indian Civil Service, later on became leaders of the nationalist movement in the country. The Indian National Congress founded in 1885, became the rallying ground for many nationalistic-minded youths. In succeeding years, the Congress emerged as the principal political party which started a long and strenuous struggle for greater participation of Indians in the administration. Later on it changed its objectives to obtaining complete independence of the country from British control.

As the independence movement gathered momentum during the early twentieth century, nationalists in the Congress eventually split into Hindu and Muslim factions. The initial British response to the nationalist's demand was a stiff attitude and several nationalist leaders, including Mahatma Gandhi and Jawaharlal Nehru were imprisoned for long terms. At times the nationalist movement was somewhat weakened by centrifugal forces within the Congress party, as well as due to the emergence of a new party, the Muslim League, founded in 1904. With the emergence of the Muslim League, the Congress, which had all along mobilized a broadly based nationalist support, felt hampered in dealing with the British by itself. As talks on independence advanced during the 1940s, the Muslim League leadership, spearheaded by Mohammed Ali Jinnah, opted for a separate independent "homeland" for the Muslims, to be named

Pakistan and to be carved out of the subcontinent. The goal of a united political entity for the country became unfeasible and the Indian National Congress agreed to the partition of the country in order to achieve independence from the British.

The two independent nations were finally created out of the subcontinent in 1947, principally within the geographical framework of the concentrations of Muslims and non-Muslims in the subcontinent. The initial years of independence were full of turmoil for both countries. Even before the partitioning of the subcontinent, Hindu-Muslim riots had erupted in areas which underwent division between the countries, notably the Punjab in the west and Bengal in the east. Over 12 million persons, both Hindus and Muslims, left their ancestral homes and migrated to areas of their religious affiliations across the newly-created international borders. Homes and business establishments were abandoned in the flight for life. The new international borders showed little regard for the traditional established networks of railroads, irrigation canals and manufacturing areas. This division posed severe problems of inequitable distribution of human and natural resources between the two countries. Refugee rehabilitation for millions was, however, the immediate problem before any programs of economic development and administrative establishment could be instituted. Fortunately, well-trained Indian and Pakistani officers quickly brought civil order in the two new countries in the aftermath of partition.

## The New India

The partitioning of the subcontinent signaled the end of a long chapter of foreign domination and the beginning of a new era in Indian history. The new nation inherited from the British a mixed bag of assets and liabilities. The struggle for independence was long, and the freedom gained was marred by bloody communal clashes between Hindus and Muslims (and also between Sikhs and Muslims). Age-old rights within communities had further widened during the British rule due, in some measure, to government's policy of preferential treatment to the selected communities. The eastern and western borders faced unfriendly neighbors. The northern borders which had never been clearly defined or demarcated on the ground were left with an uncertain, and as the Indo-China war of 1962 proved, a violent future. Over 650 princely states, anachronistic pieces of the past, split the country into a patchwork of irregularly shaped territories of varying sizes, degrees of administrative control and economic development, inimical to new India's territorial integrity. Economy, disabled by the mass migrations of millions of refugees across the international borders, was in disarray.

On the assets side, the British left behind a cadre of well-trained and efficient civil and military personnel, well-entrenched judicial and legislative systems, a large network of railroads, roadways and irrigation canals. These measures together with the introduction of English as an official language in schools, courts, and civil and military services, calculated undoubtedly to effectively control and administer the country, had brought India closer to the western and the modern scientific world. Large investments in public works, and modern communications had linked the country more closely to the British industrial requirements of Lancashire. Although India contained a modest to fair amount of mineral, agricultural and infrastructural potentials, the British did not make any effort for the diffusion of large-scale industrial growth or the export of the commodities other than tea, cotton, indigo and opium prior to the nineteenth century.

As India stepped into a new era of independence, it faced enormous challenges of reconstruction and development, political integration and social harmony. Independence clearly brought into focus new aspirations and hopes. The founding fathers quickly devised for free India's government a constitution which is federal, republican and parliamentary and came into operation on January 26, 1950. Although the federal experiment has not always functioned as the framers of the constitution intended, over the years the federal government steadily gained power at the expense of the states in dealing with the economic, political and social unrest within the country. Indira Gandhi, who served as a prime minister from 1966 to 1977 and from 1980 until her assassination on October 31, 1984, behaved in a manner that most observers considered authoritarian. The democratic foundations of the country, however, remained unshaken despite the continuous pressures from the extremist elements like the Sikhs and the Nagas. Elections were, by and large, fair and free, particularly before the 1980s, though they were occasionally marred by violence. The country did not witness any military takeover of its administration unlike several of its neighbors and other developing nations.

Among the many problems at the transfer of power from the British in 1947 was the status of over 650 princely states. After the passing of the States' Reorganization Act of 1956, a complete revamping of the country's map, based primarily but not entirely, on the linguistic distribution, was undertaken. While all the states were supposed to contain the population speaking the same language, few states like Bombay and Punjab still remained bi- or multi-linguistic. Following intense ethnic agitations in these states, a Gujarati-speaking state was created in 1960 and in 1966 the central government divided Punjab into three states, one of which contained Punjabi-speaking majorities, ostensibly to satisfy the Sikh demands for a Punjabi *suba* (state) (the other two are the Hindi-speaking areas). Since 1966 several more states were created to placate ethnic or cultural minorities, particularly along India's northeastern borders.

The Sikh community was not entirely satisfied with the creation of a Punjabi state as it left many of its members in the newly-created state of Haryana. Their agitation was increasingly hardening into a demand for an independent "homeland", named *Khalistan* (land of the "pure"). The radical elements turned to violence, bombing and shooting Indian officials, and Hindus. Determined to end these Sikh extremists' violence, India's Prime Minister, Indira Gandhi sent troops into the Golden Temple in June 1984, where most of them had taken sanctuary. Hundreds were killed in that raid. In an act of apparent vengeance, Indira Gandhi was assassinated by two of her Sikh bodyguards in October 1984. This infuriated the Hindus and as an act of reprisal, violent mobs of Hindus went on orgy of looting and killing hundreds of Sikhs in the capital. The demand for an independent *Khalistan* continued among the radical Sikhs.

Besides Sikhs, other communities also demanded homelands, and resorted to violence. Tribal groups, on India's northeastern border, among them the Nagas in particular, started agitation for an autonomous homeland. They frequently clashed with the Indian forces but after the creation of the state of Nagaland in 1963 their demand for a separate country for the Nagas lost much of its momentum.

The partitioning of the subcontinent had created an unfriendly border with Pakistan, and an uncertain frontier with China. From the beginning Indo-Pakistani relations became sore over the the two particularly thorny issues: the status of the princely state of Jammu & Kashmir, and the distribution of the Indus waters. While the Indus water dispute was resolved peaceably by an Indo-Pakistani agreement in 1960, the Kashmir issue led to armed conflicts in 1948, 1965 and 1971. In 1971, when the Pakistani army attacked India on its western borders, the Indian forces moved into East Pakistan and drove out Pakistani forces declaring the newly-liberated area to become a sovereign nation of Bangladesh. Pakistan has, in fact, made the Kashmir issue intractable. The situation grew increasingly dangerous as the demand for its secession from India gained momentum in the late 1980s and early 1990s sparking off violent protests within the state. Elsewhere, along India's northern borders territorial disputes between China and India led to a military confrontation in 1962.

On the social front the most intractable problem has been a large, and rapidly growing population which swelled from 350 million in 1947 to over 860 million in 1991. Huge increases year after year have clearly been a source of choking off nation's economic development. In 1947, a little over 80 per cent of the population lived in rural areas: three-fourths of which were illiterates, and over one-half lived in poverty. Food shortages were endemic between 1950 and 1970, and foodgrain imports ranged between 10 and 20 million tons during the critical years of poor harvests.

In 1951, India launched its first of a series of Five Year Plans which aimed at the alleviation of poverty, attaining food self-sufficiency, curbing its rapidly increasing population, and achieving a self-sustained economy. The Eighth Five Year Plan started in 1991. Although the nation has not always reached the targets, the balance-sheet is not entirely dismal. Since 1947 India has more than doubled its foodgrain production, making it nearly self-sufficient in its requirements. The economy is growing at over 5 per cent per year. Literacy levels have considerably gone up. India now manufactures automobiles, ships, electronic goods; has its own airline, and new supplies of energy including offshore oil wells and atomic plants. Its defense forces are among the largest and well-equipped in the world.

Overall, the Indian democratic structures have withstood the internal political stresses and external pressures that have constantly threatened its apparently fragile unity. The politicians at the time of

**Figure 3.2**
*Pan-Indian Empires*

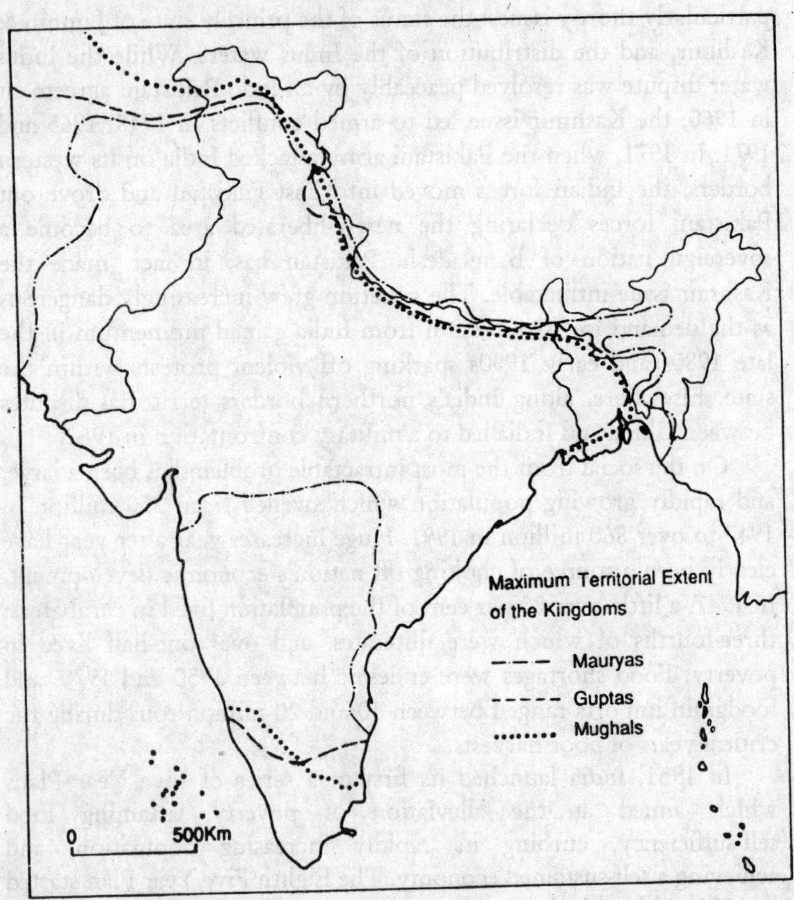


Independence had raised politics to a high level of moral idealism in order to achieve goals of national and public interests. But after Nehru's death, the political system in general got fragmented and regionalism, localism and sectional politics came to dominate the democratic political system of India with the result that out of 12 Lok Sabhas constituted between 1952-99, only seven could complete their five-year term. And since 1996, four governments have been ousted from power indicating that even coalition ministries are of no good to the federal polity.

**Political and Cultural Core Areas**

It can be clearly discerned from the preceding overview of the history of Indian civilization that a recurrent pattern of territorial instability of political bases and development of sub-regionalism in the subcontinent has been a major theme of Indian historical geography. The cherished goal of establishing a pan-Indian empire was accomplished on a few times, under Ashoka, Aurangzeb, and more recently during British rule (Figure 3.2). Even during these exceptional periods of vast empires, the southern tip of the country escaped control of the central authority with the exception of the British rule.

No less striking is the theme of the broad cultural division of the subcontinent into the Aryan North and Dravidian South. Hinduism, however, transcended the linguistic and ethnic regionalism dividing the northern and western plains of the subcontinent from its non-Aryan component, permeated into, and made an enduring stronghold in Dravidian India. Recurrent political instability and territorial fragmentation of the political units, combined with the Aryan-Dravidian dichotomy, produced cultural sub-regionalism or core areas based on local political control over ethnic and linguistic sub-groups (Figure 3.3). Even during the British rule, direct administrative control extended over three-fourths of the country, the remainder one-fourth of territory was composed of princely kingdoms of varying sizes.

The various core areas thus developed in rich agricultural lands and were, in general, isolated from each other by physical barriers like mountains, deserts or forests. Insulated from the mainstream of history for centuries, areas of refuge and isolation also developed regional cultures, as in the Kashmir valley and the Assam valley. Several smaller refuge areas are still occupied by tribal groups (Bhils, Gonds, Santhals, etc.) often speaking languages unrelated to the Indo-European or Dravidian families, and still retaining distinctive archaic cultures. Such areas are found in the jungles of Gondwana, the Chota Nagpur hills between Bihar and Bengal, the mountains surrounding the Assam valley, and in the Cardamon hills in southern India, and many hilly locales in central India.

The most persistent external boundary of the country has been

**Figure 3.3**

*Historical Boundaries and Nuclear Regions*

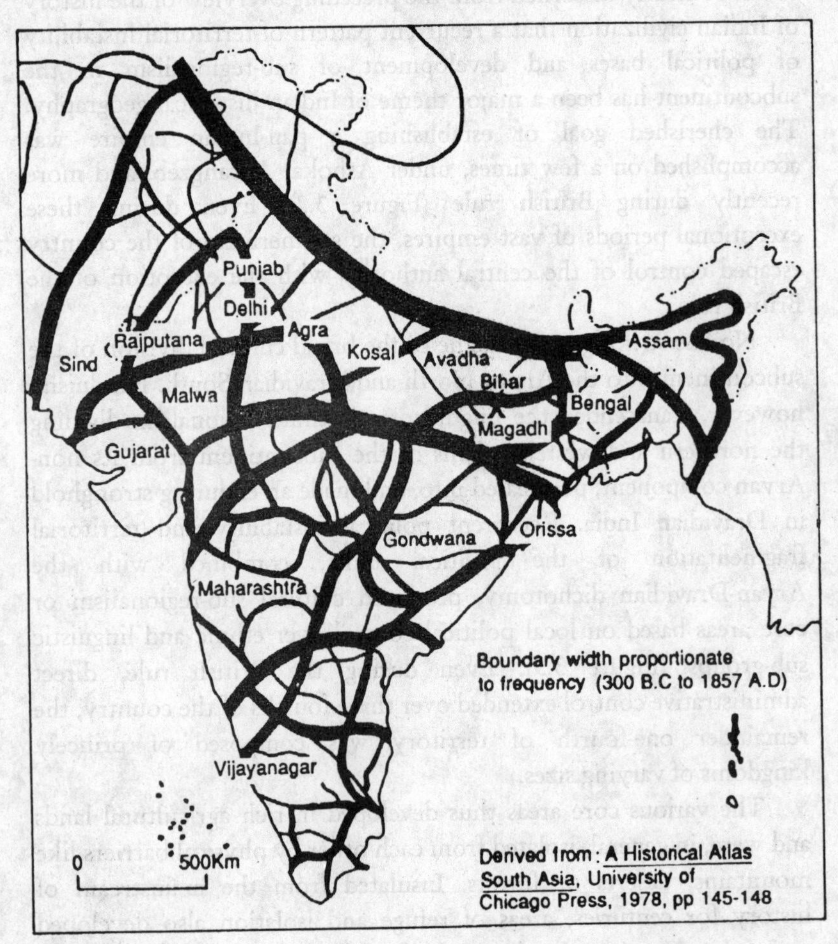


one to the north and west of the Indo-Ganga plains, historically separating these plains from Nepal, Tibet and Afghanistan. In the northeast, Assam has also remained for long periods outside India's central authority. The most stable of the internal political boundaries within the subcontinent has been along the Narmada-Son Mikir line, separating the Indo-Aryan Gangetic plains from the peninsula (with the exception of Maharashtra) essentially Dravidian India. Within this broad Indo-Aryan Dravidian supra-regional framework, either core (nuclear) regions crystallized which remained relatively stable, and submerged temporarily and recurrently under national (pan-Indian) authority. Only in modern times did the British superimpose their authority over the entire subcontinent.

### *References*

Basham, A.L., *The Wonder That Was India*. New York, 1954.
Chatterjee, A.C., *A Short History of India*. New York, 1957.
Dodwell, Ed., *Cambridge Shorter History of India*. Cambridge, 1934.
Dutt, Ashok K., *Atlas of South Asia*. Boulder, Colo., 1987.
Irfan, Habib, *An Atlas of the Mughal Empire*. Delhi, 1982.
Kumar, Ravinder, *The Social History of Modern India*. Delhi, 1983.
Rawlinson, H.G., *India: A Short Cultural History*. London, 1952.
Singhal, D.P., *India and World Civilization*. Vol. 2, London, 1972.
Spear, P., *A History of India*. vol. 2, Harmondsworth, 1966.
Schwartzberg, J.E., *A Historical Atlas of South Asia*. Chicago, 1978.
Thapar, R., *A History of India*. Vol. 1, Harmondsworth, 1966.
Watson, F., *A History of India*. New York, 1975.
Wolpert, S., *A New History of India*. 4th ed., New York, 1993.
Wolpert, S., *India*. Berkeley, 1991.

# II

# Society : Geographical Basis

# 4

# Cultural Patterns

India's cultural landscape can be likened to a kaleidoscope revealing the interplay of several religions, sects, creeds, languages, races and ethnic groups. The diversity and complexity of these culture groups remain virtually unmatched by any other major country. Cultural impact from a variety of sources—Hindu, Buddhist, Greek, Persian, Islamic—and British have left their distinctive imprints and transformed the cultural landscape.

## Spatial Patterns of Religions

Of the diverse cultural forces affecting the cultural and political life of most Indians, religion is unquestionably the most dominant. It permeates virtually all aspects of their daily personal and family life.

Geographically, the religious distributions may be studied in several spatial contexts. At the level of the village, religion is a pervasive force. Religious beliefs sanction caste restrictions on types of labor, and prejudice against certain cropping systems and veneration of cows have inhibited technological and agricultural progress. Religious groups and caste divisions have often generated political disharmony in villages. Social stratification, based on religion and caste, has been manifested in residential segregation and has restricted cooperative and collective village efforts. In urban centers which are well-linked among themselves but not well-linked with a vast rural society, there is a greater diversity of religious adherence which has compounded the spatial interactions developing among the various communities. The urban world affords greater opportunities for the various groups to solidify their political forces into political parties. Sharp cleavages along social, linguistic and religious lines fuel inter-community riots, occasionally some of which are masterminded by religion-based political parties. Indian history, during the British occupancy as well as contemporaneously, is replete with examples of Hindu-Muslim riots, which were actively supported by political groups.

Another spatial level of geographic enquiry consists of groups of districts (administrative divisions) containing a religious community with a numerical majority. Such compact areas tend to exercise their political influence on state and union administrations. During the closing days of the British rule agitation for a separate homeland by Muslims based on their numerical distributions in the western section of the country and the Ganga-Brahmaputra delta region was motivated by religious considerations. Their demand proved successful in 1947 when an independent country of Pakistan was formed by the union of territories in which Muslims had a majority status. In 1966, the political demand for a "linguistic homeland" (Punjabi-speaking state) within India by Sikhs was fulfilled when, within the newly created state of Punjabi speech, they attained a majority status.

## Religious Communities

Table 4.1 shows the distribution of religious communities in 1991. Two major communities, Hindus and Muslims, together comprised nearly 94 per cent of India's population in 1991. Hindus numbered 687.6 million or 82 per cent of the total population, and Muslims formed 12.1 per cent of the population. Before partition, the Muslim population was 92 million, accounting for 24 per cent of the country's population. In 1947, they numbered 36 million, and the Muslim community has now grown to over 101.6 million, making India as one of the major Muslim countries in the world. In addition, there are substantial numbers of Sikhs, Christians, Buddhists and Jains as well as the Zoroastrians and groups belonging to tribal faiths.

**Table 4.1**

*Major Religious Communities, 1991*

| | *Population (in millions)* | *Per cent of Total Population* |
|---|---|---|
| Hindus | 687.6 | 82.0 |
| Muslims | 101.6 | 12.1 |
| Christians | 19.6 | 2.3 |
| Sikhs | 16.4 | 1.9 |
| Buddhists | 6.4 | 0.8 |
| Jains | 3.4 | 0.4 |
| Scheduled Castes | 139.0 | 16.5 |
| Scheduled Tribes | 67.7 | 8 |
| Others | 3.2 | 0.4 |

*Source:* *India: A Reference Annual, 1997*, New Delhi.

Hindus form majorities everywhere except in a few districts (Figures 4.1 and 4.2), where Christians or Sikhs predominate. In fact, only in few areas do they form less than 84 per cent of the population: (a) along the borders of Bangladesh and Pakistan; (b) the state of Jammu & Kashmir where Muslims account for 65 per cent of the population; (c) areas of Christianity and tribal religions in Assam and Nagaland along the Tibet and Burmese borders; (d) Kerala state; and (e) scattered areas in Deccan and the Ganga plains where large scale conversions of natives to Islam took place between the twelfth and eighteenth centuries. The spatial near ubiquity and predominance of Hindus has had a significant impact on the cultural and political geography of India. Other communities are, in general, dispersed among Hindu majorities, and have been Hinduized in varying degrees. Many caste practices, characteristic of Hindu society, have been retained by non-Hindus, most of whom are the descendants of converts from Hinduism. Hindus registered a decline in per cent of the total population, from 75.1 per cent in 1881 to 74.2 per cent in 1901, to about 66 per cent in 1941. After partition in 1947 (with most Muslim majority areas gone to Pakistan and the massive migration of Hindu refugees into India), the proportion of Hindus increased to 84.1 per cent of the population. By 1981 it had dropped again by 1.4 points. This declining trend is attributed to a lower fertility among

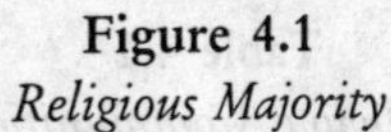
Figure 4.1
*Religious Majority*

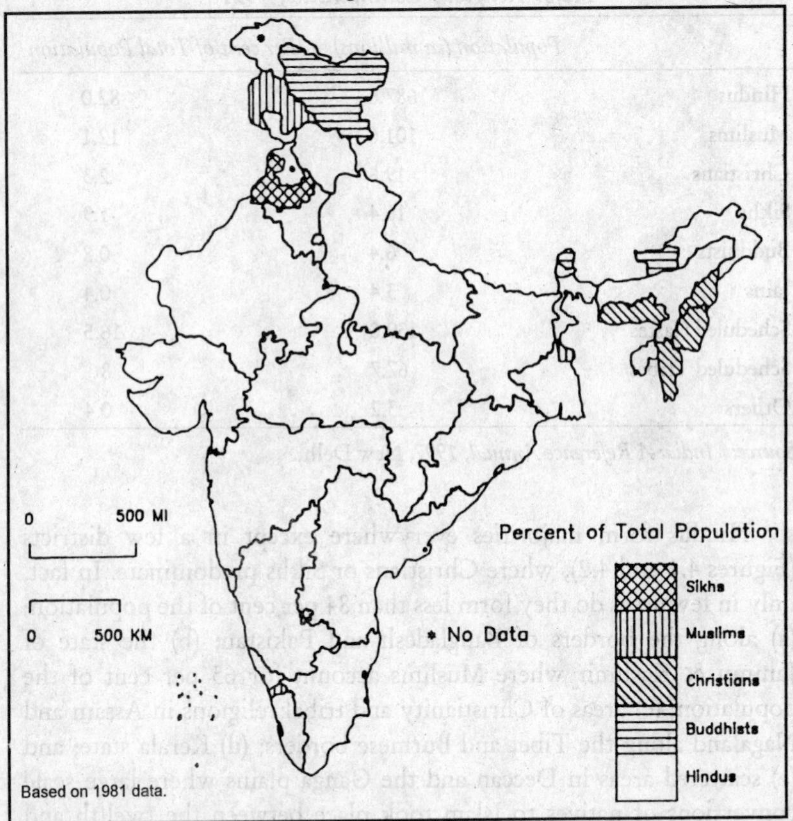


Hindus and loss through conversions to other religions. In the 1950s significant numbers of untouchables were converted to Buddhism. Nearly 140 million Hindus (16.5 per cent of the country's population in 1991) were in the category of the scheduled castes or untouchables, now known as the *Dalits* (literally the depressed castes).

Except in Jammu & Kashmir, and one district each in Kerala and West Bengal, Muslims are nowhere in a majority status. Along the Bangladesh border in West Bengal, 20 to 50 per cent of the population in several districts is Muslim. Muslims are also concentrated in and around Delhi, and in 14 districts of Uttar Pradesh. Delhi was the seat of Muslim rule between the twelfth and nineteenth centuries and Uttar Pradesh, was for centuries, the hub of Muslim rule. The share of

**Figure 4.2**
*Distribution of Hindus*

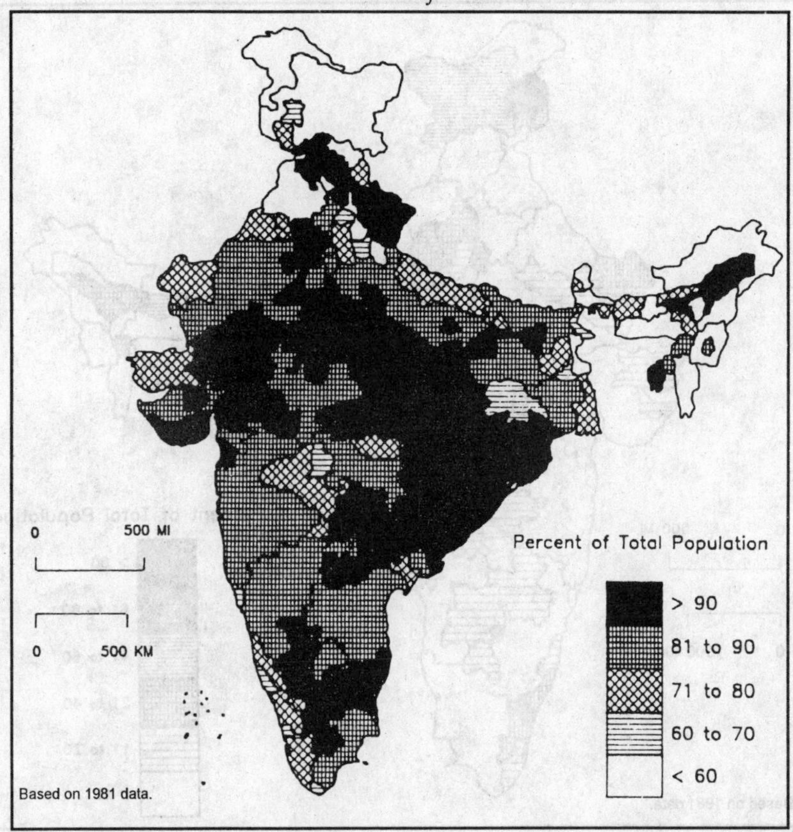


Muslims in India's total population increased steadily, from 20 per cent in 1881 to 24.2 per cent in 1941. Since Independence they have formed 10 to 11 per cent (11.67 in 1991) of the population. The higher fertility rate among Muslims results from their religious attitudes, especially from a greater tolerance of widow remarriage. Muslims have been less literate and less urbanized as a group than Hindus. They had fallen into a position of inferiority to Hindus in government services during the British rule. Their demand for a separate homeland derived partly from the prospect of advancing social and economic power for the Muslim majority within a self-governing territory. Figure 4.3 gives the geographical distribution of Muslims.

**Figure 4.3**
*Distribution of Muslims*

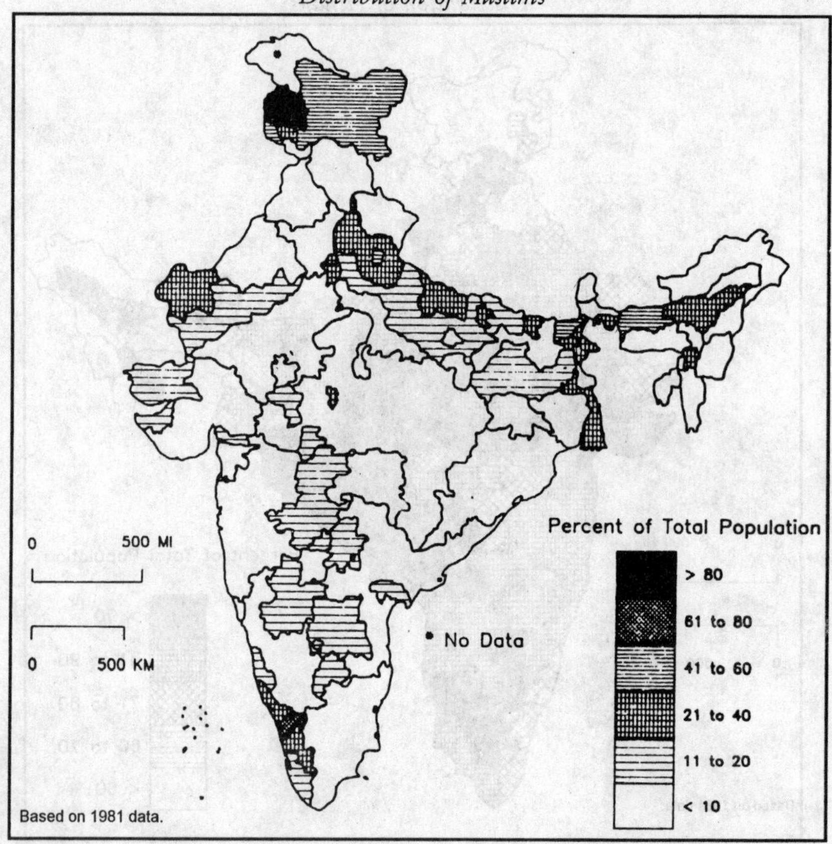


The Christian population (19.6 million or one-fifth of India's total population) is dispersed mostly in the rural areas. Its main concentration lies along the western coast around Goa, and in Kerala where Christians represent 20-24 per cent of the population. These areas have had historical contacts with Christianity (Figure 4.4). According to a legend such contacts in Kerala were established as early as the first century, although regular contacts were developed after Vasco de Gama's arrival in 1498. Most Christian missionary activity in India was, however, intensified as the British firmly gained control in the eighteenth century. The Christian population is relatively numerous in the tribal hill areas of Assam, Meghalaya, Mizoram,

**Figure 4.4**
*Distribution of Christians*

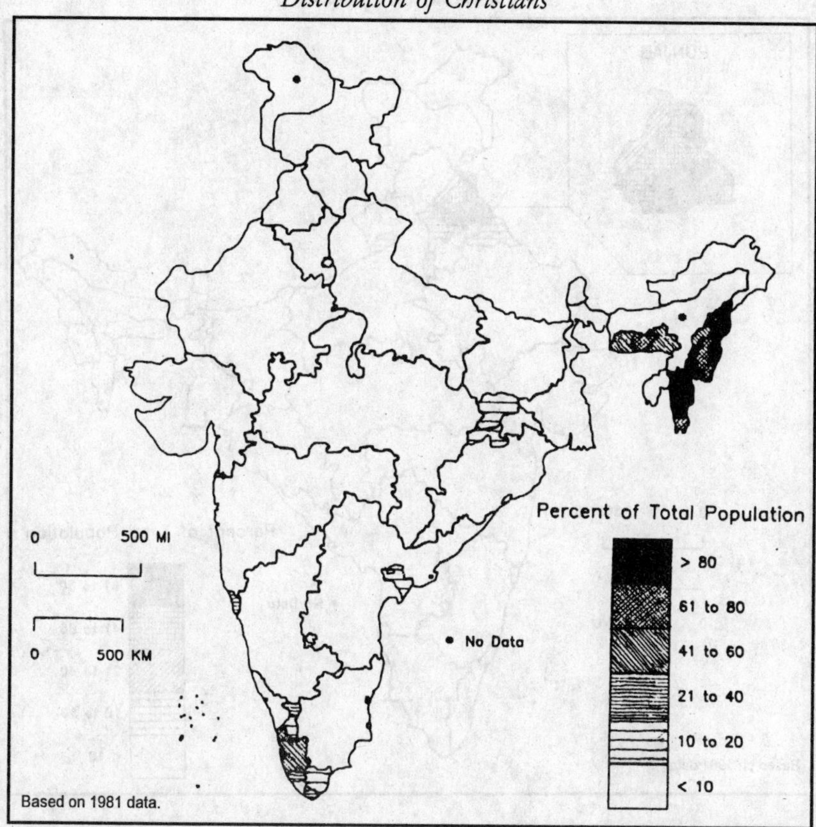


Assam excluded.

Nagaland and Chota Nagpur where Christian missionaries have been active, many tribals were converted to Christianity during the British rule. A demand for a cultural and political homeland for the tribal population was fulfilled when the state of Nagaland was formed in 1963 in which Christian groups have a majority status. Christians gained in their share of the country's population, from 0.71 per cent in 1881 to 2.6 per cent in 1971 but declined to 2.4 per cent in 1981. Among the Christian population are the Anglo-Indians—off-springs of mixed European and Indian parents. Anglo-Indians adopted a European lifestyle and were preferred in such public services as the railways under the British rule.

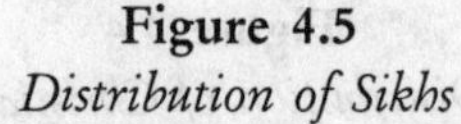

**Figure 4.5**
*Distribution of Sikhs*

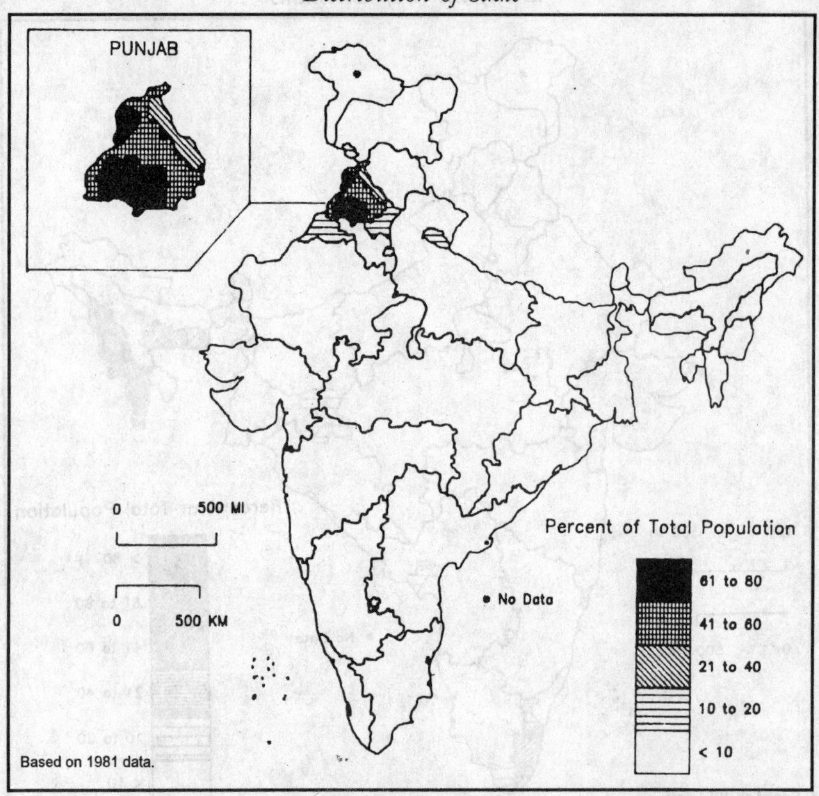


Sikhs, numbering 16.4 million in 1991, are mostly concentrated in a small section of the country in the state of Punjab—their original homeland. Nearly 90 per cent of all the Sikh population lives in a critical area bordering Pakistan (Punjab, Haryana), and in metropolitan Delhi (Figure 4.5). An offshoot of Hinduism, the Sikh religion retains many Hindu practices, and many Sikhs intermarry with Hindus. In the seventeenth century, Sikhs were persecuted by Muslim rulers, and assumed a militant posture against Muslims. The partitions of India in 1947 cut directly through their territory, and about 1.5 million Sikhs who found themselves in Pakistani territory moved to India mainly to the states of Punjab and Delhi. Since then, Sikhs have been politically the most active religious group, and successfully agitated in favor of a Sikh-dominated state, which was created in 1966. Sikhs have experienced a remarkable growth in numbers since 1881 (from 1.8 million to about 13.1 million in 1981), and improved their share in the country's population from 0.75 per cent in 1881 (in pre-partitioned India) to 1.9 per cent in 1981. They registered a growth rate of 33 per cent between 1961-71 and a little over 26 per cent during 1971-81. Their growth rates are among the highest in the country. They are also more literate than either Hindus or Muslims. Sikhs account for 60-65 per cent of the population in most districts of Punjab and form minorities in parts of Haryana, Himachal Pradesh, Jammu, Rajasthan and western Uttar Pradesh.

Jains and Buddhists are two older communities that originated in the early sixth century B.C. as a reaction against the caste system and ceremonial ritualism of Hinduism. However, their religious philosophies retain the basic Hindu beliefs of *karma*, transmigration of soul and *ahimsa* (non-violence). Jains (population a little over 3.4 million) are among the highly literate groups, and are mostly concentrated in the urban areas of Rajasthan, Gujarat and Maharashtra. Buddhists number about 5 million, with 90 per cent of the population concentrated in the state of Maharashtra (another major area in Kashmir) where large-scale conversion of low-caste Hindus took place following the example of Ambedkar, their famous leader (Figure 4.6). Parsi population is less than 100,000 concentrated mostly in the city of Mumbai. The most literate and urbanized of all the communities, Parsis have played significant entrepreneurial roles in the business and commerce of Mumbai, and have excelled in numerous professions.

The Indian census of 1991 enumerated tribal population of over 67.8 million, forming nearly 7.8 per cent of the country's population, listed under the comprehensive category of "Scheduled Tribes". In 1991 they numbered 53.8 million or 7.8 per cent of the population. They are the descendants of the aboriginal population and are known as *Adivasis* (literally, "indigenous"). These tribes have long retained their distinctive languages and religious beliefs. These tribes are geographically scattered. The main areas of their concentration (Figure 4.7) are generally hilly, forested and inaccessible: (a) Assam and Arunachal Pradesh, Mizoram, and Manipur along the Tibet and Burmese borders; (b) the central Indian hills and plateaus; (c) northeastern sections of Himachal Pradesh; and (d) to a lesser degree

**Figure 4.6**
*Distribution of Buddhists*

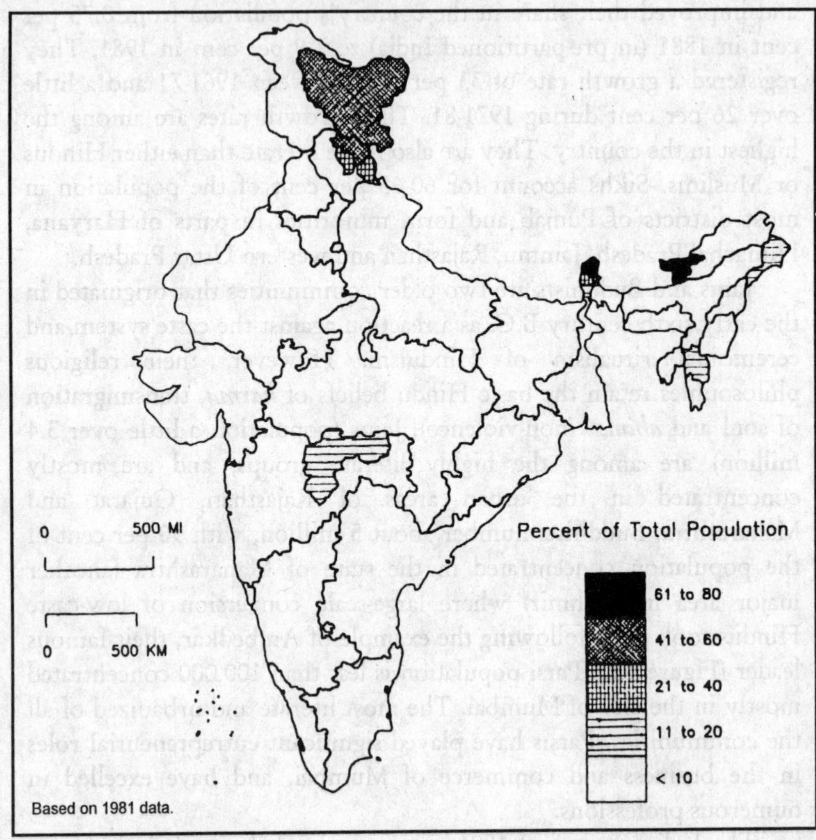


in the hilly regions of Andhra Pradesh and Tamil Nadu. Relative geographic inaccessibility has no doubt imposed on them cultural, environmental and economic isolation from the mainstream of Indian society. Their isolation and comparatively socially autonomous status have posed administrative and economic problems of integration of their territory into a unified national political system. These problems have become particularly critical in areas where they have developed marked cultural and political consciousness. During the British rule, missionary activity among them extended not only literacy but also a sense of tribal nationalism. The Nagas along the Burmese border belong to such a category. They actively agitated for the formation of

**Figure 4.7**
*Distribution of Scheduled Tribes*

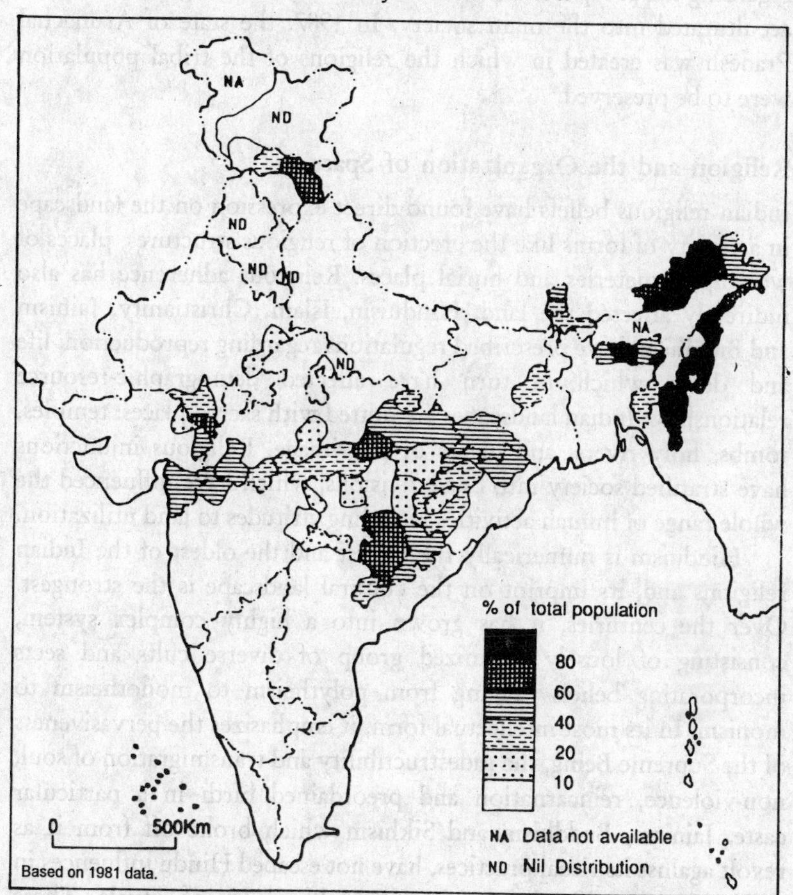


a state in which they could further their cultural and religious interests. Nagaland, a separate state, was created in 1963.

Most other tribal groups, such as the Bhils and Gonds in central India, the Santals in Bengal-Bihar and several minor tribal groups in south India remain largely poor, illiterate, and culturally remote. With increasing contact with the outside world, they are gradually being assimilated into Hindu society. Their lot is roughly analogous to the tribal Indian groups in the United States of America. On the one hand, there is the problem of their integration into the mainstream of society, on the other there is a genuine concern regarding the prospective loss of their identity and religion as they are acculturated into the main society. In 1987, the state of Arunachal Pradesh was created in which the religions of the tribal population were to be preserved.

## Religion and the Organization of Space

Indian religious beliefs have found direct expression on the landscape in a variety of forms like the erection of religious structures, places of worship, cemeteries and burial places. Religious adherence has also indirectly affected the land. Hinduism, Islam, Christianity, Jainism and Buddhism have prescribed regulations regarding reproduction, life and death which in turn have affected demographic-resource relationships. Indian landscapes are dotted with sacred places: temples, tombs, holy rivers, and places of pilgrimage. Religious injunctions have stratified society into caste divisions, which have influenced the whole range of human activities including attitudes to land utilization.

Hinduism is numerically the largest and the oldest of the Indian religions and, its imprint on the cultural landscape is the strongest. Over the centuries, it has grown into a highly complex system, consisting of loosely organized group of diverse cults and sects incorporating beliefs ranging from polytheism to monotheism to monism. In its most intellectual form, it emphasizes the pervasiveness of the Supreme Being, the indestructibility and transmigration of soul, non-violence, reincarnation and preordained birth in a particular caste. Jainism, Buddhism and Sikhism which broke off from it as revolt against its ritual practices, have not escaped Hindu influences in their social stratification, and in the observance of rituals. These religious groups have, however, maintained their distinctive places of worship and their sacred shrines.

Formal expression of Indian religions on the landscape is manifested in their sacred structures, use of cemeteries, and assemblages of plants and animals for religious purposes. Streets, parks, bridges, trees, and rivers are dotted with the statues or abstract figures of gods, which constantly receive the the propitiations of the passers by. Sacred structures are widely and conspicuously distributed all over the country, ranging from minor inconspicuous village shrines (or even a small idol embedded in a roadside wall) to large

Hindu temples, monumental mosques or ornately designed cathedrals in large metropolitan areas. Whether it is a Hindu temple, or a Sikh *gurdwara*, or a Muslim mosque, or a Buddhist monastery, such communal sacred buildings differ in size, form, space use and density, depending on the ideological and organizational requirements of the religious order. All, however, perform a basic religious function.

A distinction between structures, housing a god or those meant for congregation may be made. Hindu temples invariably enshrine a statue or a symbol of a deity, whereas mosques, *gurdwaras* or churches are basically designed for religious congregations. In addition to their basic religious functions, such structures tend to acquire secular attributes as well, and often maintain guest rooms for pilgrims and other visitors. Business and political conventions, folk festivals, and recreational activities are also arranged in areas specifically demarcated for such purposes within these religious structures. Although many of the Hindu holy places established at remote places (Badrinath, Kedarnath), most religious structures are located where adherents can regularly attend the services, their spacing and density parallel that of the hierarchy of settlements. Large Hindu temples are maintained by the various caste groups, wealthy people or charitable endowments. Over the years, a vast amount of wealth has been offered to the temple gods and priest. The total wealth of Hindu temples in the form of jewelry and property alone is estimated to be in billions of dollars. Hindu temples often house a number of gods, since many Hindus are eclectic and propitiate not one, but several gods, believing that all gods are merely manifestations of one Supreme entity.

A typical Hindu temple is distinctive in layout and architectural style. Unlike a mosque or church, it does not necessarily require a large closed interior space for congregation where prayers are held. The statue of the chief deity is usually sheltered in an inner shrine. Surrounding it are the corridors (pavilions) for ritual circulation (*parikrama*) by the public. Elaborate temple gateways direct one's entry into the building. Jain and Buddhist temples also adhere to this basic plan. Sikhs do not enshrine a statue, and Sikh temples tend to adopt the congregational aspect of Islam and Christianity; space is allocated for the purpose. Islam and Christianity emphasize the congregational aspect of prayer. The focus of activity in their places of worship is congregational prayer space for which a hall or a compound is specifically allocated within a church or *masjid* or *jami*. In Islam, community worship is scripturally prescribed and universally practiced. Within the precincts of a mosque, an open space, usually rectangular in shape and enclosed by walls, is maintained.

The relative impress of religious structures on the land depends on the frequency of their distribution, the number of clients of the religious order, the wealth and enthusiasm of the religion's clientele and the ecclesiastic requirements of the various religions. In Goa, Kerala, parts of Assam and large urban centers, Christianity's impress of churches, mission houses and Bible Centers is pronounced. Areas controlled by Muslim rulers between the thirteenth and eighteenth centuries, especially urban centers like Delhi, Agra, Allahabad, Hyderabad and Lucknow, clearly exemplify the strong Islamic impress on the cultural landscape. Mosques like *Jama Masjid* in Delhi and *Shahi Masjid* in Agra (both built by Shahjahan in the seventeenth century) are among the noble examples of monumental structures which dot most large Indian cities.

Religious functions can also be performed outside temples of worship, often in a home or even in an open space. Virtually every household—Hindu, Sikh, or Buddhist—has a family shrine and place reserved for worship. Many Hindus keep household gods, symbolized by statues or paintings, located at a designated place in their homes. Ritual prayers of Muslims (*namaz*) can be said anywhere at appropriate times, even in a moving train.

Cemeteries, burial places, and cremation grounds constitute other direct expressions of religious affiliation on the land. The Hindu, Buddhist, Jain and Sikh tradition of cremation of the dead is very different from the Muslim and Christian institution of burial in a community ground and thus leaves a different cultural imprint on the land. Burial grounds, for example, impose a squeeze on useful land within cities.

Religions occasionally bestow on certain plants or animals a degree of sanctity or ritual function. Such plants and animals are kept in religious structures or dispersed along pilgrims' routes. *Tulsi* and *pipal* trees, though of limited economic or decorative value, are sanctified by Hindus and Buddhists. The turmeric plant's sanctity is recognized because its pigment is used in Hindu rituals. Sandalwood's dye is considered sacred and is used in the ritual marks on the forehead by religious Hindus. Rice is a sacred plant because it is used in most Hindu and Buddhist ceremonies and ritual observances. Coconut and *ghee* (clarified butter) are also used in ritual ceremonies.

Associations between religions and places are developed in many other ways. The distribution of holy places and place names honoring thousands of Hindu gods and Muslim and Christian saints, indicate the direction and history of religious movements. Hundreds of places of pilgrimages express the direct formal expression of religion on the land. The geography of Indian holy places (Hindu, Jain, Buddhist, Sikh, and Muslim) is rendered complex by the large pantheon of Hindu gods, Muslim saints, Jain *tirthankars* (preceptors) and Sikh *gurus*. River sources such as Manasarowar Lake (in Tibet) and Badrinath in the Himalayas, river confluences (Allahabad), physiographic breaks in the course of rivers (Hardwar, Srirangam), mountain peaks (Mt. Abu), lakes (Pushkar) and caves (Amarnath) have all been sanctified by several religions. Large temples have been established in many of these locations, and these, in turn, have favored urbanization. Amritsar located at a spring lake (a holy place of the Sikhs) is another example. Many present-day large urban centers originated with a predominantly religious function as sacred places of pilgrimages like Varanasi, Madurai, Tirupati, Allahabad and Mathura. In addition to their normal sanctified character as pilgrimage sites, many of these attract regular periodic assemblages, e.g., *Kumbha Mela* every 12 years.

Pilgrimage has been an important mechanism of religious circulation (*tirthayatra*) involving million of Hindus, Jains, Buddhists, Muslims, Sikhs, and to a lesser extent Christians each year, a number increased manifold during *Kumbha Mela*. These periodic pilgrimages afford a continuing forum for the exchange of religious ideas, often cutting across linguistic and sometimes religious barriers. Pilgrims also generate cultural exchange, social mixing, and trade and contribute to political integration. Pilgrim circulation, however, can also diffuse epidemic diseases. An informal hierarchy of the distribution of pilgrim centers may be recognizable at the national, regional and local levels, each level maintaining its religious hinterlands (Bhardwaj, 1989: 225-228).

Religions have prescribed or encouraged taboos on work and

food, which may interact with the cultural landscape. General taboos against the killing of animals and meat eating among Hindus, Buddhist and Jains has resulted in a lack of development of the dairy and beef industry. The taboo on beef eating by Hindus, and extension of the idea of *ahimsa* (non-violence), has resulted in the accumulation of an enormous cattle population, a large part of which is aged, diseased and unproductive. It is a pity that very little attention has been paid to efficient stock breeding in India. However, future breeding should be controlled and the stock improved for the extension of a dairy industry within the framework of the ideals of *ahimsa* (Sopher, 1967: 40).

Religious mandates largely regulate human occupations. Hindu society is scripturally stratified into caste categories with associated occupations (first expounded in the Laws of Manu). Basic notions of ritual purity among high-caste Hindus have affected their occupations. Occupations like fishing and leather tanning are downgraded as low-caste occupations. Cultivation is a lowly profession and high castes are reluctant to take up farming, which has traditionally suffered from neglect by castes who possess capital and managerial skills. Non-Hindus also display similar occupational traits. Jainism prohibits any form of agricultural occupations since plowing must destroy some insect life. Jains have, therefore, taken to trading and banking in large numbers, and are concentrated in the cities.

Among Hindus, different castes display different food consumption traits as sanctioned by scriptures. Avoidance of meat eating is very strict among Brahmans, the highest caste. Lower castes, like the untouchables, who are considered ritually "unclean", do not avoid meat eating. Most upper caste Hindus abstain from fish and fishing as this occupation is accorded low social status. The fishing industry, in general, has not prospered in India. The development of high class liquor manufacturing industry has also lagged, as most Indian religions forbid or discourage the consumption of alcohol. Religions also mandate food habits. Muslims are forbidden to eat pork. The Sikh religion proscribes the use of tobacco and thus discourages Sikhs from entering into any tobacco-related business.

The rhythm of farming activities is regulated by a ritual calendar affecting the quality of production. Religious festivals must be attended even if growing crops need attention. Work in the fields is usually suspended at prayer time (devout Muslims observe prayers five times a day).

Religious mandates have indirectly affected population-resource ratios. High and middle caste Hindus and Jains forbid widow remarriage, resulting in lower fertility rates among them as compared to the higher rates for Sikhs and Muslims. The Hindu desire for a child early in life encourages early marriage and high maternity rates among young females. Another scripturally sanctioned desire widely prevalent among Hindus is to have at least one male child (who is responsible for most ritual functions for the parents).

Religious beliefs also affect social and economic behavior. Based on the Hindu-Buddhist-Jain value systems, indifference to worldly gain has, in general, inhibited the formation of capital investment and labor input. In this respect, Sikhs have proved to be more liberal and successful in farming and Parsis in business than have Hindus. The Islamic practice of segregating females deprives them of many economic functions (working in the field or in an office or factory).

Religious sentiments often dominate one's political behavior. The partition of the Indian subcontinent into two countries in 1947 was a successful realization of a goal of the Indian Muslim League, a strong political party inspired by religious motives. At several levels—local, district and state—religion-inspired political parties identified on the basis of their religious affiliation fight elections with the ostensible purpose of winning concessions and privileges for their religious groups.

### The Caste System

Caste has been one of the oldest and most distinctive features of Indian society. It is a major force in the socio-economic and political systems of the country. Theoretically, Sikhs, Buddhists, Muslims and Christians are immune to caste distinctions. But most non-Hindus in India are the descendants of converts from Hinduism; they interact with the traditional caste-based Hindu society regularly and have maintained the caste traditions of Hindu society. Despite the weakening of caste structures in the cities, the lives of over 600 million villagers are still profoundly shaped by the institution of caste.

A precise definition of caste is difficult, but the concept includes at lest two levels of comprehension. One is the philosophical and broad level comprehending caste as a hierarchical division of society on the basis of *varna* (literally "color") into Brahmins, Kshatriyas, Vaishyas and Shudras. From the original *varna* categories, the Hindu society evolved into thousands of castes or *jatis*. Any visitor to India would soon discover that caste as a functional organic system is more complex than this classification of society into these four broad groups. The second level of caste comprehension is the segmentation of society into thousands or *jatis*, or castes, each internally bound by marriage, lineage, and with characteristic occupational, religious, ritualistic, and social, roles. With increasing urbanization, democratization and spatial mobility, the occupational affiliations of the castes have been undergoing social and structural transformation. The functional division of society into castes of *jatis* is a unique feature of India's human geography, and is of greater socio-economic relevance than the four broad *varna* categories.

Within the broad *varna* framework most *jatis* are traditionally stratified on the basis of their social status, ritual marriage ties and common descent. The most simplistic explanation of the origin of castes may be traced to the Aryan settlement and expansion in India sometime in the second millennium. The Aryans, it is widely believed, were pastoral nomads in Central Asia, from where they migrated into northwest India and pressed into the Indo-Ganga plains. In the early days of their settlement, society is thought to have been divided on the basis of race into fair-skinned Aryans and the dark-colored natives. Vedic literature (1500 B.C.-500 B.C.) suggests such a simple division of society. Gradually, as Aryans spread to other parts of the country, increasing contact with and assimilation of the natives, and the crystallization of the functional roles of the various groups, society was divided into a hierarchical strata of social classes. First, the four *varna* divisions emerged; later thousands of caste groups or *jatis*, (distinctive in kinship, lineage, customs, and social taboos) grew up within the varna framework. The precise procedures of this societal segmentation, and the formalization of ritual, lineage and social codes are not properly understood.

Caste mechanics, although grounded in antiquity are, in general, well known. Caste membership is hereditary and usually immutable. One's caste affiliation, at birth, is a function of one's *karma* (reward or punishment for actions performed in a previous earthly existence). To play a desirable role in life, according to *karma*, is to lead a life prescribed by *dharma* (duty), the appropriate rule of religious and social obligations to one's caste.

Caste territories are spatially demarcated on the basis of marriage links along endogamous lines. Physical contact with other castes is circumscribed by restrictions on contact, associations, dining or eating food cooked by other castes. Caste membership is emphasized by one's *jati* name, by the individual's identification with his caste in the eyes of the community, by one's conformity to the customs of one's caste and by one's subjection to *jati* government. Ritual observances of the caste are strictly maintained. *Jati* solidarity is kept by traditional occupations. *Jati* sanctions prescribe what forms of work one may or may not undertake, whom one should marry, what, how and with whom one may eat, and what rituals one may perform. Caste affiliations have a major effect on local and state politics as well.

### *Polarization and Interdependence of Castes*

In a traditional rural society, developed around a largely self-sufficient village community, various castes have performed their well-regulated functions. In ancient and medieval times spatial mobility and linkages were restricted to a few pilgrimages to sacred places, and to trips associated with marriages. During the last fifty years, with the establishment of an increasing number of spatial links of villages to urban centers, caste structure is undergoing change and its ritual role is declining.

In *varna* hierarchy, Brahmans stand at the top. Principle castes of this *varna* have been those of priests, teachers, custodians of sacred ritual practices and the arbiters of correct social and moral behavior. Geographically, they are the most ubiquitous since they must officiate in a variety of rituals. By virtue of their traditional prestige, they have been able to collect large amounts of land through grants by local rulers and patrons. Ritually barred from cultivation, they are now a prominent land owning and money lending class. their services are constantly in demand by other castes for major ritual functions at births, marriages and deaths. Some castes, especially the *Harijans* (scheduled caste) utilize the services of non-Brahman priests. Generally, Brahman households are few in a typical village, but

Brahmans command a wide variety of services from other caste households as they own a large part of the agricultural land. Brahmans are generally better educated than the other groups. With increasing urbanization, they have shown a tendency to gravitate to cities for service in clerical and administrative positions. In many parts of South India, especially Tamil Nadu and Kerala where they have historically played an unusually important role, anti-Brahman feelings have been building up over the last fifty years, and non-Brahmans have become active in the once Brahman-dominated administrative services. Many Brahmans who have adopted clerical and administrative occupations continue to perform ritual functions of other classes.

The Kshatriyas, next in *varna* ranking, are also very powerful. The traditional Kshatriya (warrior caste) role of defense is now largely submerged under the role of land ownership. When castes form large majorities in a village, the village is usually identified by the *jati* name of the Kshatriyas, for example a Rajput village (in Punjab), *Jat* village (in Uttar Pradesh), Thakur village (in Himalayan regions), and Nair village in Kerala. Vaishyas rank below Kshatriyas, but fall within the ambit of ritually high *varna*. Major castes within this *varna* were those who were engaged in occupations of farming and retail trade. Several prominent Vaishya groups have established successful monolithic businesses all over the country. Shudras, ranking below Vaishyas, belong to the lower category and are debarred from several ritual privileges. Shudras are mostly engaged in cultivation, and in a wide variety of artisan services, such as carpentry, metal work and basket weaving. Shudras form the bulk of the country's population.

At the bottom of the social ladder are the *niravasita* meaning "excluded", or the "exterior" castes, to so-called "casteless", officially "scheduled castes." Since the Government of India Act of 1935 they have been listed in the special official schedules for administrative and representational purposes. Gandhi addressed them as *Harijans* (children of God). Their occupations are "unclean" (to the higher castes): disposal of the dead, flaying of dead animals, menial work, cleaning latrines etc. To the upper castes they are "untouchables". The practice of untouchability started with their original "unclean" occupations, and physical contact by the upper castes with them was prohibited. A complex code of contact, ritual purity and prejudice slowly grew up as society was formalizing codes of conduct during the early crystallization of the castes. Untouchables were denied admission to temples, certain roads, schools and certain collective village social activities, and were also banned from the practice of occupations reserved for the high castes. Caste stratification can be seen in its worst form in south India. Until recently *Harijans* were not permitted the use of sandals, residence in brick houses, or the wearing of an upper garment on their body in Tamil Nadu and Andhra Pradesh. By tradition, these were privileges enjoyed by the "clean" higher *varnas*.

The Hindu social system's organic functioning, though segmented into various castes (*jatis*), was based on the principle of caste interdependence. In theory, each caste has traditionally defined duties and obligations relative to its status and ritual purity in conformity with the needs of the village community. A unique system of hereditary service and patronage among the castes (and families) thus grew up. This system is known as the *jajmani* system. The *jajman* or the hereditary patron receives the services of a hereditary artisan or servant and pays in goods perquisites, or cash on a regular basis for the performance of such services. For example, village landowning castes (usually upper castes) received ploughs, pots, and leather goods, as well as services connected with cleaning and repairing their households, from the lower castes (*Harijans* or Shudras); and pay for these in grains, vegetables and even land use privileges. In certain ritual functions, the lower castes also act as *jajmans*, the patrons, as for example the Brahman performance of ritual (priestly) functions for the lower castes. Brahmans act as the keepers of village shrines. In lieu of these services, they obtain cash and other gifts on a regular basis. In essence, the *jajmani* practice sets up a social network of alliances among groups of people who exchange goods and services. Payment for goods and services is traditionally fixed and related to the amount of service rendered, the relative status of the caste involved in an exchange, and the needs of the receiving family. Payments by the *jajman* are also made in return for special services at ceremonial occasions such as birth or marriage.

In a traditional, largely self-sufficient, caste-structured society, the need for such an arrangement is crucial for those castes whose ritual behavior keeps them from doing manual and menial jobs. Brahmans, for example, traditionally do not till the land, and Shudras or lower castes perform these functions for them in exchange for grain and/or cash payments. Several castes perform the functions of craftsmen, such as carpentry, pottery, weaving, barbering, shoe making and the laundering of clothes of their *jajmans*, and receive regular fixed payments. Different castes carry specific social and religious obligations. In planning a marriage, or building a new house, or arranging entertainment, the householder must call upon the services of a wide range of castes in connection with sweeping the floors, performance of rites and procurement of goods needed for the occasion. Different castes thus perform traditionally defined economic, social and ceremonial roles within the village. The upper castes depend on the lower ones for labor, artisan goods and menial services; the lower castes look upon the upper ones for financial rewards and loans in times of need. Historically, this interdependence has preserved and strengthened the caste structure. Harmonious inter-caste relations lead to economic cooperation. It is advantageous for a landowning farmer to be on good terms with a carpenter, blacksmith, weaver or barber. The traditional cash payments to them were meant to be a hedge against the inflation of an urban commercial society, although inflation has clearly been eroding into its real value. Through centuries, this system has enabled a large, stratified and heterogeneous population to live in peaceful coexistence. Despite segmentation and polarization, the caste institution has persisted as a remarkably cohesive force in a rural, self-contained society. Clearly, the impact of the caste system is most clearly discernible in the rural areas, where it is impossible to trespass the caste boundaries.

### *Spatial Dimensions of Caste Distributions*

No enumeration of the castes (except for "scheduled castes") has been made since the Indian census of 1931 which listed 2378 "main" castes, or *jatis*, for the country. Several of these claimed million of followers. The present-day caste distribution in the country probably follows closely the patterns of the 1931 census. Although Brahmans and the "untouchables" (also listed as "depressed" classes) perform wide-ranging services, and are more pervasively distributed, they still tend to show degrees of regional concentration. Other castes vary greatly in their regional distributions. At the micro level of a village, only a few of the castes are usually represented. Within a village the number and size of castes depends on the size of the village, its regional location, its economic structure and above all its historical experience. A tiny settlement of a few hundred farmers may contain only a few castes (farmers, shepherds). Neighboring larger villages provide it with ritual and other services. A typical average village contains a mixture of 5-15 main castes, a few landowners (e.g., *Brahmans, Jats, Rajputs*), a few members of the lower castes performing menial jobs, and a large number of households of one or more farming castes.

Within the village, caste hierarchy is revealed by residential segregation of the castes, the size of houses and the quality of materials with which they are constructed. Better houses typically belong to the higher castes and are usually located in the center, the other castes inhabit the peripheral parts. Residences of the untouchables often occupy areas outside the main village. Muslims and other lower castes usually restrict their housing to one side of the village, but are not necessarily segregated in areas outside the village.

The predominant caste may not necessarily be the dominant one. A "dominant" caste is the one which is economically and politically the most powerful. If a numerically weak caste holds most of the land, business, and public offices, it usually controls the politics and administration of the village as well and emerges as the dominant caste. Often a dominant case possesses high societal ranking.

Above the village level, two spatial levels of caste distributions can be identified: the exogamous caste region incorporating those villages which are functionally interconnected by marriage ties; and a wider region grouping several hundred or even thousands of villages of a predominant caste.

Caste regions based on the marriage alliance are recognizable almost universally in north and west India. According to the rule of exogamy, one is not supposed to find a wife within one's own *gotra* (ancestral lineage traced to a common ancestor). Marriage alliances are developed with the outside villages, usually in a 4-12 mile range. A few studies indicate that in Punjab the greatest clustering of marriage links lies within 8 miles, and in most parts of Uttar Pradesh between 8 and 12 miles. Usually no marriages are arranged within 2 miles of the residence. An average distance of marriage ties in Uttar Pradesh is 10 miles. A typical marriage circle may sometimes cover about 300 villages, with about 80 villages accounting for half of all the marriage linkages. Spatially inter-connective directions of the flow of girls in marriage ties are usually established and maintained. Spatially oriented marriage links also determine social ranks. "High" villages obtain the girls, and "low" ones supply them. Two sets of villages are therefore spatially organized; one set usually receives wives, and the other sends wives. Financial penalties may be imposed on the families of their castes if territorial affinal boundaries are transgressed. The idea of hypergamy, i.e., men of high caste taking wives from the caste below them, and that of exogamy, helps to perpetuate the establishment of "high" and "low" villages, although this practice is not so widespread. In Punjab and Haryana, three marriage regions based on marriage ties have been identified: *Majha*, *Doaba* and *Malwa*. In south India, however, marriage alliances are encouraged within the same village because of the marriage preference for cross cousins and nieces.

Schwartzberg (1978) has identified caste regions of north India by grouping several districts (civil divisions) on the basis of predominant castes, caste dominance and rate of increase in the number of castes with increasing village size. He recognizes five distinctive regions for north India. One region includes the Punjab plains and a contiguous area in Rajasthan. The predominant caste in this region is the farming caste of *Jats* (mainly Sikhs, but many Hindus as well). Whatever other types of castes they may contain, most are inhabited by one single caste, most often *Jats*. In some villages Rajputs (non-cultivating land-owning caste of high ritual status), Sainis (gardeners), Kambohs and Ahluwalias (cultivators) lead in villages. In a typical *Jat* village there is usually not a single household of Rajputs, emphasizing the mutual exclusivity of the two castes. Other castes well represented in the region are: Chamars (leather workers) and Mazhbis (Sikh sweepers). In some urban areas several Vaishya (merchant) castes are locally predominant. Lying to the east is a region which includes Punjab, Delhi and Uttar Pradesh. The caste structure in this region is diversified with two or three almost equally numerous castes coexisting, as well as many less numerous castes. In some parts, the scheduled castes are the most numerous ones. Several Muslim castes are also well represented. Chamars, Jats, Baniyas (merchant caste), Rajputs, Ahirs (milkmen-cultivators) and Kurmis (cultivators) are numerically strong castes. In the lower Ganga plain in Bihar Ahirs are predominant, with Brahmans and Rajputs making up only 10 per cent of the population. Bhumihar, a cultivating Brahman caste, is also locally significant in several areas. Another region stretches from the eastern Nepal border to the Bay of Bengal. Sheikhs (Muslim caste) predominate numerically. Mahishya (cultivators) and Kayastha castes are locally significant as landlords. The region surrounding the Calcutta metropolitan area but also extending south to Bihar, contains Brahmans, Mahishyas and Sheikhs as important castes. Tribal groups like Santals, predominate in several villages, and are being gradually Hinduized (Schwartzberg, 1965).

### *The Changing Caste System*

The caste system has persisted for perhaps three thousand years. Throughout these years it has been continually expanding, with the assimilation of new elements, to form new castes. Essentially, it has worked reasonably well in a rural, self-contained, conservative society, each social stratum performing its assigned functional role. But it is a patently undemocratic system based on inequalities in social relationships. Moreover, it presupposes a fixed environment of unchanging societal norms and needs.

Although the caste system is pervasive and deeply entrenched, it finds itself increasingly pitted against the forces of literacy and constitutional democracy as society becomes more urbanized and moves towards modernization. Traditional restrictions and sanctions are slowly being sloughed off in the cities. Caste inequalities are being eroded perceptibly. City life inhibits the observance of caste rituals in the matter of interdining, and polluting physical contact in public places like cinema houses, shopping areas, city transportation and restaurants, etc. Spatial diffusion of these social amenities to the rural areas is still limited, but is on the increase. Since Independence, transport facilities have been extended to tens of thousands of villages and the traditional self-sufficiency of the villages has been reduced by the penetration of a money economy, and by a progressive disappearance of handicrafts and thus of the caste-related *jajmani* system. Chamars (a lower caste of shoemakers) are, for example, now lured to becoming milkmen to earn more money ever since factory-made shoes from the cities are being increasingly sold in the villages (Srinivas, 1960).

The three main forces responsible for bringing structural and institutional changes into the caste system are: secularization, westernization and sanskritization. *Secularization* may be defined as a process by which the scriptural and ritual behavior and symbols cease to impart the traditional mandate. Purity becomes a matter of personal hygiene rather than a caste-prescribed ritual social code. For example, "pollution" arising of interdining with lower castes is being replaced by interdining via washed teacups and dishes. Many educated pilgrims are now more concerned with the polluting drains flowing into the river Ganga than with the river's holiness. There is a gradual erosion of priestly authority. Ritual customs at weddings are becoming lax. All these are the outcome of secular processes (Srinivas, 1967).

*Sanskritization* refers to the adoption of the lifestyles, rituals, and beliefs traditionally associated with the upper castes by the lower ones in order to achieve social elevation. Although an old ongoing process, the opening up of economic opportunities, the spread of education, and the concept of equality before law have all contributed to this process of vertical social mobility among the lower castes. In Karnataka, parts of Tamil Nadu, Punjab, and several other parts of the country, the *Harijans* are renouncing the consumption of liquor, meat eating and widow remarriage in order to claim higher social status by the adoption of the traditional mores of the upper castes.

*Westernization* is a process of adoption of the lifestyle of the western world. It gained initial popularity among the upper castes, who westernized their life pattern by the adoption of British social mores during and since British rule. The educated Indian elites (usually the upper castes) acquired such forms of British social behavior as drinking, and adoption of western dress and forms of greetings during their contact with the British in clubs, officers' messes, civil departments and courts. Many rituals related to caste purity and physical pollutions were abandoned. The three processes mentioned above have made a greater, impact in the cities where the vast, congested, mobile, secular, and somewhat westernized society would tend to diminish the original roles of the castes.

On the other hand, several forces have tended to perpetuate the caste institution, especially in the cities. Since Independence caste groups have been politicizing during the British rule. The growth of caste hostels, caste-based religious schools, public dispensaries and restaurants has fostered the interests of the individual castes. Caste lobbies have sprung up in the state legislatures. In the villages, the dominant castes have been tightening their hold on the village councils, despite the seeming democratization of the village councils (*panchayats*).

The most basic cleavage in society, however, despite its apparent segmentation, remains in the form of its polarization into two basic groups: the "clean" and the "unclean" (polluting castes). The "unclean" castes are the untouchables, or the *Harijans*, who numbered 138.2 million in 1991 (excluding Jammu & Kashmir where 1991 census did not take place). Generally ubiquitous, they are concentrated in larger proportions in rural than in urban areas. Historically, they have been oppressed, but they have functioned as an integral link in the socio-economic fabric of the society. Although they generally reside in segregated quarters from the main settlement nucleus, they work in the homes and fields of the upper castes, and may still receive *jajmani* payment from them. Geographically, the North Indian Plains contain about one-half of their population, although in no part of the country are they altogether absent.

Untouchability has been proscribed by India's constitutional mandate, and its practice made a criminal offense since 1955. But these legal sanctions have proved fruitless. Social pressures against untouchability are still quite feeble. The untouchables increased from 50 million in 1951 to 138.2 million in 1991, registering a higher rate of increase than that of the general population. Figure 4.8 gives the distribution of the scheduled castes. Although universally distributed, their concentration in the North Indian Plains, Kerala and Tamil Nadu coastal areas is particularly noteworthy where they account for more than 20 per cent of the population. About 40 per cent of the untouchables are tenant cultivators, 35 per cent work as hired agricultural laborers, and most of the remainder work in household services and cottage industries. In rural areas, approximately two-thirds of them are in debt to the village landlords. Illiteracy is nearly universal (90 per cent) among them (compared to 70 per cent for the total population). The Indian government is keenly aware of their backward status and has, since 1950, adopted several measures to improve their condition. Special quotas of reserved seats in the state and federal legislatures, public services, and educational institutions, as

**Figure 4.8**
*Distribution of Scheduled Castes*

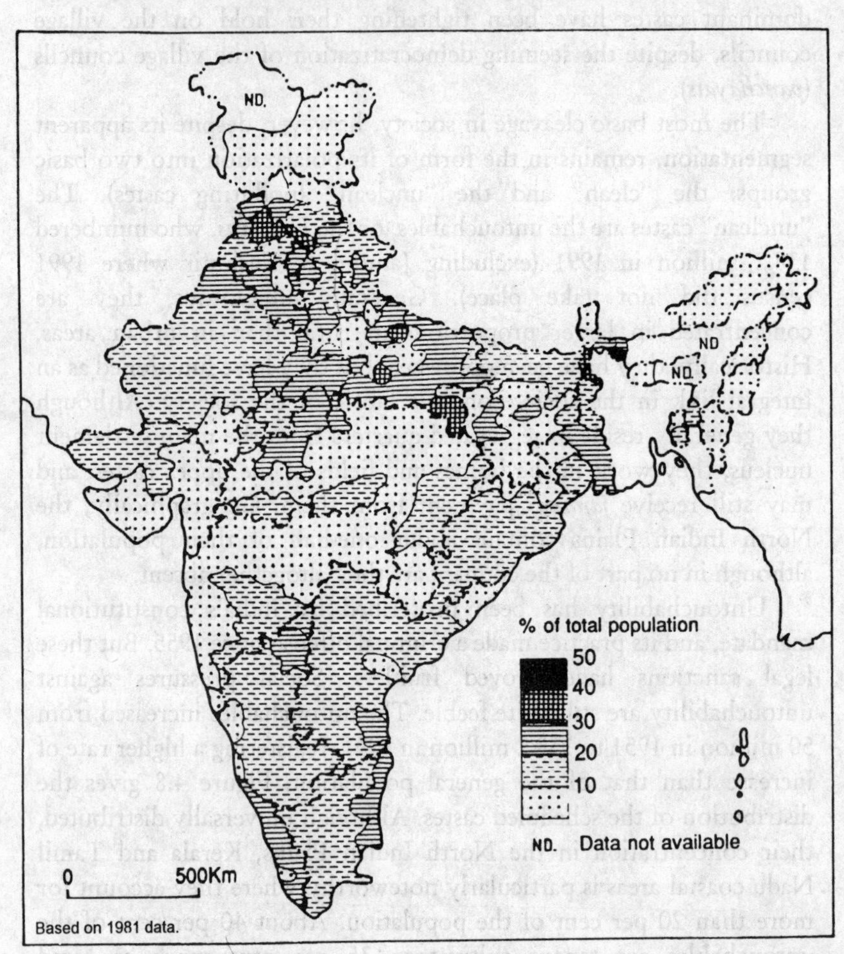


called for in the Constitution of India, have been set aside for them. Such legislative measures to compensate the *Harijans* for historic injustices have ironically encouraged their separateness from the rest of the community, and has even bred hatred leading to serious civil riots. Rather than accelerating their socio-economic advancement by protective legislation, social cleavages has, in some respects widened. Meanwhile, exploitation of the *Harijans* continues by the more powerful high caste groups. A report of the central government in

1969 indicated that very little improvement in the status of *Harijans* has been achieved. The institution of caste, the report points out, is still deeply rooted. The report recommended several institutional and structural changes to root out untouchability. These include abolition of hereditary priesthood, biological assimilation of the social groups through intermarriage, and stricter enforcement of legal equality.

The architects of India's independence had dreamt of a "casteless and classless society". The constitution forbids untouchability, and in theory the scheduled castes were granted privileges through legislation, like free schooling, reservation of seats in universities and colleges, positions in civil service, and seats in *Lok Sabha* (parliament's lower house). In reality, however, there is little evidence of significant change. Occasionally riots break out between the untouchables and the upper castes over government's preferential legislation for the scheduled castes. There were violent riots in Gujarat between the landed castes and the scheduled castes in 1985 and 1986, and similar riots in Bihar in the 1980s. In 1990 student riots erupted in several cities like Chandigarh, Delhi, Jaipur and Chennai, demanding abrogation of the proposed legislation granting substantial quotas to the scheduled castes as recommended by Mandal Commission. The future of untouchability is linked to modernization and urbanization. Undeniably untouchability is disappearing in urban schools, temples, hospitals and public transport. In the villages where illiteracy and ignorance are more prevalent and where the traditional social taboos are deeply entrenched, untouchability remains a fact of life.

## Linguistic Divisions

The linguistic configuration of India is unique among nations, and comparisons with other linguistically plural countries like Canada, Switzerland and Belgium are misleading. Although like them the distribution of diverse languages has profoundly affected its political, social and economic structures, India's linguistic divisions run deeper and are infinitely more complex creating a multidimensional problem as a result of several forces: a constant interplay between local, regional and all-India languages; an assertion of language-related territorial claims for the various states; interminable demands of the language elites seeking administrative privileges; the overall problem of a national language for the country; and the intractable relationship between languages, religion and political aspirations.

The *Linguistic Survey of India* listed 179 languages and 544 dialects spoken within the country, while the census of 1961 recorded 1,018 different languages and dialects. The Constitution of India recognizes 17 major languages, most of which (with the exception of Sanskrit) claim million of speakers (Table 4.2) Twelve of these find territorial identification in large political units (states).

**Table 4.2**
*Languages of the Eighth Schedule, 1991*

| *S.No.* | *Languages* | *Speakers* | *Percentage of All-India Population* |
|---|---|---|---|
| 1 | Hindi | 337,272,114 | 40.22 |
| 2 | Bengali | 69,595,738 | 8.30 |
| 3 | Telugu | 66,017,615 | 7.87 |
| 4 | Marathi | 62,481,681 | 7.45 |
| 5 | Tamil | 53,006,368 | 6.32 |
| 6 | Urdu | 43,406,932 | 5.18 |
| 7 | Gujarati | 40,673,814 | 4.85 |
| 8 | Kannada | 32,753,676 | 3.91 |
| 9 | Malayalam | 30,377,176 | 3.62 |
| 10 | Oriya | 28,061,313 | 3.35 |
| 11 | Punjabi | 23,378,744 | 2.79 |
| 12 | Assamese | 13,079,696 | 1.56 |
| 13 | Sindhi | 2,122,848 | 0.25 |
| 14 | Nepali | 2,076,645 | 0.25 |
| 15 | Konkani | 1,760,607 | 0.21 |
| 16 | Manipuri | 1,270,216 | 0.15 |
| 17 | Kashmiri* | 56,693 | 0.01 |
| 18 | Sanskrit | 49,736 | 0.01 |
|  | Total** | 807441612 | 96.29 |

*Excluding Kashmiri speakers in Jammu and Kashmir.

**Speakers of the Eighth Schedule languages excluding Jammu and Kashmir.

*Source:* Census of India, 1991, Series-1, India, Paper No.1 of 1997, Table C-7, *Language.*

**Figure 4.9**
*Language Familes and Major Languages*

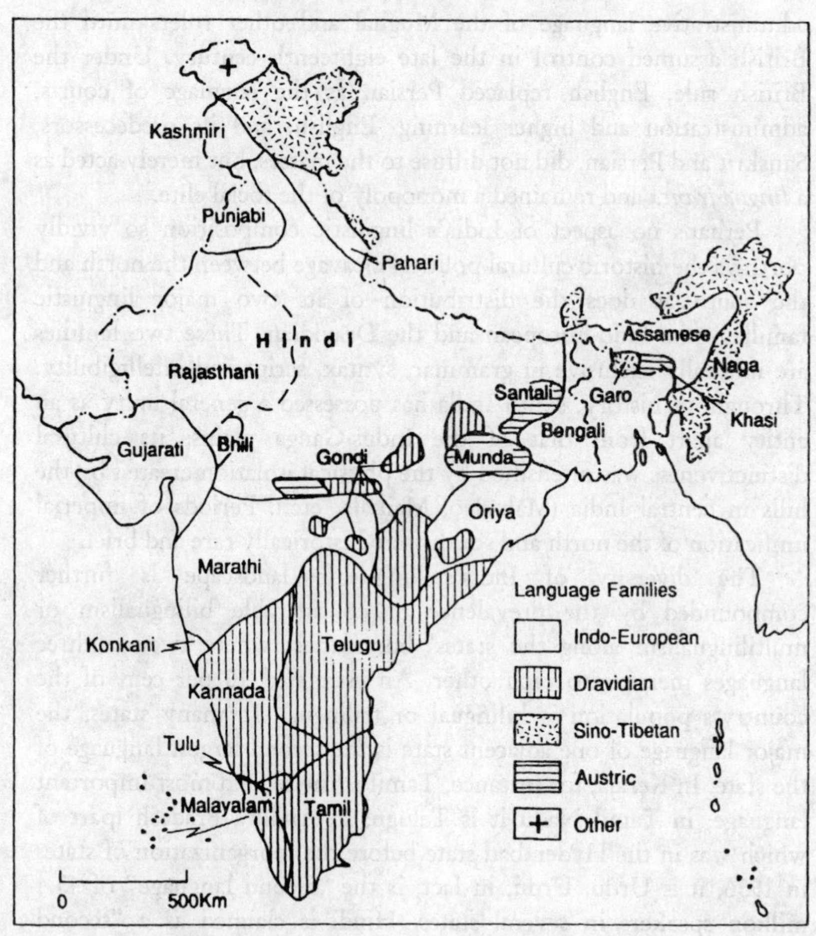


Despite the seeming linguistic diversity, Indian languages and dialects can be grouped into four language families: Indo-European, Dravidian, Sino-Tibetan and Austro-Asiatic. Secondly, 98 per cent of the country's population belong to the first two large families. Thirdly, with the exception of minor spatial scattering of the tribal languages, the major Indian languages are localized in compact territorial blocks (Figure 4.9). And finally, only a handful of the languages have historically been the vehicles of political hegemony.

Until the ninth century, Sanskrit was the language for administration and of the cultural elite. Persian (later Urdu) was a court and administrative language of the Mughal and other rulers until the British assumed control in the late eighteenth century. Under the British rule, English replaced Persian as the language of courts, administration and higher learning. English, and its predecessors, Sanskrit and Persian, did not diffuse to the masses, but merely acted as a *lingua franca* and remained a monopoly of the social elite.

Perhaps no aspect of India's linguistic composition so vividly portrays the historic cultural-political cleavage between the north and the south as does the distribution of its two major linguistic families—the Indo-European and the Dravidian. These two families are mutually exclusive in grammar, syntax, script and intelligibility. Throughout history, south India has possessed a general unity as an entity apart from that of the Indus-Ganga plains; its cultural distinctiveness was intensified by the physical isolation created by the hills in central India (Mahadeo, Maikala, etc.). Periods of imperial unification of the north and south were historically rare and brief.

The diversity of India's linguistic landscape is further compounded by the prevalence of considerable bilingualism or multilingualism along the states' boundaries, where two or three languages merge into each other. An estimated 10 per cent of the country's population is bilingual or trilingual. In many states, the major language of one adjacent state is the second largest language of the state. In Kerala, for instance, Tamil is the second most important language; in Tamil Nadu it is Telugu; in Andhra Pradesh (part of which was in the Hyderabad state before the reorganization of states in 1956, it is Urdu. Urdu, in fact, is the "second language" of 43.4 million speakers in several states. Hindi is claimed as a "second language" by 30 million speakers residing outside the "Hindi" region (Uttar Pradesh, Madhya Pradesh, Bihar, Haryana, Himachal Pradesh and Rajasthan). Nearly 10 per cent of the Telugu-speakers live outside the Telugu-speaking state of Andhra Pradesh. Cosmopolitan cities like Mumbai and New Delhi contain significant numbers of speakers of several languages.

Within the Indo-European family, Hindi claimed the largest number of speakers or nearly one-fourth of India's population according to the 1991 census. Including its variants, it ranks

**Figure 4.10**
*Hindi Speakers (including Rajasthani and Bihari)*

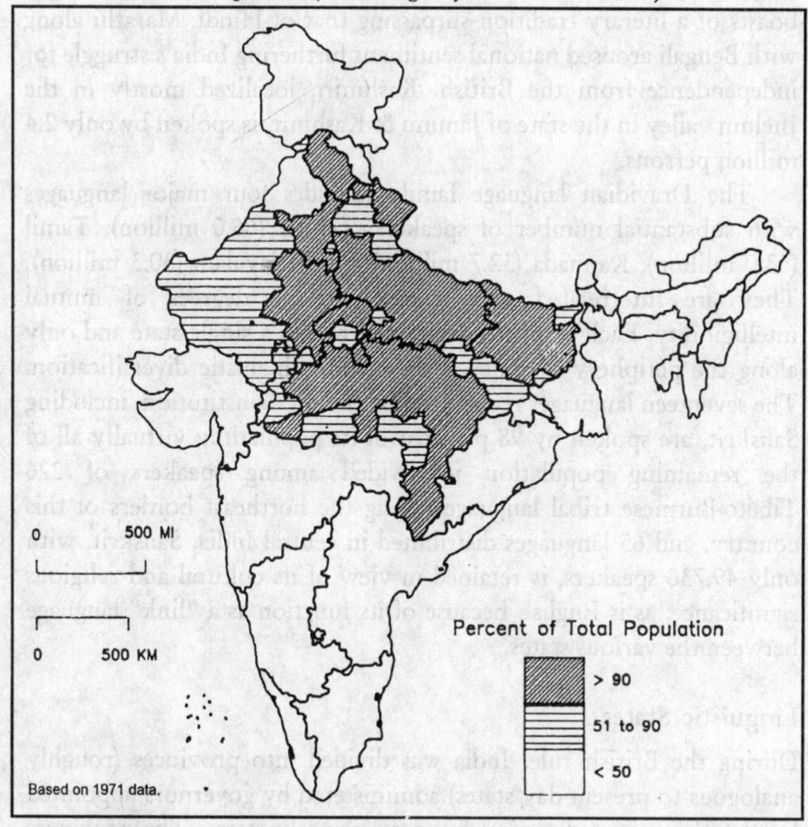


numerically as one of the five major languages of the world, after Chinese, English, Russian and Spanish. Territorially, it is spread over the states of Uttar Pradesh, Bihar, Madhya Pradesh, Haryana, Delhi and Rajasthan. Languages of Bihari, Rajsthani and Marwari, Urdu (distinguished from Hindi in script and in vocabulary of Persio-Arabic origin, with 43.4 million speakers), and Punjabi (23.3 million speakers) along with Magadhi, Chhatisgarhi, Maithili and Pahari make the Hindi-related language group the largest language with nearly 337.27 million speakers of 40 per cent of the country's population (Figure 4.10). Other major Indo-Aryan languages are: Bengali (69.5 million speakers), Marathi (62.4 million speakers), Gujarati (40.6 million speakers), Oriya (28.1 million speakers) and Assamese (13.1 million speakers), each contained mainly within a single state. Each of these languages possesses a mature literature and associated culture. Bengali boasts of a literary tradition surpassing that of Hindi. Marathi along with Bengali aroused national sentiment furthering India's struggle for independence from the British. Kashmiri, localized mostly in the Jhelum valley in the state of Jammu & Kashmir, is spoken by only 2.4 million persons.

The Dravidian language family includes four major languages with substantial number of speakers: Telugu (66.0 million), Tamil (53.0 million), Kannada (32.7 million) and Malayalam (30.3 million). They are interrelated and possess varying degrees of mutual intelligibility. Each is mostly contained within a single state and only along the periphery of states is there some linguistic diversification. The seventeen languages recognized by India's Constitution, including Sanskrit, are spoken by 98 per cent of its population; virtually all of the remaining population is divided among speakers of 226 Tibeto-Burmese tribal languages along the northeast borders of this country, and 65 languages distributed in central India. Sanskrit, with only 49,736 speakers, is retained in view of its cultural and religious significance; as is English because of its function as a "link" language between the various states.

## Linguistic States

During the British rule, India was divided into provinces (roughly analogous to present-day states) administered by governors appointed by the Viceroy, and several hundred princely states. The provinces were created haphazardly in conformity with the accidental course of British annexations and often included territories of two or three linguistic communities. The province of Bombay, for example, included Gujarati and Marathi speakers' territory and territory of several other languages. The idea of linguistic states was, however, repeatedly espoused by the Indian National Congress, the major political party in the struggle for Indian's Independence. The proponents of linguistic patriotism pointed out that the creation of linguistic states would release energies which had previously been consumed by struggles between linguistic groups within the multilinguistic provinces.

Independence in 1947 thus became a signal for agitation favoring realignment of state boundaries. After resisting demands for linguistic states for six years, reasoning that centrifugal forces would gain at the expense of national unity, the central government yielded in 1953 to the demand for a separate state of Andhra for the Telugu speakers of the northern part of Madras province. This opened the gates to other similar demands, and the central government quickly appointed a States' Reorganization Commission to look into the question of formation of linguistic states. Following the Commission's Report in 1955, the government carried out in 1956 a large-scale reorganization of the internal political boundaries within the country. This resulted in the dismantling of some of the former provinces, the emergence of new states, and a sweeping modification of the boundaries of most of the others. In effect, the government yielded to nearly all of the demands by major linguistic groups. India was thus reorganized into 14 states, each of which, with the major exceptions of the state of Bombay, and Punjab, contained a clearly dominant language. Agitation for further changes, however, persisted. Several revisions of state boundaries between 1956 and 1976 increased the number of states from 14 to 27. The last of the major linguistic states to be thus created was Punjab, which was created to placate Sikh demands seeking political identification for their religious language. Boundary revisions since 1966 have been largely on ethnic considerations, created mostly by splitting the existing states along India's northeastern frontier. The nation is currently composed of 27 states and seven union territories. As a consequence of the boundary revisions most present day states reasonably coincident with the linguistic regions.

## National Language and National Unity

The assertion of a national language for a polyglot country like India, undoubtedly poses a real and serious problem. Since Independence, Hindi, spoken by the largest number of people and covering several states, has been championed as the national language by the Indian National Congress, the party which, for 43 years (until 1990) without interruption, was in control of the union government in New Delhi. The Indian Constitution bestows upon 15 of the country's major languages the status of official languages (Table 4.2), with a pride of place for Hindi (in *Devanagari* script) which is recognized as the "official" (the term "national" was avoided) language for all-India communication at the central level in New Delhi. The Constitution of India also provided that English would continue to be used for all purposes at the central level for a period of 15 years (1950-65). At state level, the regional languages are recognized as official. English is accepted as a special language for its role in central government's transactions, in parliamentary debate, in the courts and as an instructional medium for specialized subjects for higher education, as well as for its value as a "link" language between various states.

The constitutional mandate that Hindi should become the central language by 1965 always met with great resistance. Hindi has a less developed literary tradition than several regional languages, such as Bengali, Marathi, Telugu and Tamil. Moreover, as officially promoted it has been steadily assuming a strange form as a result of its being infused with a new Sanskritic vocabulary. The main focus of resistance to Hindi has been in the Dravidian language states in south India. Hindi was accorded the status of optional administrative language at the level of the central government and in the Public Service Commission examinations which was considered by many states to be a discriminatory imposition of "Hindi imperialism". To allay the fears of non-Hindi speaking states, the parliament passed in 1963 a language bill providing that English would continue to be used after 1965, in addition to Hindi. Nevertheless, as 1965 neared, there was apprehension the government might decide to keep Hindi as the sole official language at the federal level. Riots, based on linguistic nationalism, erupted in Madras (now Chennai), Calcutta and other places. Central government swiftly and wisely responded in 1965 by allowing the continued use of English "as an associate official language" until all the non-Hindi states have consented to the use of Hindi, thus in effect retaining English virtually indefinitely. The feelings of the Dravidian states, who were at one time contemplating a secession of Dravidian areas from India to form an independent country of "Dravidistan" were particularly assuaged. English, therefore, remains an official associate language of the central government, along with Hindi, and is serving as a link between the different states.

In effect, Hindi has clearly failed to elicit universal popularity. In literary and intellectual appeal, it is seriously rivaled by such highly developed languages as Bengali, Marathi, Tamil and Telugu. Hindi enthusiasts have injected into it a strong Sanskrit-based vocabulary and have only increased the distance between the language and the masses. Furthermore, speakers of closely related languages such as Urdu and Punjabi find themselves drifting away from the newly developed Sanskrit-based Hindi. These three closely related languages, namely, Hindi, Punjabi and Urdu are mutually intelligible to a large extent, but are divided in scripts, which are favored by different religious communities. For Sikhs, Gurmukhi is the religious (therefore the favored) script for the Punjabi language. Urdu derives its script from Perso-Arabic characters and is espoused by Muslims. Hindus favor the usage of *Devanagari* script for all these languages. In this context, the Sikh demand for a Punjabi-speaking state was interpreted as communal idea seeking a homeland for the Sikh community. In 1966, the demand was fulfilled when the state of Punjab was partitioned into two states, the Hindi-speaking Haryana state, and the mainly Punjabi-speaking state in which Sikhs possessed a narrow majority.

The important question of the future of English still remains unresolved. Undeniably, it has played a major role in India's national consolidation and development. During the British rule it was the official language at the provincial and national levels, universally used in civil and military services, in the judiciary, in politics and in the all-India struggle for independence. It had created an intellectual elite on an all-India basis. It provided independent India with a modernizing leadership, and its place in the educational system became well established. Despite its wide-ranging influence, however, the masses remained virtually insulated from it. This created a deep dichotomy between the elites trained in English and the uneducated masses.

Since Independence, the role of English has been declining. The states have been advancing their regional languages as the primary media of instruction in the schools and public services at the expense of English. Circulation of regional language newspapers and films have tended further to weaken the usage of English as a *lingua franca*. Many educated Indians, however, consider the gradual disappearance of English as an intellectual tragedy. Already, deterioration in the educational standards in the universities is widespread in several states, a situation explained largely by the replacement of English as a medium of instruction in higher learning, as well as by problems faced by graduates moving to other states where higher education is in a different regional language. These tendencies have once again underscored the need for the retention of English at the federal and state levels, in public services, and in higher education. Advocates of English further point out that such a retention would contribute to stronger links between the states, and internal administrative cohesion. Critics of English, however, are quick to point out that an alien language should never be accepted as an official language. One interesting suggestion favors the idea of several plural official languages, with the retention of English as an associate language in the government and higher education. The language problem is clearly putting India's overall structure of national unity to a severe test.

### *References*

Bhardwaj, S.M., *Hindu Places of Pilgrimage in India*. Berkeley, 1989.

Brown, W.N., *The U.S. and India, Pakistan, Bangladesh*. Cambridge, 1972.

Brush, John E., "Distribution of Religious Communities in India," *Annals of the Association of American Geographers*, 1948, pp. 81-96.

Bouton, M.M. (ed.), *India Briefing*, Boulder Colorado, 1989.

Chatterji, S.K., *Language and the Linguistic Problem*. London,1945.

Dasgupta, J., *Language Conflict and National Development*. Berkeley, 1970.

Dumont, L. *Homo Hierarchicus*. Chicago, 1970.

Dutt, A.K., "Religious Pattern of India," *Geo Journal*, Vol. 3 (2), 1979, pp. 201-204.

Harrison, S.S., *India, The Most Dangerous Decades*. Princeton, 1960.

*India: A Country Study*. U.S. Government Printing Office, Washington, D.C. 4th ed., 1985.

*India, 1997: A Reference Annual*. New Delhi, 1997.

Muthiah, S., *et al.* (eds.), *A Social and Economic Atlas of India*. New Delhi, 1987.

*Report of the Committee on Untouchability, Economic and Educational Development of the Scheduled Castes and Connected Documents*. Government of India, New Delhi, 1969.

Schermerhorn, R.A., *Ethnic Plurality in India*. Tucson, 1978.

Schwartzberg, J.E., "The Distribution of Selected Castes in the North Indian Plains," *Geographical Review*, Vol. 55, 1965, pp. 477-495, and map plates.
Schwartzberg, J.E. (ed.), *A Historical Atlas of South Asia*. Chicago, The University Press, 1978.
Sopher, D.E., *The Geography of Religions*. Englewood Cliffs, N.J., 1967.
Ṣrinivas, M.N., (ed.), *India's Villages*. Bombay, 2nd ed., 1960.
Srinivas, M.N., *Social Change in Modern India*. Berkeley, 1967.
Wolpert, S., *India*. Berkeley, 1991.

# 5

# Population Dimension

Having crossed the one billion mark on 11 May 2000, India is one of the most populated countries in the world, second only to China. Containing nearly 16 per cent of the world's population on only 2.4 per cent of its area, its population far exceeds the combined populations of the United States and the Commonwealth of Independent Russian Countries (the former U.S.S.R.). No other country of comparable size even approaches its present population density of about 370 persons per square kilometer (over 675 persons per square mile)—a figure well over five times the world's average. Its current annual rate of growth of 2 per cent (*World Population Data Sheet, 1997*), though not among the highest in the world, is nearly twice as high as it was 40 years ago. Translated in absolute terms, the country is adding 17 million persons—more than Australia's total population—to its existing population every year. If the current rate of increase remains unrestrained, the country's population will double itself in 43 years, and is estimated to surpass that of China's population in 2010.

Large population increments in the last forty years left the country until recently with food shortages and dependent on other countries for foodgrain imports. Government is cognizant of these problems and has launched family planning programs. A real breakthrough in this is not yet in sight. Although considerable gains toward achieving foodgrain self-sufficiency have been made, but the results may be temporary. The population explosion and foodgrain inadequacy are still among the major problems facing the country. This chapter deals with India's demographic composition and selected aspects of its population problem. Five areas of the population problem have been identified for brief discussion. These are: problems of illiteracy, foodgrain inadequacy and population explosion, manpower and unemployment, programs dealing with reduction in human fertility, and the problem of public health and disease prevalence.

## Growth Patterns

India has been one of the early centers of population concentration in the world. As far back as two to three millennia B.C., a sophisticated urban civilization flourished in the Indus basin. Several sites (Harappa, Mohenjodaro, etc.) in the Indus basin have revealed the existence of well-planned cities, domestication of agricultural plants and animals, a wide circulation of artifacts prepared of copper, gold, silver, and bronze and the development of manufacturing crafts like weaving and metal work. Early Aryans who settled in northern India from 3000-4000 years ago also possessed a technology capable of supporting a fairly dense population. During the period of Aryan settlement (Vedic India), over 20 towns flourished in north India (Schwartzberg, 1978, Plate III. A.1). By the fourth century B.C., the North Indian Plains contained large concentrations of populations. Greek and Indian records of this time indicate that the Aryan civilization of northern India compared favorably with that of early Europe. Chandragupta (c. 321-297 B.C.), the first pan-Indian ruler, is believed to have maintained a large standing army of 700,000 men. During Ashoka's rule (c. 274-236 B.C.), Indian civilization attained a high level of development in administration, commerce, agricultural techniques and the utilization of metals. Indian population during this time has been estimated at 110-150 million. For about 2000 years following Ashoka's rule population fluctuated within a narrow range. Population gains of peaceful times were canceled by periods of high mortality resulting from wars, epidemics and famines. The estimates of population of the area ruled by Akbar (c. 1556-1605 A.D.) range between 87-117 million (Nath, 1929).

Rapid gains in population numbers began to occur only after the early part of the twentieth century, when natural checks on population growth were diminished by the virtual elimination of several diseases (plague, malaria), the introduction of better public health standards, the spread of law and order and a consequent reduction in banditry and crime, and a reduction in the frequency and severity of famines and consequent increases in life expectancy. The first complete census of the country in 1881 yielded a count of 256 million (including present Pakistan and Bangladesh)—almost twice the population of the sixteenth century. Since then a comprehensive decennial census has been maintained. Table 5.1 gives the results of the total population count since 1901.

**Table 5.1**

*Growth of India's Population, 1901-91*

| *Census* | *Population (in Millions)* | *Change in Millions* | *Percentage of Change From Preceding Decade* |
|---|---|---|---|
| 1901 | 236.3 | - | - |
| 1911 | 252.1 | +15.8 | +5.7 |
| 1921 | 251.4 | -0.7 | -0.3 |
| 1931 | 279.0 | +27.6 | +11.0 |
| 1941 | 318.7 | +39.7 | +14.2 |
| 1951 | 361.1 | +42.4 | +13.3 |
| 1961 | 439.2 | +78.1 | +21.5 |
| 1971 | 548.1 | +108.9 | +24.8 |
| 1981 | 683.3 | +135.1 | +24.7 |
| 1991 | 846.3 | +163.0 | +23.8 |

*Sources:* 1. *Census of India, 1991*, New Delhi.
2. *Census of India, 1961*, Paper No. 1 of 1962.
3. Kingsley Davis, *The Population of India and Pakistan*, p. 27.

*Note:* Before 1951 figures are adjusted for areas of post-independence India.

In the first two decades of the twentieth century, population growth was slow and sporadic. Since 1921, however, population has shown a continuous trend of accelerating growth. The contrasting patterns, before and since 1921, are apparent, and are related to the mortality situation of the two periods. Before 1921 death rates were estimated at 42-48 per 1,000 persons per year, thus nearly canceling increases resulting from high birth rates. Since then death rates have declined markedly (Table 5.2) but only after 1950 did birth rates fall below 40 per 1,000 per year. These demographic factors account for the doubling of the rate of increase from 1 to 2.6 per cent between 1921 and 1971. In absolute terms, between 1961 and 1971, 304.7 million people were added to the total population—a figure greater than the current total population of the United States. In a single decade of 1981-91, population registered an increase of 160.6 million.

**Table 5.2**

*Estimated Crude Birth Rate, Crude Death Rate and Natural Increase in India, 1901-97*

| | *Crude Birth Rate (per 1000)* | *Crude Death Rate (per 1000)* | *Natural Increase in per cent Per Year* |
|---|---|---|---|
| 1901-11 | 48.1 | n.a. | n.a. |
| 1911-21 | 49.2 | n.a. | n.a. |
| 1921-31 | 46.4 | 36.2 | 1.0 |
| 1931-41 | 45.2 | 31.2 | 1.4 |
| 1941-51 | 39.9 | 27.4 | 1.2 |
| 1951-61 | 41.7 | 22.8 | 1.8 |
| 1961-71 | 39.0 | 12.4 | 2.6 |
| 1971-76 | 38.0 | 13.0 | 2.5 |
| 1988 (estimated) | 31.3 | 10.9 | 2.1 |
| 1991 | 31.0 | 10.0 | 2.1 |
| 1997 | 25.9 | 9.6 | 1.6 |

*Sources:* 1. *Growth of Population in India, Ministry of Health*, Government of India, New Delhi, 1962.
2. *Census of India*, 1941, 1951, 1961, 1971, 1991.
3. *1991 World Population Data Sheet*, Washington, 1991.
4. *India 1990: A Reference Annual*, New Delhi, 1997.

The average density of population of the country is 791 persons per square mile (306 per sq km). Within the country, however, there is a wide variation in the distribution of population. In Figure 5.1, the regional variations are categorized into six levels, which can be generalized into three broad spatial types: (a) high density (more than 400 persons per square kilometer), (b) low density (less than 100

**Figure 5.1**
*Population Density*

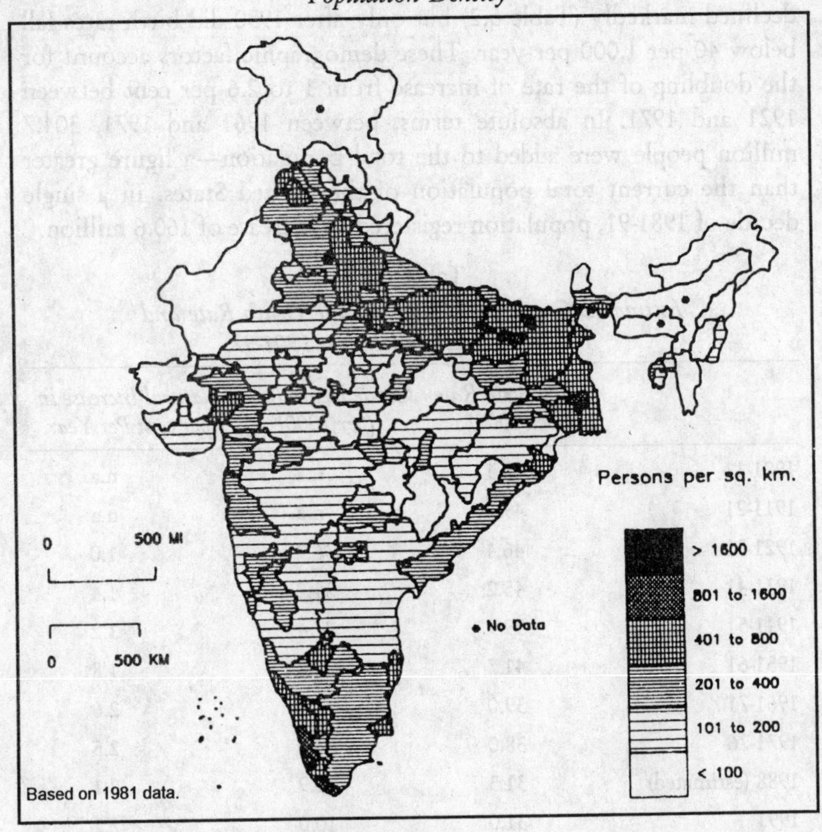


persons per square kilometer), and (c) medium density (100-400 persons per square kilometer).

Areas of high density are mainly peripheral to Peninsular India, and to the arid and semi-arid parts of the Thar desert (west Rajasthan). These high density areas include the Ganga plains, north Punjab plains in north India; southern Gujarat, Kerala (Malabar coast) along the western coast; Tamil Nadu (especially the Kaveri delta) in the extreme southeast; and the eastern littorals of Orissa and Andhra Pradesh (especially the river deltas). All these areas are well-watered, arable lowlands of fertile soils and easy transportation, and contain old settlements. The single most populous section within the belt of high density is the Ganga plains and the adjoining areas of north Punjab—a vast territory of over 400 million persons. Average nutritional densities (number of persons per unit area of cultivated land) of over 3,600 persons per kilometer exist in several areas within the densely populated Ganga plains, and part of West Bengal, north Bihar and eastern Uttar Pradesh, where some 80 per cent of the population is engaged in agricultural activities. Gujarat has high density in its alluvial plains of Sabarmati river. Increased urbanization, consequent upon the industrial and commercial development of the last sixty years, has raised the density of population. Along the western coast the plains of Kerala are well-watered and composed of fertile soils. The Kaveri delta in Tamil Nadu along the eastern coast has also been a densely populated area of fertile alluvium for over two millennia. The urbanized districts of Calcutta, Delhi, Mumbai, Chennai, Hyderabad, Chandigarh, Kanpur and Bangalore contain very high densities (over 1,600 persons per square kilometer).

Areas of low population density are the Thar desert (western Rajasthan), the hill/mountainous areas of northeast India (Meghalaya, Nagaland, Mizoram, Arunachal Pradesh) and the western Himalayan region, and the district of Bastar in Madhya Pradesh. These areas contain poor soils and rugged terrain, and are handicapped by a serious shortage of water supply for any large-scale agricultural activity. In parts of the Thar desert population density is below 20 persons per square kilometer. The remaining parts of the country contain only medium densities 200-400 persons per square kilometer. Most of the Deccan plateau, south Bihar, central India, eastern Rajasthan, west Orissa, and most of Assam belong to this category.

These patterns of population distribution and density have developed over long periods of time in response to physical and cultural forces. Some of the physical factors have been pointed out above. Cultural factors have also interacted to produce these patterns. Settlement during the ancient and medieval times was based on the original consideration of defense, water supply and caste affiliations, etc. The development of urban centers as foci of administrative and commercial functions, the development of roads, growth of railroads, the development of canal irrigation facilities, growth of ports, and the opening up of new areas for settlement during the British occupation are some other factors that led to the differential rates of population growth in the various parts of the country.

Despite the differential growth of the various parts of the country, however, the overall population patterns, as shown in Figure 5.1, have not changed noticeably except in Assam and in Punjab during the last fifty years, although within this framework (areas of high and low) densities have been consistently rising. Growth rates between 1981 and 1991, however, suggest that population increases have occurred at lower rates in the high density areas (Uttar Pradesh, Bihar, Tamil Nadu)—a situation suggestive of their probably having approached a saturation level in terms of supporting further increases. Punjab, Haryana and Rajasthan continued to show higher increase in densities resulting mainly from intensification of irrigational facilities.

Figure 5.2 depicts population growth during the decade 1981-91. Areas having a high rate of growth (over 35 per cent) include: (1) the northeastern states of Nagaland, Mizoram, parts of Manipur, and northern and southeastern sections of Arunachal Pradesh where the Indian government invested large amounts of funds for development programs such as road construction, and village industries during the last thirty years in order to pacify several tribal groups inhabiting the area; (2) the urbanized districts of Delhi, Mumbai, Hyderabad, Bhopal, Bangalore, Jaipur and Lucknow; and (3) the districts of Jaisalmer and Bikaner in Rajasthan, Sidhi and Jhabua in Madhya Pradesh, Aurangabad in Maharashtra, and Surat and Gandhinagar in Gujarat experienced increased urbanization and in several instances industrialization. The islands of Andamans also fall in the category of high growth.

In the category of low growth (below 20 per cent in 1981-91) lie most of Tamil Nadu, southern Kerala, western Karnataka coastal plain, southern coastal plain of Maharashtra, western Gujarat coastal plain, areas generally containing high densities, with a few exceptions, where saturation levels of population may have been reached. Elsewhere, the relatively underdeveloped districts scattered all over India registered low growth rates such as the district of Gumta in Bihar, Bolangir in Orissa, the northern hilly districts in Himachal Pradesh and Uttar Pradesh. Districts along the Pakistan border in the strife-ridden Punjab also registered growth rates below 20 per cent during this period (1981-91).

**Figure 5.2**
*Population Growth, 1971-1981*

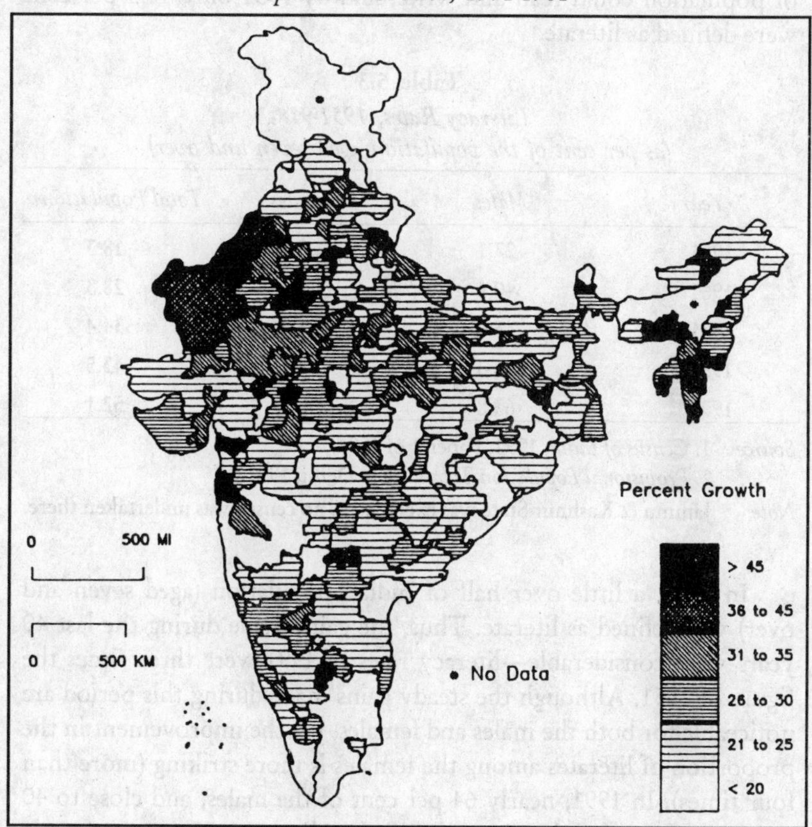


## Literacy

India has the largest population of illiterates in the world. Despite improvements in the literacy levels during the recent years, only 52.21 per cent of its population aged seven years and above could read and write in 1991. Although statistical increases in literacy are substantial —net gains of 118 million literates from 234 to 352 million in 1991. In the 1991 census, 482 million were recorded as illiterate, which excluded children below seven years of age (who would not normally know how to read and write). This fact of a vast illiterate humanity overshadows the significant gains in literacy since 1951 (Table 5.3). As a matter of fact, the actual number of illiterates increased by 76 million between 1981 and 1991 (from 406 million to 482 million).

Prior to 1951 literacy levels were very low. In 1901 only 5.4 per cent of population could read and write, and by 1951 only 18.3 per cent were defined as literate.

**Table 5.3**

*Literacy Rates, 1951-91*

*(as per cent of the population aged seven and over)*

| *Year* | *Males* | *Females* | *Total Population* |
|---|---|---|---|
| 1951 | 27.1 | 8.9 | 18.3 |
| 1961 | 40.4 | 15.3 | 28.3 |
| 1971 | 45.9 | 21.9 | 34.4 |
| 1981 | 56.3 | 29.7 | 43.5 |
| 1991 | 63.8 | 39.4 | 52.1 |

*Sources:* 1. *Census of India*, 1991, Paper 1 of 1991.
2. *Provisional Population Totals*, New Delhi, 1991.

*Note:* Jammu & Kashmir State was excluded as no census was undertaken there.

In 1991, a little over half of India's population (aged seven and over) was defined as literate. Thus, the gains made during the last 40 years were considerable—literacy rates in 1991 were three times the figure of 1951. Although the steady gains made during this period are noticeable for both the males and females, yet the improvement in the proportion of literates among the females is more striking (more than four times). In 1991, nearly 64 per cent of the males, and close to 40 per cent of the females were classified as literates. However, female literacy remains very low. In 1991, only 131 million females as compared to 230 million males were classified as literates (male-female ratio being 1.8:1).

In India, as in many other countries, females form the backbone of tradition and conservatism. Traditionally, literacy has had little relevance for females in the rural areas. Their prime functions in the rural areas are to cook, to feed the household members, to work as agricultural laborers, and to rear children. Since 1951, however, gains in female literacy have been higher than among males, especially in rural areas, largely due to increasing awareness among the masses as evidenced by increasing female enrollment in the rural schools. Government legislative measures encouraging compulsory and free

**Figure 5.3**
*Literacy, 1981*

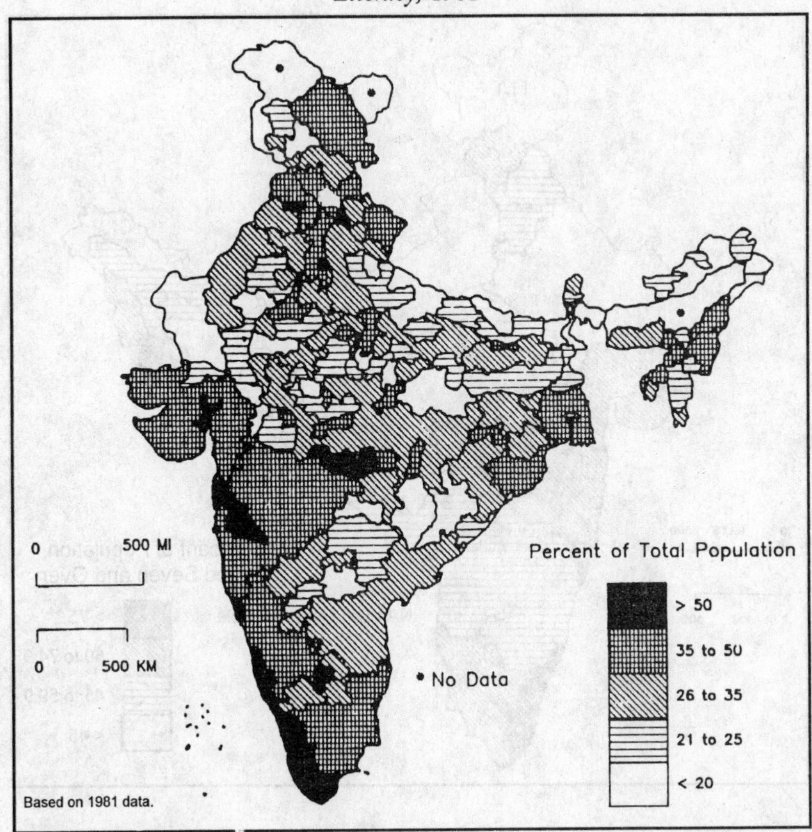


education up to the primary school level have helped create public opinion favoring female education.

Literacy differences between rural and urban populations have been striking, but are slowly narrowing. Urban literacy was twice as high as for the rural population (56.7 per cent vs. 28.6 per cent in 1981). Although female literacy is universally lower, urban male-female differentials are smaller than those in the rural areas. In 1981 male and female literacy rates in rural population were 39.3 and 17.3 per cent respectively. In urban areas male literacy rate was 64.9 per cent while only 47.6 per cent of the females were literates.

Figure 5.3 clearly brings out the broad regional disparities in the literacy levels across the country. In general, areas of higher literacy

**Figure 5.4**
*Literates, 1991*

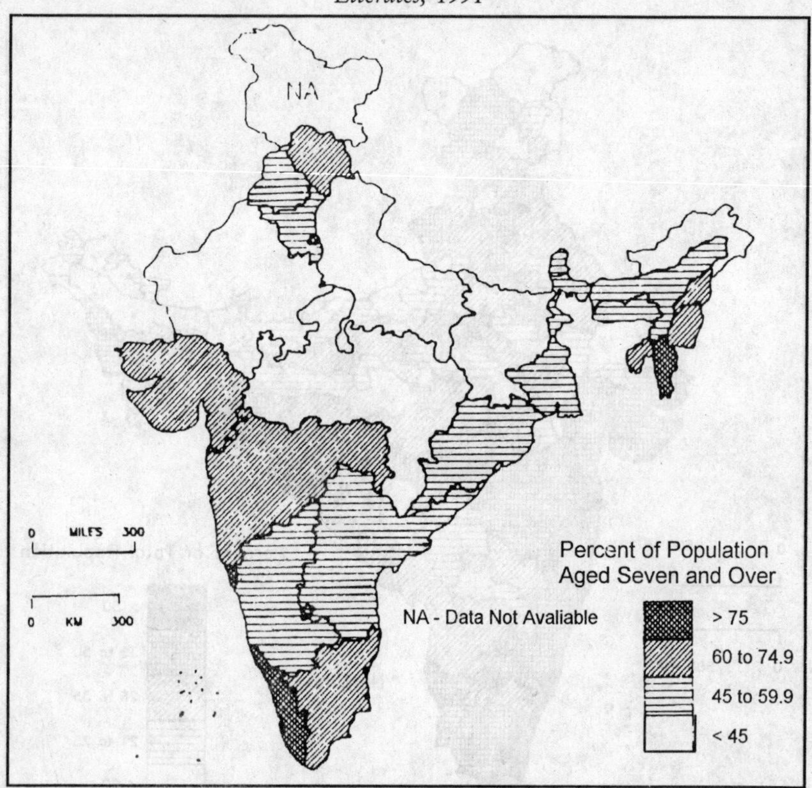


(over 50 per cent of the population) are concentrated in the urbanized districts of Delhi, Mumbai, Calcutta, Chennai, Pondicherry, Chandigarh, Hyderabad, Ahmedabad, Goa, Bhopal, Jalandhar, Shimla, Bangalore, and the entire state of Kerala. About one-third of Kerala's population is Christian. Intensive Christian missionary activity, which was responsible for opening parochial schools during the British rule, also accounts for higher literacy rate for Kerala and all the above mentioned districts. Christian missionary activity was also strong among tribal groups in Nagaland, Meghalaya and Mizoram where literacy rates in 1981 were above 35. Figure 5.3 may be compared with Figure 5.4 where the latter shows the distribution of literacy levels in the states in 1991. The broad similarity of patterns between the two figures is easily noticeable. Kerala in 1991 claimed a near universal literacy (94.4 per cent among males, and 86.9 per cent among females). Mizoram (81.2 per cent), Chandigarh (78.7 per cent), Goa (76.9 per cent), Delhi (76 per cent) and Pondicherry (74.9 per cent) were next below Kerala. Chandigarh, Goa, Delhi and Pondicherry are "city-states" (union territories) with high urbanization. Maharashtra, Tamil Nadu, Gujarat, Manipur, Tripura, Himachal Pradesh and Nagaland had literacy levels above 60 per cent mark as compared to the national average of 52.1 per cent.

In 1966, the government set up the goal of imparting free and compulsory education to all children in government schools up to the primary grade. Furthermore, thousands of new schools, teacher training and technical institutions, libraries, vocational schools and degree colleges have been opened since 1951. Large outlays of funds (over a billion dollars for the five-year period of the Fourth Plan 1966-71, and several billions during the previous plan periods) have been spent on such measures. The impact of these efforts is only slowly manifesting itself, but the goal of a completely literate society remains as yet a planner's dream. Large annual increases in population make the problem of raising literacy levels even more difficult.

## Population and the Food Problem

In order to meet with the critical foodgrain deficiency during the three decades after Independence, India spent large amounts of scarce foreign exchange on importing foodgrains. Foodgrain imports during this period averaged 3 to 5 million tons a year, rising to a peak of 10.8 million tons during 1966-67, when the United States alone supplied 8 million tons. In 1971, for the first time after Independence, the government announced that the country had achieved foodgrain sufficiency. Whether this status can be maintained in the future or whether demographic increases will outstrip foodgrain production is still problematic. Furthermore, the nutritional requirements of the people would have to be met. On the average, the caloric intake of an Indian is 2,300 per day as compared to 3,700 that of an American. Possibly famines and large-scale foodgrain shortages which plagued the country from time to time will be averted in the future. But it will be a long time before the nutritional requirements of the people can be achieved.

About 80 per cent of the total cultivated area in the country is under foodgrain crops such as rice, wheat, millets and maize. Increasing their production is chiefly a question of increasing their output yield per hectare since a possible extension of the productive areas is limited. It is estimated that 20 per cent of India's total area is in the form of culturable wastes (usually unused infertile patches full of weeds), much of which has been lying fallow since medieval times. Only a small part of the so-called "culturable waste" can be reclaimed, and they generally show little promise for future development.

Between 1947 and 1966, foodgrain production increased about 60 per cent, from 50 million tons to 80 million tons. The addition of another 10.5 million tons was recorded between 1966 and 1970. by mid-1970s production had risen to 129 million tons making India self-sufficient in her foodgrain requirements. It was officially claimed that whereas population could be expected to grow at a rate of 2 to 2.5 per cent a year, foodgrain production could be expected to grow at 4 to 5 per cent a year for several years. Officials claimed that India might never again have to depend on foodgrain imports. Official claims must, however, be interpreted with caution. By its own admission, the government report warned that the spectacular growth of 1970-71 in food production cannot be indefinitely maintained. In fact, during 1971-72, the country experienced a severe drought which forced the administration to import approximately 2 million tons of foodgrains from Canada, Australia and France. Between 1976 and 1988 foodgrain production continued to climb, except for the years of drought (1977, 1982, 1987). Production oscillated between 120 million tons for poor years and 170 million tons for 1988-89. The Seventh Plan had targeted production of 178-183 million tons for the period 1985 to 1990. The country continued to import small amounts (2 to 4 million tons annually), as population continued to grow at rates of nearly 2 per cent a year, in order to tide over emergencies as a buffer stock. A critical factor in maintaining a steady increase in production is the proverbial vagaries of the Indian monsoon. Targets are difficult to be met during the drought years, when the target-achievement gap becomes 15 to 20 million tons, and the country relapses into foodgrain shortages. The country appears now to have virtually attained foodgrain sufficiency. Future outlook, therefore, depends on several factors, such as diffusion of new techniques, population increases and natural hazards (the drought years), the last of which is of great importance.

Even the productive land of the middle Ganga plains (Uttar Pradesh and Bihar) and lower Ganga plains (Bengal) are deficient in foodgrain supply. In these areas, population densities are very high, and despite large-scale foodgrain production, the requirements remain unmet. Among the factors creating this situation are over-cropping, soil exhaustion, lack of fertilizers, backward farming methods, fragmented holdings, and the extreme poverty of the *ryot* (cultivator). These areas were also the hardest hit during the famine of 1966-67. Other areas of high population densities and foodgrain deficiency are: Kerala, western Tamil Nadu, western Andhra Pradesh, parts of Karnataka and eastern Madhya Pradesh. These areas suffer from poor rainfall, rugged terrain, infertile soils and/or limited irrigation facilities.

Reports of recent farming booms in many parts of the country, especially in the selected districts where the "Green Revolution" brought increases in wheat and rice production (e.g., Ludhiana district in Punjab, Thanjavur district in Tamil Nadu) have given some hope for the future. The Green Revolution is largely attributed to the utilization of hybrid seeds of rice and wheat, better fertilizers, and pesticides. However, the impact of the Green Revolution is yet to be felt in most areas of foodgrain deficiency. Regional deficiencies in foodgrain production continue to persist. Several studies have pointed out that the areas of foodgrain deficiency were hit hard during epidemics and famines. Movement of grain to these areas was hampered by poor transportation, inadequate storage conditions, faulty distribution facilities and administrative controls exercised by the states with surpluses. Administrative planning should be focused on narrowing the regional foodgrain production disparities. The national problem of foodgrain self-sufficiency is in reality a regional problem. Planners should realize and deal with the problems faced by farmers in the various regions (Chakravarti, 1970). Efforts should be made to universalize the diffusion of Green Revolution techniques, to augment farm activities under the existing Community Development Programs, and to enforce cultivator-assisting tenancy laws. In addition, the agricultural systems, such as the growing of alternative high-yielding food crops like cassava, sweet potatoes or bananas in areas of lesser fertility have also been recommended.

## Diseases and Public Health

India has an endemic home of several diseases, notably smallpox, tuberculosis, cholera, malaria, diarrhea/dysentery, pneumonia and typhoid. Occasionally recurrent epidemics of plague also brought death and destruction to millions especially in the urban areas and along routes of trade and pilgrimage throughout historical times. Besides, cancer also accounts for a significant number of deaths in the country.

Until 1955, malaria was the major cause of mortality, claiming about 2 million deaths a year. Its chief foci lie in the Tarai areas and in the riverine plains and rainy plains and rainy slopes of the coastal *ghats* (hills). It has been prevalent in an epidemic form in the Ganga plains, the coastal plains and in the interior parts of the country it has been endemic. There were between 60,000-65,000 yearly deaths from malaria in the 1960s, but since then there have been fewer cases. The virtual eradication of malaria had apparently been achieved by the large-scale spraying of DDT all over the country under the malaria control program aided by the World Health Organization (WHO) during the 1950s. During the 1970s, however, DDT-resistant strains of mosquitoes surfaced and the incidence of malaria has increased dramatically during recent times.

Cholera incidence and mortality have also been significantly reduced during the last sixty years primarily as a result of a large-scale inoculation program. However, it is still responsible for large numbers of deaths during religious pilgrimages. It spreads explosively along the pilgrim routes and in the pilgrim centers, especially during the mammoth gatherings of millions of pilgrims in Allahabad and Hardwar every twelfth year, and at other places at recurrent times of religious pilgrimages. Its normal and less violent occurrence zone lies in the Ganga-Brahmaputra delta, from where it generally fans upstream, often accompanied by violent outbreaks at places with poor drinking water facilities, and insanitary environment conditions.

Smallpox has been one of the major epidemic diseases in India, accounting for 30,000 to 40,000 deaths annually until 1979 when the WHO claimed that it had been virtually eradicated.

A wide-ranging campaign of inoculation and pilot projects to reduce TB incidence are currently underway. Bronchitis and related pneumonia are among the important.' causes of death, particularly in north India during winter, and among the poor in urban slums. Dysentery/diarrhea is another major health problem and is the cause of hundreds of thousand deaths. Improved sanitation and personal hygiene, and better standards of living are slowly reducing mortality due to dysentery and diarrhea.

Morbidity control in India has rapidly reduced mortality in the last fifty years. The crude death rate has fallen from 36 per 1000 a year in 1921 to 9.1 per 1,000 in 1996. The average life expectancy of an Indian has risen from 27 years in 1931 to 63 years in 1997. Maternal and infant mortality are still very serious problems, with their rates among the highest in the world. Excessive child-bearing, poor nutrition, low standard of living, inadequate maternity care in general, and inadequate sanitation and public hygiene are largely responsible for high infant mortality.

The decline in mortality in India has been sex, status and residence selective. Men are comparatively healthier, and less prone to diseases than females. Persons enjoying a higher socio-economic status are healthier, less prone to diseases, and have lower mortality rates than the average. The mortality rate is lower for urban areas than in the countryside, where public health facilities and sanitation are poorer. Despite recent reductions in mortality levels, they are still very high (10 per 1,000 population in 1997 as compared to 9 per 1,000 for the world, as reported in *World Population Data Sheet, 1997*). Further reductions will depend on several factors, chief among which are: government health policies, spread of literacy, and changes in the standards of living and improvements in infant health care. India's difficulties in this regard are heightened by the vast size and poverty of its public and inadequate financial help afforded by the government. Public health was accorded low priority in the first three Five Year Plans. The doctor-population ratio remained at a low level of 18 to 19 per 100,000 persons during the forty years of Indian planning from 1951 to 1991. An allocation of a little over $500 million for public health programs for the Sixth and Seventh Plans (1980-85; 1985-90) works out about 2 per cent of the total outlay for each plan, or over 5 per cent if the allocations for water supply and sanitation programs were added. Although this represents a substantial increase in total outlay of public expenditure from the previous plans, it is insignificant on a per capita basis (amounting to less than Rs 90 or $3.20 annually per capita for public health, medical service and water sanitation).

## Manpower and Employment

About 600 million persons belonged to the working age group of 15-59 in 1991; of which only 306 million were working. Only 37.5 per cent were counted as unemployed or partially employed. To provide for this vast unemployed labor force is one of the greatest challenges to the country. Whatever statistics on current employment are available, show that more than 15 per cent of the labor force were seeking jobs in 1990. In addition, 37.4 per cent of the labor force was estimated to be unemployed. Nearly a half of the unemployed category were educated beyond high school.

A large proportion of unemployment among the literates and the educated resides in the urban areas. Nearly 50 per cent of the urban illiterates are unemployed, as compared to the rural illiterate—employed figure of about 40 per cent among the labor force. The steady stream of rural population composed mostly of illiterates, pouring into urban centers in search of jobs in the competitive, commercial-industrial labor market, will probably increase. This urbanward migration seriously aggravates the unemployment situation.

The employment situation in the rural areas appears equally grim. Although its exact size remains unascertained, a sizable portion of the population engaged in agriculture perpetually remains underemployed and seasonally idle. During the 1990s there will be an addition of 65 million people to the rural areas who will be faced with finding work on the farms. The creation of such a large underutilized and unutilized reservoir of labor will undoubtedly intensify migration to the cities, which themselves will be feeling the crunch of inadequate additional job opportunities. By a conservative estimate, nearly 150 million additional jobs will have to be created by the end of this century. Obviously, drastic measures will be required to permit the absorption of the unemployed into the employment pool.

The problem is not merely one of idle manpower, underemployment or unemployment, but also that of raising workers' productivity. Tradition and sentiment emphasize the measurement of productivity per unit of land, or in industry by capital returns, and not according to the output per worker. Labor productivity in India is among the lowest in the world. Vast opportunities exist in the introduction of new, scientific and progressive processes of production in which the vast human resources of the country can be mobilized. Government's response to the employment problem has been to encourage cottage industries, subsidized by public funds, and protected by regulated markets within the framework of the Five Year Plans. Cottage industries, however, have failed to absorb a significant portion of the idle labor force. Private factory production has exceeded the government ventures in efficiency, consumer orientation and quality of production. The solution, therefore, lies not only in the development of cottage industries only, but in the development of progressively directed new rural industrial estates and the funneling of rural surplus labor into these agro-industrial centers, spatially organized around the varying size-groups of villages and small cities. In order to establish functioning market centers, a vast amount of rural reconstruction is also required, such as building irrigation ditches, *bunding*, (i.e., damming) well-digging, rural road construction, building of grain warehouses and building of public facilities. The idle labor force should be encouraged to adopt more productive, efficient and rewarding measures, like the introduction of labor intensive techniques of cultivation, increased utilization of forests and underground water (Johnson, 1965). The newly developed agro-industrial market centers near the villages will absorb much of the rural migratory streams which are currently moving to the larger cities. The principle bottle-neck in the implementation of these plans is the scarcity of required capital.

## Demographic Structure and Population Explosion

Perhaps, the most disturbing aspect of India's population is its demographic structure. Birth rates, now estimated at 31 per 1,000 per year, have remained high, between 40 and 45, during the first half of the twentieth century. Death rates, on the other hand, fell markedly during the same period, from 36.2 to 22.8 per 1,000 population. Infant mortality experienced a particularly noticeable decline, as life expectancy at birth rose from 32 years in 1951 to 57 years in 1991. A combination of these forces pushed the rate of population increase from 1.3 per cent a year in 1941 to nearly 2.1 per cent in 1991.

Virtually universal and early marriages, low divorce rates, and a lack of recognition and practice of birth control measures in a tradition-bound society have contributed to high fertility rates. About 36 per cent of the population is under 15 years of age, causing the working population to carry a large dependency load. A demographic representation such as the age sex pyramid dramatizes India's population problems. In it, each except the last few five-year age groups (known as a cohort) is larger than the one above it. Such a trend would push the current population of 860 million to nearly 1.2 billion by the year 2000. In sum, the total population is already large, fertility is high, death rates and infant mortality are declining, the dependency load is large and rates of population increase at nearly 2 per cent annually.

A careful analysis of the growth of India's population suggests that even in the event of significant reductions of birth rates, India's population would rise to 986 million at the turn of the century. Another study's estimates based on the assumption that the Indian women will beget only an average of 2.4 children (as compared to the figure of 3.9 in 1990) and these birth rates are, reduced by one-half as prevailing in 1990, India's population will still reach to a billion mark. Other scenarios, based on assumptions of higher birth rates, are more grim. The Seventh Plan estimates that the number of females in the reproductive age group 15-44 will go up from 183 million in 1990 to 228 million in 2000. Although the fertility levels in India during the last seventy years have declined, the total number of births have continued to climb, and population continues to soar.

## Family Planning Experiment

The population problem, or more specifically the problem of high birth rates, was one of the major pressing domestic issues confronting the Indian government after independence in 1947. When, in 1951, the government formulated its First Plan (1951-56), family planning programs were included in it, and India became the first major nation among the developing countries to adopt a public policy aimed at achieving reductions in birth rates. Specific targets to achieve reductions in birth rates from 45 to 32 per 1,000 persons annually for 1969 were fixed for the First Plan, and revised later to a figure of 25 per 1,000 for 1974. Initial government efforts were half-hearted and slow. Policy was primarily designed to diffuse information on birth control among the illiterate masses. During the Second Plan (1956-61), there was a shifting of emphasis from mere publicity advocating voluntary abstinence to more tangible and effective measures, such as a large-scale male sterilization (vasectomy) program, and free supply and servicing of the IUD (intra-uterine device) to females. Several states even offered such inducements as monetary awards and three days' paid vacation to go with each vasectomy operation. During the Third Plan (1961-66), this drive was further intensified. However, practical results were not forthcoming. The basic reason was that the lifestyle and traditional beliefs of the people remained unchanged. By the end of the Third Plan, it was clear that the achievement of the targets in birth rates was merely a demographic pipe dream.

Allocations of men, materials and services were increased several-fold during the Fourth Plan (1966-71). The union government's administrative set-up was tightened up and coordinated with those of the various states. Special recognition of the family planning activity was provided by creating a new division in the Ministry of Health to achieve greater efficiency. A federal "task force" of doctors was newly constituted to train at least one doctor specializing in family planning servicing for every 20,000 persons. Chief among the more recent innovations was an effort to elicit the help of the private advertising media. It was argued that the use of Indian manufactured condoms known as "Nirodh" (literally meaning "prevention") might increase if a campaign were to be conducted by well-known private companies. Condoms could be sold all over the country, over the counters of tobacco, tea, soap and flashlight shops for a unit price of one-half a cent.

Yet, major changes were not forthcoming despite pious wishes, enactment of budgetary allocations, and propagandizing of new methods. By the mid-1970s it was widely acknowledged that government family planning programs had failed. Government policy had indeed been, so far, ill-designed, and its implementation was regionally selective. Government efforts had been characterized by vacillation and half-heartedness. Backed by the government, a dramatic drive during the Emergency (1975-77) helped push sterilization (vasectomies) programs to an unprecedented level of 8.2 million in 1976-77 (many performed on involuntary clients) as

**Figure 5.5**
*Age Group*

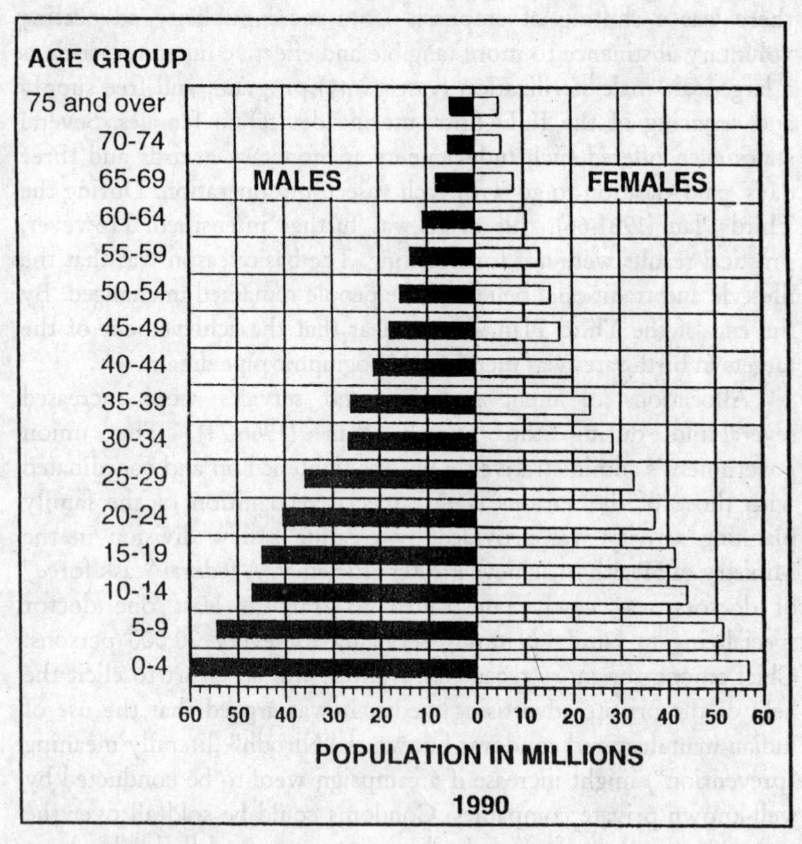


*Source:* Based on Information from Eduard Bos, My T.Vu, Ann Levin, and Rodolfo A. Bulatao, *World Population Projections, 1992-93 Edition*, Baltimore, 1992, 266.

compared to a figure of 2.61 million for the previous year.

During the 1980s the government policies shifted to voluntary sterilizations, and programmes aiming at improvement in family health particularly that of females and infants. There was a general consensus that couples could be effectively motivated to limit the size of their families only if those already living had a better chance of reaching maturity. The texture and tone of the family planning programmes were drastically altered as the emphasis was focused on maternal and childcare. The Sixth and Seventh Plans (1980-85; 1985-90) revised the targets set by the previous plans of birth rate at 29.7 by 1991 and 23.7 by 2001. The previous plans had envisaged to reduce the birth rates to 25 by 1981. The Department of Family Planning was reorganized to form the Department of Health and Family Welfare. The Seventh Plan recognized the importance of universal literacy and extension of educational facilities, and the involvement of voluntary and official agencies in order to bring about a real change in the motivation of the people to reduce the size of families. Results, however, were not noteworthy. Performance fell substantially short of targets. The Sixth Plan had targeted 24 million sterilizations, and 8 million IUD insertions to be accomplished during the five-year period, of which 17 million sterilizations and 7 million IUD insertions had been carried out. It is estimated that only 8.2 million persons were using contraceptives in 1985. To achieve the targets of Seventh Plan 50 to 60 million people must regularly use contraceptives.

The expenditure outlays for family planning during the Plans have remained at nearly Rs 25 per head per reproductive age couple, a figure which is hopelessly inadequate to effectively deal with population problems. In addition to inadequate resources, there is the question of managerial skills, and technical know-how to initiate improvements in the diffusion of family planning programs among the illiterate masses which the government teams lack. Public resistance to new gadgets and innovative techniques is also bound to be strong among the tradition-loving people, particularly when large families are prized as help in agricultural production and as social security in old age.

## *References*

Agarwal, S.N., "Family Planning Program in India", *Asian Survey*, Vol. 7 (120), 1957, p. 859.

Bhardwaj, R.C., *Employment and Unemployment in India*. New York, 1969.

Bose, A., *India's Urbanization, 1901-2001*. New Delhi, 1978.

Cassen, R.H., *India: Population, Economy, and Society*. London, 1978.

Chakravarti, A.K., "Foodgrain Sufficiency Patterns in India", *Geographical Review*, Vol. 60,1970, pp. 208-228.

Coale, A.J. and Hoover, E.M., *Population Growth and Economic Development in Low-Income Countries*. Princeton, 1958.

Davis, Kingsley, *The Population of India and Pakistan*. Princeton, 1951.

*India: Ready or Not, Here They Come*. Population Bulletin, Vol. 25 (5), 1970, specifically subsection by John P. Lewis on "Population Control in India".

*India: A Reference Annual*. Publications Division, Government of India, 1971, 1973, 1977-78, 1990.

Johnson, E.A.J., *The Organization of Space in Developing Countries*. Cambridge, Mass., 1970.

Learmonth, A.T.A., *Health in the Indian Sub-Continent*. Canberra, 1965.

Lewis, J.P., *Quiet Crisis in India*. Washington, D.C., 1962.

Muthiah, S. *et al.* (eds.), *A Social and Economic Atlas of India*. New Delhi, 1987.

Mishra, R.P., *Medical Geography of India*. New Delhi, 1970.

Nath, Pran, "A Study of the Economic Conditions of Ancient India", *Proceedings of the Royal Asiatic Society*. London, 1929, Chapter 5.

Schwartzberg, J.E. (ed.), *A Historical Atlas of South Asia*. Chicago, 1978.

*The Seventh Five Year Plan, 1985-90, Vols. I and II*, Government of India, Planning Commission, New Delhi, 1985.

Spate, O.H.K. and Learmonth, A.T.A., *India and Pakistan*. London, 3rd ed., 1967.

*World Population Data Sheet, 1997*, Population Reference Bureau, Washington, D.C., 1997.

# 6

# Rural India

A majority of Indians live in villages. According to the 1991 census, India had nearly 580,000 villages and nearly 67 per cent of these contained less than 1,000 inhabitants. Ubiquitously dotting the landscape, except where topographic, soil and climatic conditions are unfavorable, these villages are interspersed with towns and relatively few large cities. In 1996, there were only 3,768 urban places in India, two-thirds of which contained less than 20,000 inhabitants, as compared to 557,100 villages of much smaller size. Table 6.1, showing the number and size categories of rural places, indicates that there is an inadequate development of a graded hierarchy or rural settlement sizes in the country.

A functionally integrated and graded spatial system of settlements acting as central places (i.e., serving areas around them) and linked with higher level settlements within their travel range, has yet to evolve in the country. Most villages have grown relatively self-sufficient, isolated, and not effectively linked with higher level settlements even within their travel range and are inaccessible to the larger market centers. Large market centers, i.e., centers for the exchange of goods and services, are only sparsely scattered along a few linearly developed railroads or paved roads. Spatial dispersion, of small villages resulted over a long time from such historical factors as the movement of caste and religious groups due to local variations in soil fertility or due to availability of water. Villages emerged as

**Table 6.1**

*Distribution of Rural Settlements, 1981*

| *Population Range* | *No. of Villages (1000's)* | *Per cent of Villages in this Range to Total Number of Villages* | *Population in Millions* | *Per cent of Population in this Range to Total Population* |
|---|---|---|---|---|
| Less than 200 | 120.0 | 21.6 | 12.2 | 1.4 |
| 200-499 | 150.7 | 27.0 | 51.2 | 10.1 |
| 500-999 | 135.9 | 24.4 | 97.1 | 19.2 |
| 1000-1999 | 94.5 | 17.0 | 131.5 | 25.9 |
| 2000-4999 | 46.9 | 8.2 | 137.3 | 27.0 |
| 5000-9999 | 7.2 | 1.4 | 47.1 | 9.3 |
| 10,000+ | 1.8 | 0.4 | 31.2 | 6.1 |
| Total | 557.1 | 180.0 | 507.6 | 100.0 |

*Notes:* 1. These figures exclude Assam, which in 1971 had 21,995 villages.
2. Percentage figures are rounded off.

*Sources:* 1. *Statistical Outline of India, 1989-90*, Bombay, 1989, p.44.
2. *Census of India, 1981, Part II A(i)*, pp. 704-707.

self-contained socio-economic units, and their functional linkage (commercial links) with higher ranking settlements did not develop. Commercial and industrial enterprises were concentrated in the few, larger urban nuclei, which developed at places where the local chieftains established their administrative headquarters. A pyramidal centralization of administrative and commercial functions (and in some instances of religious activities) in a few urban centers was thus established in ancient and medieval times. Under the British rule (latter half of nineteenth and early twentieth centuries), a linear pattern of railroads linking these urban centers was established, especially in the North Indian Plains, without a similar development of transverse feeding sub-branches to feed the smaller towns away from a few linear axes. With the passage of time, ancient and medieval central places such as Patna, Allahabad, Delhi, Agra, Madurai and Thanjavur greatly increased in size as a result of railroad linkage and enlargement of their market functions. As modern highways were built, they tended to parallel the railroads and further polarized the settlement pattern. During this extension of the railroad and road network by the British great amounts of funds were spent to link the large cities and the hinterland with the ports, and there was "entirely too little allocated to build a network of roads" that would have helped to commercialize the rural landscape (Johnson, 1970: 157). A systematic, spatially graded central place development, containing interconnected transport and communication facilities, thus failed to evolve.

The movement of goods, capital, people and entrepreneurship to cities, rather than their allocation in various size-categories of central places which are spatially dispersed in a regular manner, has resulted in an increase in the size of larger cities.

Thus, a structural dichotomy of dualistic economies evolved which is represented by the tradition-bound, relatively inaccessible, self-contained village of small size on the one hand and a large, over-crowded, urban center in which trade, industry and professions are concentrated, on the other (Johnson, 1970: 152). Such a dualistic pattern with concentration of industry, trade and profession in large urban centers and rural subsistence economy in the villages has divided society into two distinct worlds—urban and rural. Between these two worlds, there is a striking deficiency of various size-categories of central places. Johnson calls it "central place deficiency", which is measured in terms of the ratio between villages on one hand and the urban centers on the other (Johnson, 1970: 174). Developed countries contain relatively low ratios, suggestive of a lower "central place deficiency". India's high ratio indicates a highly deficient development of central places.

Government planning has largely neglected development of a central place hierarchy of settlements, and has allocated only meager funds for this purpose in the Five Year Plans (1951-83). The establishment of medium-sized and smaller cities acting as market for the countryside has not aroused sufficient enthusiasm, and the virtues of the village as a self-contained unit have been extolled. Most new development has been located in the proximity of existing large cities. Of the 397 new industrial estates established in India during the first three Five Year Plans (1951-66), only a few were located near the medium-sized cities. This policy of developing large cities and locating most new industry close to them has only accentuated the regional differences. What is needed is the development of a network of coordinated and graded market places between the villages and the large cities which could commercialize the stagnant agrarian society and relieve the unplanned migration of surplus rural labor to the cities. Industries also need to be dispersed in the countryside in the form, of a. well-planned spatial clustering of related industrial enterprises in or near small cities that "will supply the necessary frame for the whole network of development sequence, linkages, and feedbacks upon which the successful transformation of...the countryside largely depends" (Johnson, 1970: 169). However, Johnson's controversial ideas have been criticized particularly on grounds of initial heavy investments and the time-lag between the investments and anticipated returns.

## Village Social Order

Studies of traditional Indian society generally begin with the village as a functioning socio-economic unit. Approximately 80 per cent of India's population resides in over half-a-million villages. About 70 per cent of the population depends directly on agriculture for a living. The innate conservatism of the society is reflected in its stubborn adherence to the caste system, backward agricultural practices, and an economic order little affected by modern technology. Large metropolitan areas (of over 100,000 population) modified by modern technology and performing industrial-commercial functions account for only about 10 per cent of the population. Villages and metropolitan areas present two different worlds—two levels of experience. Linked poorly by transport and communications, and with a restricted exchange of goods and services, each world has crystallized into a distinct spatial reality.

Rural India remained largely untouched by outside influences. Within it, a complex system of socio-economic interdependence among the people has grown over the centuries. The various groups and castes are in varying degrees bound together in landlord-tenant (*zamindari*), patron-client (*jajmani*) and creditor-debtor relationships, in caste and ritual observances, and in a loose form of traditional self-government (*panchayat*). The British rule did not attempt to break up the traditional economy of the village.

A typical village consists of a cluster of houses or huts, with agricultural fields around it, except where the settlement pattern is "dispersed" as in parts of Bengal and Kerala. Different groups in a village are bound together in social and economic relationships. The landlords, remnants of the feudal class structure (either by inheritance or as creations of British colonial policy), typically form the apex of a social hierarchy. This group enjoys a powerful political, social and economic position in the village since it owns most of the land. Landlords lease land to tenant-cultivators for agricultural work (usually as sharecroppers). A section of this landowning group, the non-cultivating, absentee landlord section, is generally the one which can also acquire power from outside. The resident non-cultivating landowning section is generally poorer than these landlords. Below these groups are the ordinary peasants and sharecroppers, many of whom occasionally have to accept wage labor to augment their real incomes as cultivators. Finally, in such a socio-economic set-up, the landless necessarily occupy the lowest ranks.

This division of the village population into the basic categories of peasant landlords, absentee landlords, other peasants, tenants and agricultural workers is paralleled by social rankings as well. The highest order of social esteem is generally obtained by owning land; those who perform manual labor are placed at the lower level. Supervisory work carries with it a traditional social status. High social esteem is gained by abstaining from work. Sharecroppers and agricultural laborers are held in low esteem. The lowest status, reserved for the landless agricultural laborer, arises from three factors: his landlessness, his performance of manual labor and the stigma attached to manual work under the regular supervision and control of another person. The hired landless laborer may be economically better off than the sharecropper, but he stands at a lower social status.

Most village land is owned by the high castes (e.g., Brahmans, Kshatriyas), regarded as "cultivating" castes. Brahmans are ritually and traditionally discouraged from tilling. They generally engage sharecroppers to do this work for them. Many middle ranking castes (e.g., Jats) are, however, traditionally farming castes, too. Lower castes including the lowest of all, the former "untouchables" or "outcastes" are usually landless and are engaged as agricultural laborers. Each person is born into his caste and place in the village hierarchy and possibilities for upward movement are limited. Land reforms since independence have resulted in some land redistribution to the landless.

Since villages have remained relatively self-contained, their economic links with the outside world consisted of the transfer of agricultural surpluses to outside. Their basic needs for foodstuffs, fuel, cloth, sugar and a few manufactured items such as agricultural implements and leather materials were filled within the village. Castes traditionally looked after their time-honored occupations. Village castes, for example, included Dhobi (washerman), Teli (oilpresser), and Chamar (worker in leather craft). A complex system of exchange of goods between such specialized castes and the cultivating castes in return for payments in kind and services the *jajmani* system grew up.

Agricultural surpluses of the village which enter the market are largely controlled by village landlords and moneylenders. The cultivators find it difficult to circumvent the powerful landlord and offer their surplus produce in the open market, since in time of need, they have to turn to the landlord for credit. The landlords seek to monopolize the sale of surplus to their economic advantage. Such socio-economic parasitism of the privileged group has been perpetuated over centuries. Many cultivators are, therefore, usually disinclined to produce large surpluses.

Another crucial factor in raising land productivity and agricultural surpluses is the traditional tenurial system. The sharecropper is not enthused about attempting new improvements in agriculture, such as the adoption of modern techniques, and the use of fertilizers and pesticides, because of the insecurity of his tenure, his lack of capital and, more importantly, his fear that the landlord would be the recipient of a major portion of the produce without working for it. Landowners are indifferent to investing more capital for improvement since their receipts from rents and crop-shares are adequate for their needs. The stigma attached to labor further impedes intensification of agriculture. Many would prefer semi-starvation on a tiny plot of land to higher income possible from wage employment. One basic reason for the disinclination of the farmer to introduce improvements in agriculture is that traditionally agricultural production has been organized to maintain a conventional mode of living in a caste-structured society rather than as a means of acquiring profit for reinvestment and expansion of production.

## Role of Agriculture in the Villages

Since agriculture forms a major component of India's economic activity and is basic to country's rural structure, its role in the socio-economic geography must be reviewed. Broadly speaking, there are two main types of traditional cropping systems: sedentary farming (irrigated or dry) and shifting cultivation. Each type is associated with a characteristic social and economic organization. In areas where shifting cultivation is predominant, as in the tribal territories of Assam and Arunachal Pradesh, the economic system remains largely self-contained and subsistence-oriented. In zones of sedentary cultivation, some surplus beyond the family's need is normally produced and transferred to the local market or to the landlord in return for receipt of customary services and credit facilities from him and some goes into the households who provide goods and services within the *jajmani* system.

Customarily, throughout ancient and medieval times, each peasant family had a right to till the land for its own use and to dispose of the harvest. This family could be evicted only by a community decision. Authority to allocate the wasteland or reserve land surrounding the village in times of need rested with the village community. The village community was not only a geographic clustering of houses but also a social unit. Those who were born in it shared its religion and customs. The system of land tenure in which the community held proprietary rights prevailed in most parts of the country, although in principle ultimate government ownership of land prevailed as far back as the Mauryan times.

Under the British rule several modifications in the tenurial system were introduced. Right of ownership was firmly transferred from the community to the government. Landlords who traditionally received rents from the peasants became collectors of revenue for the government, and gained rights to evict the peasant from the land or increase the rent if revenue was not readily forthcoming. The land of the evicted was then added to the landlord's private estate. A large class of private landowners, uninhibited by traditional tenurial rules, thus grew up. Landlords also received from tenants traditional "gifts" at specified occasions, in addition to the rents or land. Gradually, landlords grew in power through the annexation of vast amounts of land by peasant evictions. They hired agricultural laborers to work on their land and supervisory staff to oversee their land as well as for the collection of rents. Many lived in cities and left the working of land to the sharecroppers or hired tenants. Rent collection was delegated to local agents, who formed a growing cadre of intermediary rent collectors.

Often, the peasants, fearing eviction for non-payment of rent or delay in rent payment, would borrow money at exorbitant rates from village moneylenders, who in many cases would be the landlord himself. Landless peasants also needed loans for fulfilling customary obligations of children's, especially, daughters' weddings and other ceremonial functions. This indebtedness often remained unpaid. Indebtedness was even passed on from father to son. The landlords continued to profit from this situation and exploited the peasant proprietors over the years. The power of moneylenders in the traditional rural structure was further buttressed by their ancillary roles as merchants.

During the nineteenth and twentieth centuries some other changes were occurring on the rural landscape. Population was rising, first slowly, since 1921 at an increasing rate, without a compensating expansion in the cultivated area. In the absence of the laws of primogeniture among Hindus and Muslims, land was shared equally among all male heirs, resulting in sub-division and fragmentation of land under cultivation. Eviction of small delinquent peasants from their land, combined with the fragmentation of land holdings, contributed to the growth of a big landowner class, as well as that of the landless. The lot of a large peasantry of agricultural workers holding units of land too small to provide them with an adequate livelihood.

Over a period of time, many landholdings became very difficult to operate. Excessive sub-division of land interfered with the irrigational facilities and utilization of new farming techniques. Small scattered holdings have multiplied the labor involved in the application of fertilizers, and the use of improved techniques. All these factors have contributed to the present agricultural inefficiency and low productivity.

Traditional social systems have also affected the utilization of agricultural labor. Absentee landlordism has removed a large portion of the agricultural population a useful source of productive work.

Hereditary caste groups like Brahmans and Rajputs have resisted manual labor. Female participation ratios have been limited due to religious and social taboos among Muslims and Hindus. The agricultural work year is short and seasonal underemployment is estimated to be four to six months per year for most cultivators. Customs and religion claim much time for festivals, holidays and ceremonial occasions.

## Socio-Economic Aspects of a Village

This section deals with three selected aspects of a typical Indian village, namely, the joint family system, village *panchayat* (council) and village handicrafts.

The *joint family* or *extended family* system is widely adhered to in Indian villages. In a joint family, men (fathers, brothers, sons), their children, and wives (several nuclear families) are grouped together as a single consuming unit, and often as a single producing unit. These families reside together and share a common kitchen. The cultivating joint family also works together in the fields. When members of a joint family migrate to cities, they may find the urban environment not particularly conducive to the continuance of the family structure. Brothers among urban immigrants may find different employment in different parts of the city, and may experience transport and housing difficulties. The joint extended family tends to disintegrate, as the splintered (once extended) family finds residence in different parts of the city.

The *panchayat* or village council traditionally of five members (*panch* means five) but in practice often more, is another characteristic feature of rural society. Since villages have, in general, remained relatively self-sufficient, the traditional council or self-government composed mainly of hereditary village elders (each known as a *panch*) still persists. Such a local council looks after the fiscal needs of the village, is an arbiter of family and group disputes and has been traditionally headed by a member of a powerful local caste. The position of a village headman has largely been hereditary, passing from father to son. The village headman helps the local village officials, such as the *patwari* in the collection of taxes. The village *panchayat* has traditionally strengthened the political power of the local landlords, high caste persons and the moneylenders, who usually dominate the village council. Since 1950, the government has tried to democratize the election process for the formulation of village councils in several ways, especially by providing for a secret ballot, allowing or even requiring some membership by lower castes. Real power, however, is still wielded by the social elites, the self-selected member class and the former hereditary title holders as the headmen. The election process based on adult franchise is intended to reduce the power of the elite.

One other aspect of traditional rural life currently undergoing significant change is the gradual disappearance of the village crafts, as spatial links connecting the villages and the outside world are being slowly developed. Urban influences are percolating to those villages which lie within the travel range of the cinemas, hotels, courts, military camps and schools of the nearby cities. Educated villagers, aspiring for clerical jobs in towns and cities, have now started to commute to newly-found urban occupations. Lower caste agricultural laborers are increasingly lured to work in the nearby towns to improve their economic status. Government-instituted land tenancy reforms regarding ceilings on landholdings, abolition of absentee landlordism, democratization of *panchayats* and the opening of cooperative credit banks for granting loans to the peasants, are factors contributing to a loosening of the traditional *jajmani* system, and a shift in the agricultural methods and the eventual replacement of village crafts by modern manufactured goods from urban areas. New development programs during the Five Year Plans and the construction of transport links are increasingly bringing villages closer to the urban markets. Relative village self-sufficiency is being gradually eroded by a decline in village industries such as cloth, tea, salt, sugar, oil, soap, *beedies* (native cigarettes), bicycles, fertilizers and agricultural implements. Occupational shifts among the castes in the villages are occurring, too. Lower caste groups such as Chamar (shoemaker), Teli (oilman) and Dhobi (washerman), often find their traditional caste-based occupations no longer remunerative in the village, and are leaving for the cities in search of better prospects. This is leading to the decay of village crafts, and services.

However, modernization of the rural socio-economic structure is still far off. Increase in agricultural productivity appears to be a primary essential in this regard. Agriculturalists suffer from a lack of capital and their purchasing power for undertaking improvements is extremely limited. Private or government entrepreneurship, backed by capital through private or government resources, is generally lacking. Secondly, improvements in agriculture can be made if existing irrigation facilities are expanded, a more equitable distribution of land is carried out, and new fertilizers, better seeds, and more efficient agricultural implements are used. Thirdly, a better labor utilization can be procured by dislodging the powerful landowning class from idleness. Under the traditional caste system some caste groups are discouraged from contributing labor and energy to effectuate improvements. In sum, modernization of the rural socio-economic system cannot be achieved unless innovations are introduced and social value systems are changed.

## Rural Settlement Patterns

This section deals with the spatial dispersion, site and morphological plan of the rural settlements. While the sites of Indian villages have generally been influenced by a number of physical factors (relief, geology, drainage, exposure to wind, rainfall and sun rays, etc.) their internal forms and house plans have mostly been affected by a number of cultural factors, e.g., customs, caste and ethnic distributions, and cropping patterns. The great variety of Indian landscapes and their rich historical heritage are such that settlement patterns range from the most nulceated (compact) to highly dispersed (scattered). In general, settlements tend to be nucleated or compact in the dry farming areas of Punjab, western Uttar Pradesh, Rajasthan, northwest Maharashtra, Gujarat and relatively dispersed in areas of higher rainfall and wet farming in Bengal, Assam, Bihar and Orissa. Highly dispersed settlements are also characteristic of the coastal areas of Kerala, the Ganga delta and in the Himalayas. Four variations of these two broad categories may be identified as major rural settlement types of the country: nucleated, clustered and hamlet mixed, fragmented and dispersed (Ahmed, 1952: 139).

*Nucleated* settlements, characteristic of the western parts of the Indo-Gangetic Plain, central India, northwestern Maharashtra, Gujarat and Rajasthan, consist of a compact village surrounded by agricultural fields. Dry farming in areas of comparative aridity (annual rainfall less than 40 inches) imposes conservation of water in the village (2 or 3 wells or tanks) at selected points for efficient distribution to residences and irrigation channels for fields. Residences belonging to the lowest castes (untouchables) are often located in a separate section of the village or are not contiguous to the main village. A small street or vacant land usually separates their residences from those of the other castes. A combination of several factors has contributed to the nucleation of rural settlements in these areas. These are: uniformity of relief, soil fertility, depth of water table, relative insecurity due to recurrent invasions during medieval times, and caste interdependence. Nucleated settlements also reflect the characteristic caste or clannish solidarity among the villagers. In the *tarai* areas of the Himalayas and in the North Indian Plains where rainfall is comparatively higher, nucleated villages are located on safe and higher sites in response to the regular seasonal inundations.

*Cluster and hamlets*, the second settlement type, is common in the transitional zone between the nucleated and dispersed types in eastern Uttar Pradesh and western Bihar—the areas of adequate rainfall. Village buildings are generally located in one main nucleation, usually containing the main caste groups. This is surrounded by a few scattered hamlets, often belonging to minor caste groups of the village.

The third type, that of *fragmented* settlements consisting of clusters of residences spatially scattered in the village, is characteristic in regions where a number of lower castes predominate. Villages in this case do not, as a rule, contain any caste group with a clear majority. Such settlements predominate in western Bihar, northern parts of West Bengal and Orissa. Such hamletted villages also occur in hill tracts.

The fourth type, the *dispersed* settlement, is common in the Himalayas (Himachal Pradesh, mountainous areas of northwestern Uttar Pradesh), West Bengal and Kerala. An abundance of surface water, a multiplicity of castes and relative security from invaders during medieval and modern times have contributed to this pattern of settlements.

In the Himalayas, disposition of physical features, such as slope, soils and exposure to sun and wind, have inhibited nucleation of settlements. A variation of the dispersed type, a linear arrangement of dispersed residences, is common in Kerala, where residents are generally strung along a street backed by fields of paddy and other crops. In the Ganga delta region of West Bengal, residences are

Figure 6.1
*Samiala*

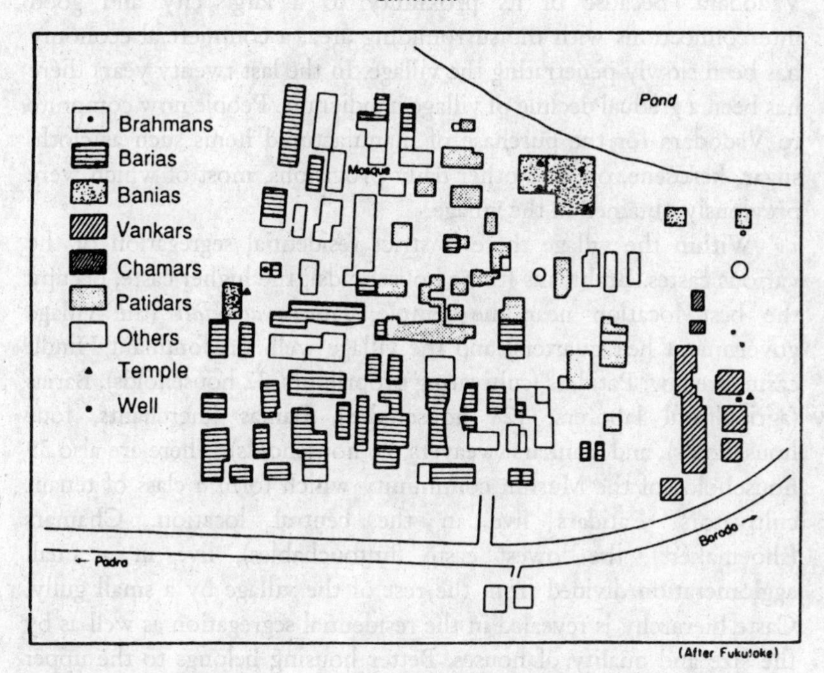


scattered in the fields, resulting in an extreme dispersal of settlement. The main reason for this is a plentiful water supply.

## Internal Structure of Rural Settlements: A Study

While there are many variations in the internal structure of rural settlement, three representative examples, a nucleated village in the dry farming area of Gujarat, a dispersed village in the paddy farming area of West Bengal and a village in Maharashtra which has undergone considerable transformation during the last fifty years are selected for study.

The village of *Samiala*, population 1,400 (Figure 6.1) is located 13 kms southwest of the commercial and industrial city of Vadodara (population 300,000) in the dry farming area of Gujarat state. It is connected by a paved road with Vadodara and with Padra—a small city located 3 kms away. There is regular bus service between Samiala and Padra. A narrow-gauge railroad line also links the village with Vadodara. Because of its proximity to a large city and good interconnections with the surrounding area, a commercial economy has been slowly penetrating the village. In the last twenty years there has been a gradual decline of village handicrafts. People now commute to Vadodara for the purchase of manufactured items such as cloth, sugar, kerosene, oil, and other daily provisions, most of which were previously obtained in the village.

Within the village there is strict residential segregation of the various castes. Brahmans (three households), the higher caste, occupy the best location near the temple, *Panchayat Hall* (the village government headquarters) and the village well. Predominant Hindu castes are few: Patidars (cultivating proprietors, 32 households), Barias (agricultural laborers, 128 households), Banias (merchants, four households), and Vankars (weavers, 26 households). There are also 39 households of the Muslim community which form a class of tenant cultivators. Patidars live in the central location. Chamars (shoemakers), the lowest caste (untouchables), live in a small agglomeration divided from the rest of the village by a small gully. Caste hierarchy is revealed in the residential segregation as well as by the size and quality of houses. Better housing belongs to the upper castes of Brahmans and Patidars. Caste solidarity is strong. Chamars (an untouchable caste) are not allowed to use the main well of the village, despite legal sanctions against untouchability. The untouchables are also denied participation in the village festivals. They have their own well and a religious shrine. The landless laborers' lot is altogether miserable, and they are increasingly seeking jobs in Vadodara.

*Supur* village (Figure 6.2), population 580, lies in West Bengal, and is an example of a dispersed village characteristic of a wet farming area. The village is about 7 kms from the nearest railroad station at Shantiniketan, and is connected with Bolpur by an uneven dirt track. Residences (117 households) are dispersed over a large area, lying close to the several ponds and tanks which are surrounded by the fields. Wide spaces between one house lot and the other are occupied by palm trees, brushes and grasses. Rainfall is over 70 inches annually and there is plenty of surface water, which is utilized in paddy fields and by the houses around the ponds. There is a greater diversity of caste

**Figure 6.2**
*Supur*

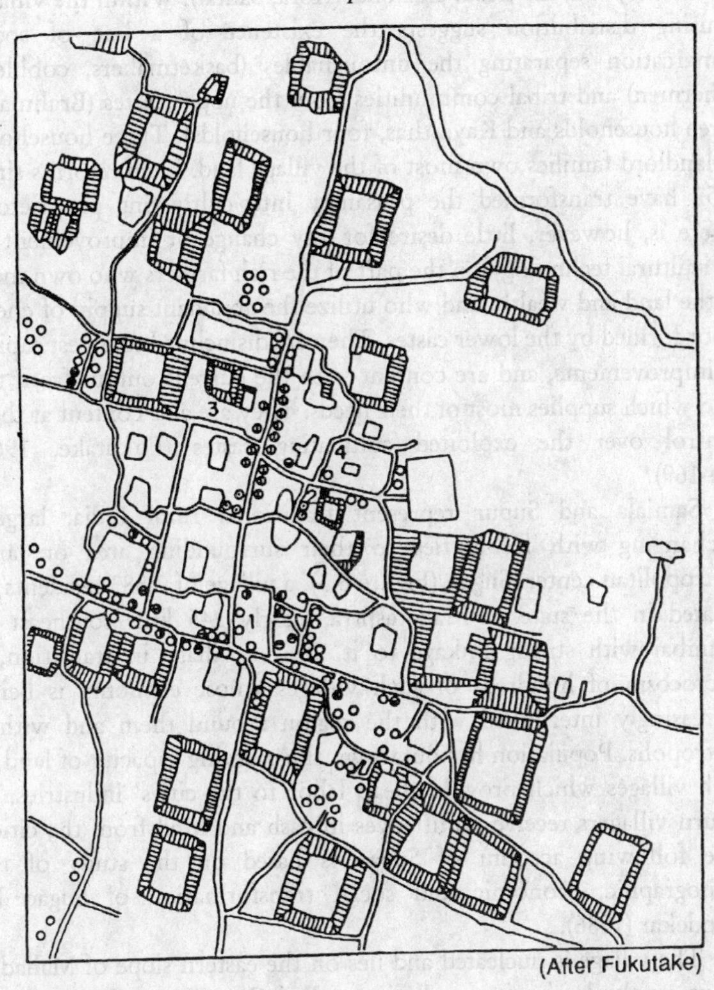


- Brahmin
- Baidya
- Kayastha
- Vaisha Saha
- Dhopa
- Suri
- Goala
- Baishnab
- Others
- Farms around ponds.

1 - Panchayat Office, 2 - Zamindar Office,
3 - Goswami Temple 4 - Zamindar Temple

groups (about 30) than in Samiala, in addition to the Muslim community and the tribal elements (Kora, Santal). Within the village, housing distribution suggests the existence of a line of social demarcation separating the untouchables (basketmakers, cobblers, fishermen) and tribal communities from the upper castes (Brahmans, three households and Kayasthas, four households). Three households of landlord families own most of the village land. Land reforms since 1952 have transformed the peasantry into cultivating proprietors. There is, however, little desire for any change or improvement in agricultural technology on the part of the rich farmers who own most of the land and wealth, and who utilize the abundant supply of cheap labor formed by the lower castes. They are disinclined to invest capital in improvements, and are content with the current output from the land which supplies most of their needs. They are also content at their control over the exploited cultivating castes (Fukutake, 1967: 164-169).

Samiala and Supur represent traditional rural India, largely unchanging with feeble ties to their surrounding area or large metropolitan center. Sugao (Figure 6.3), a village of 2,583 residents, is located in the state of Maharashtra, nearly 245 kms southeast of Mumbai with strong linkage to it. It is a village in transition, a microcosm of hundreds of such villages whose economy is being increasingly interlinked with the region around them and with a metropolis. Population has outgrown the carrying capacity of land in such villages which provide cheap labor to the cities' industries. In return villagers receive remittances in cash and kind from the cities. The following account of Sugao is based on the study of the demographic, economic and social transformation of Sugao by Dandekar (1986).

The village is nucleated and lies on the eastern slope of Mahadev range on the bank of a small river called Chandraganga. A railroad passes within a distance of 5 kms of the village of Wai, which provides Sugao with its most shopping needs in addition to holding a Monday market for the surrounding area. Sugao's inhabitants visit the shopping area and Monday market of Wai usually on bicycles. Electricity arrived in Wai in 1964, and in 1968 in Sugao. Most residents of the two villages now make use of radios and television sets, as many houses are now equipped with electricity.

**Figure 6.3**
*Sugao*

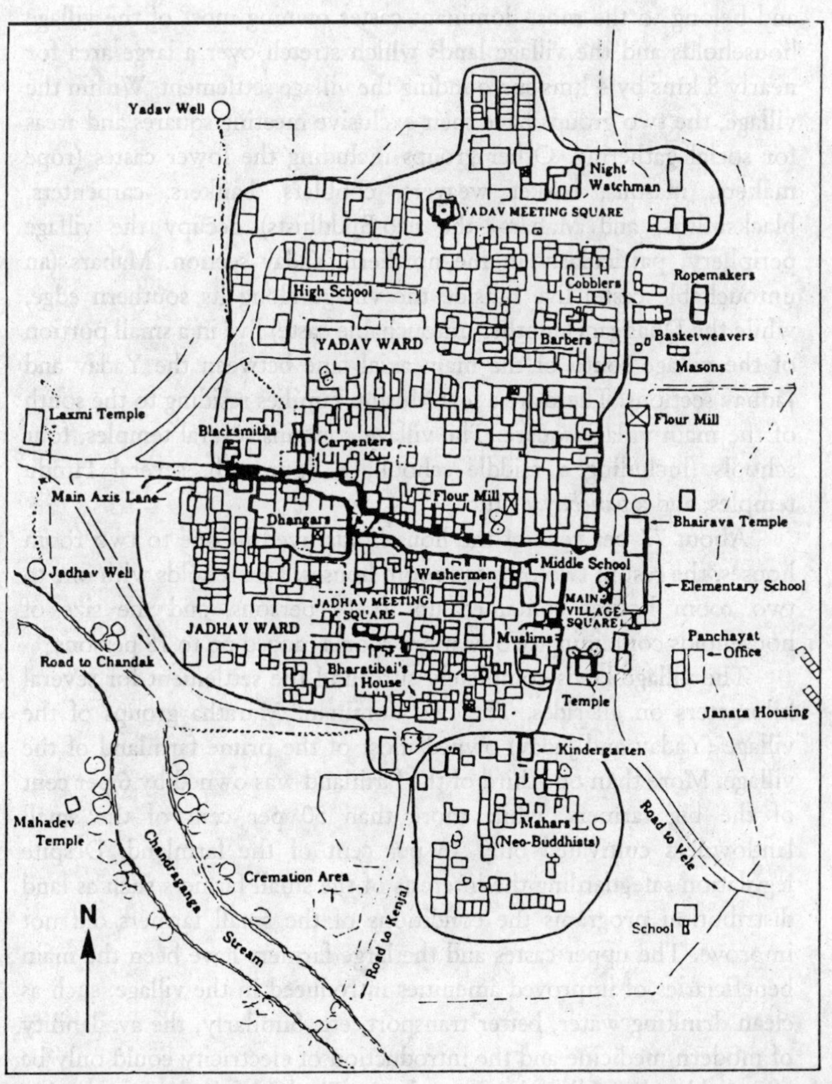


Reprinted with permission.

The village can be divided roughly into two unequal halves by an eastwest axial lane with Yadav households occupying mostly the northern section, and Jadhav groups predominating in the southern section. These two groups belong to the upper caste Maratha families, and belong to the most dominant castes owning most of the village households and the village lands which stretch over a large area for nearly 8 kms by 4 kms surrounding the village settlement. Within the village, the two groups have their exclusive meeting squares and areas for social gathering. Other groups including the lower castes (rope makers, masons, basket weavers, cobblers, bankers, carpenters, blacksmiths, and Mahars—the neo-Buddhists) occupy the village periphery, particularly in the northern Yadav section. Mahars (an untouchable caste) live outside the village along its southern edge, while the Dhangars (another untouchable caste) live in a small portion of the village south of the main axial lane between the Yadav and Jadhav sections. There are a few Muslim families residing to the south of the main village square. The village contains several temples, four schools (including a middle school), a flour mill, several Hindu temples, and a *panchayat* office.

About 77 per cent of the households lived in one to two room houses, the rest in two or more room houses. Households with one to two room houses contained up to 14 persons, and the size of households containing two or more rooms ranged up to 18 persons.

The village lands completely surround the settlement for several kilometers on all sides. The two dominant Maratha groups of the village (Yadav and Jadav) owned most of the prime farmland of the village. More than one-third of the farmland was owned by 6 per cent of the big farmers, while more than 50 per cent of the small landowners cultivated only 16 per cent of the farmland. Despite legislation safeguarding the interests of the small farmers such as land distribution programs the conditions of the small farmers did not improve. The upper castes and the large farmers have been the main beneficiaries of improved amenities introduced in the village, such as clean drinking water, better transport, etc. Similarly, the availability of modern medicine and the introduction of electricity could only be afforded by the well-to-do upper classes. The lot of the lower castes, in reality, suffered a setback, as the new industries in the village displaced the traditional artisans such as blanket weavers, rag-pickers, cobblers,

**Figure 6.4**

*Sugao Population, Resident and Migrant (by Age and Sex)*

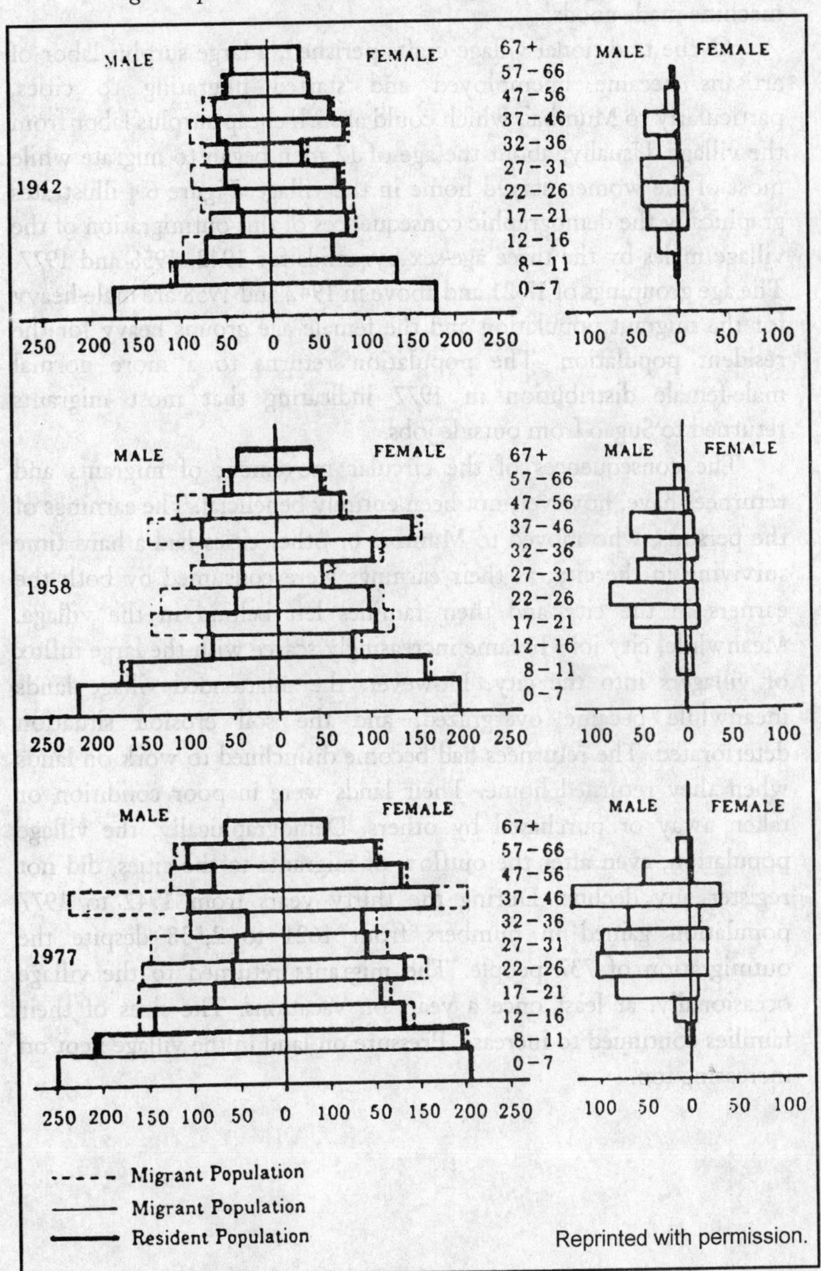


Reprinted with permission.

carpenters, and small farmers who could not enter into the new cash crop economy. Electricity displaced the village handicrafts by the machine-made goods.

As the traditional village crafts perished, a large surplus labor of artisans became unemployed and started migrating to cities, particularly to Mumbai, which could absorb cheap surplus labor from the village. Usually, about the age of 17 men began to migrate while most of the women stayed home in the village. Figure 6.4 illustrates graphically the demographic consequences of this outmigration of the village males by the three age-sex pyramids for 1942, 1958 and 1977. The age groupings of 17-21 and above in 1942 and 1958 are male-heavy for the migrant population and the female age groups heavy for the resident population. The population returns to a more normal male-female distribution in 1977 indicating that most migrants returned to Sugao from outside jobs.

The consequences of the circular movement of migrants and returnees have, however, not been entirely beneficial. The earnings of the persons, who moved to Mumbai or other cities had a hard time surviving in the city, as their earnings were consumed by both the earners in the city and their families left behind in the village. Meanwhile, city jobs became increasingly scarce with the large influx of villagers into the city. However, the unattended village lands meanwhile became overgrazed, and the soil erosion situation deteriorated. The returnees had become disinclined to work on lands when they returned home. Their lands were in poor condition or taken away or purchased by others. Demographically, the village population, even after the outflow of migrants to the cities, did not register any decline. During the thirty years from 1942 to 1977 population gained in numbers from 1621 to 2,538 despite the outmigration of 737 people. The migrants returned to the village occasionaliy, at least once a year, on vacations. The sizes of their families continued to increase. Pressure on land in the village kept on increasing too.

## References

Ahmed, E., "Rural Settlement Types in Uttar Pradesh." *Annals of the AAG*, Vol. 42,1952, pp. 223-246.

Bouton, M.M. *et al., India Briefing*. 1989. Boulder, CO, 1989.

Dandekar, H.C., *Men to Bombay, Women at Home*. Ann Arbor, 1986.

Fishlock, T., *India File*. London, 1983.

Fukutake, T., *Asian Rural Society*. Seattle, 1967.

Government of India, *India 1997: A Reference Annual*. New Delhi, 1990.

*India: A Country Study*. 4th, ed., Washington, DC, 1985.

Johnson, E.A.J., *The Organization of Space in Developing Countries*. Cambridge, Mass, 1970.

*Market Towns and Spatial Development in India*. New Delhi, National Council on Applied Research, 1970.

Muthiah, S. *et al. (eds.), A Social and Economic Atlas of India*. Delhi, 1987.

Neale, C., *Developing Rural India*. Riverdale, AM, 1986.

Owen, W., *Distance and Development: Transport and Communications in India*. Washington DC, 1968.

Singer, M., *When a Great Tradition Modernizes*. New York, 1972.

Singer, M. (ed.), *Traditional India: Structure and Change*. Philadelphia, 1969.

Srinivas, M.N. (ed.), *India's Villages*. Bombay, 1955.

*Statistical Outline of India*. 1989-90, Tata Services, Ltd., Bombay, 1989.

Schwartzberg, J.E. (ed.), *A Historical Atlas of South Asia*. Chicago, 1978.

# 7

# Urbanization

Similar to other developing nations, Indian towns and cities reflect two strikingly different faces of modernization. On the one hand, they are the repositories of economic development, higher education, and modern conveniences; on the other, they contain all the problems that follow growing modernization like the problems of crowding, growth of slums, deteriorating infrastructure and rising unemployment. Like other nations, historically, they have acted as diffusion centers of culture and economic advancement but they developed their unique traits.

Primarily containing administrative, religious and military functions during the ancient and medieval times, most cities developed as foci of economic, commercial, industrial and artistic activities. For centuries Patliputra (Patna), Varanasi, Allahabad, Agra, Ujjain, Lahore and Madurai have been well-developed cities. Several cities grew up as centers of pilgrimage (Hardwar, Gaya, Rameshwaram, Madurai), and have long been renowned for their shrines and temples. Large industrial-commercial cities in pre-British times were few in number and generally archaic. Once established as religious or administrative centers, such cities acquired commercial and industrial functions as well. Their fortunes rose and fell with the frequent movement of the seat of government from one place to another during different dynastic rules.

Modern cities based on industry and trade were, for the most part, developed as market centers during the British occupation when the Indian ports were connected by paved roads and railroads with the inland areas of agricultural production and raw materials. Some old cities languished as a result of competition with these newly emerged cities, but several adjusted to the need for a changing functional role, and acquired a modern industrial and commercial base, shaking off their traditional handicraft industries. The establishment of more modern marketing facilities for the collection and distribution of agricultural materials, and the development of railroad transport, helped quicken the growth of their commercial activities. By a process of accretion of functions, most cities became, in due course, diversified combining commercial, industrial, administrative and religious functions. Single function cities, specializing in commerce, transport, recreation, public administration or manufacturing activities, were very few. Some notable exceptions were: Jamshedpur (an iron and steel center in south Bihar), Ahmedabad (a cotton textile manufacturing center in Gujarat); Bhilainagar, Durg, Rourkela (the three steel manufacturing cities developed since Independence), and Nangal and Sindri (fertilizer industry centers). New Delhi and Chandigarh were established as centers of public administration. Many cities also served as centers of wholesale trade, transport and services.

The British also established a number of towns, including "summer capital" in the hills where most civil branches of the imperial and provincial administration would move during the hot season (mid-April to mid-September) to escape the heat of the plains. The summer capitals were connected with the plains by railroads or roads. Shimla, India's summer capital until 1947, Darjeeling and Ootacumund are some examples. During the British rule, a number of single function hill-resorts, such as Nainital, Dalhousie, Mussoorie and Mahabaleshwar also grew up.

Until 1931 urbanization was slow and halting, suggesting a relatively low rate of economic development. India's level of urbanization in 1901 was roughly comparable to that of the United States in 1830. Since 1931, the rate of urbanization has quickened (Table 7.1). At the time of the 1991 census over 215 million persons lived in India's 3,696 urban centers, an urban population exceeded only by those of the United States, and the People's Republic of China.

**Table 7.1**
*Urban Population, 1901-91*

| *Year* | *Total Urban Population (Millions)* | *Per cent Urban* | *Increase in Preceding Decade (Millions)* | *Increase in Preceding Decade (per cent)* |
|---|---|---|---|---|
| 1901 | 25.8 | 10.8 | - | - |
| 1911 | 25.9 | 10.3 | 0.1 | 0.4 |
| 1921 | 28.1 | 11.2 | 2.1 | 8.3 |
| 1931 | 33.5 | 12.0 | 5.4 | 19.1 |
| 1941 | 44.1 | 13.9 | 10.7 | 32.0 |
| 1951 | 62.4 | 17.3 | 18.3 | 41.4 |
| 1961 | 78.9 | 17.9 | 16.5 | 26.4 |
| 1971 | 109.1 | 19.9 | 30.2 | 37.8 |
| 1981 | 159.7 | 23.3 | 50.6 | 46.3 |
| 1991 | 218.0 | 25.7 | 58.3 | 26.7 |

*Source:* *Census of India*, 1961, 1971, 1981, 1991.

Urban population is defined in the Indian census as that contained in places of 5,000 population with three-fourths of its labor force engaged in non-agricultural occupations and officially accorded municipal status. In addition, places with a population of less than 25,000, a density exceeding 1,000 persons per square mile, and a dominant non-agricultural base, are classified as urban. In 1901, only 25.8 million persons (10.8 per cent of the country's population) were classified as urban dwellers. By 1991 urban population had increased some nine-fold to 218 million, representing 25.7 per cent of the country's population.

Table 7.2 also gives some statistics of urban population for 1901, 1931, 1981 and 1991. Three noteworthy trends of urban growth during the twentieth century may be observed. First, urban population has been growing at a faster rate than the country's total population. A faster rate of urban population growth indicates a net inmigration to the cities. Between 1961 and 1991, an estimated 20 million persons from rural areas moved to cities. A second trend is the remarkably higher growth rate of cities (population 100,000 and over) than of towns less than 100,000 population. Not only have cities been

**Table 7.2**

*Urban Population Growth by Size Categories*

| | *1901* | | *1931* | |
|---|---|---|---|---|
| *Size* | *Numbers* | *Population in Millions* | *Numbers* | *Population in Millions* |
| Less than 5000 | 479 | 1.6 | 509 | 1.7 |
| 5000-9999 | 744 | 5.2 | 800 | 5.7 |
| 10,000-19,999 | 391 | 5.3 | 433 | 6.0 |
| 20,000-49,999 | 130 | 4.0 | 183 | 5.5 |
| 50,000-99,999 | 43 | 2.9 | 56 | 3.8 |
| 100,00+ | 25 | 6.6 | 35 | 10.3 |
| All categories | 1,801 | 25.6 | 2,017 | 33.0 |
| | *1981* | | *1991* | |
| *Size* | *Numbers* | *Population in Millions* | *Numbers* | *Population in Millions* |
| Less than 5000 | 229 | 0.8 | 197 | 0.7 |
| 5000-9999 | 739 | 5.6 | 740 | 5.7 |
| 10,000-19,999 | 1,053 | 14.9 | 1,107 | 17.0 |
| 20,000-49,999 | 738 | 22.4 | 947 | 28.7 |
| 50,000-99,999 | 270 | 18.2 | 345 | 23.6 |
| 100,00+ | 216 | 94.5 | 300 | 139.6 |
| All categories | 3245 | 156.4 | 3696 | 215.3 |

*Sources: Census of India, Provisional Population Totals, Paper 2 of 1991*, New Delhi, p. 22, 23 and 35.

growing rapidly, their share of the urban population has also been progressively increasing. In 1901, there were 25 cities containing 23 per cent of urban population. In 1991, their number increased to 300 absorbing 65.3 per cent of the urban population of the country. The growth of "millionaire" cities also shows a parallel trend. Their number rose from 2 to 12 during the same period, and they registered over nine-fold increase in their population. Thirdly, small towns with populations less than 10,000 remained nearly constant in their absolute population (6.9 million in 1901, 7.4 million in 1931, 6.4

million in 1991), partly resulting from a changed definition of "urban" in 1961. Their share in urban population has been declining. During the Five Year Plans most large public industrial undertakings (metallurgical, transport equipment, manufacturing and chemical plants) have been located in or near the large cities.

Until 1931, the main urban concentrations were Calcutta, Bombay (now Mumbai), Madras (now Chennai) (the three major ports), and a few nuclei of modern manufacturing, commerce and services. Areas of less urbanization (25 to 40 per cent population as urban) were quite limited and were associated with those inland districts (civil divisions) which contained large commercial, industrial and administrative cities such as Lucknow, Delhi, Amritsar, Ajmer, Nagpur, Indore and Ahmedabad. By 1981, the urbanization patterns of 1931 had become intensified. Urbanization had extended along the main railroad and road links in the North Indian Plains, the industrial area around Calcutta and the heavy industrial zones around Bangalore-Mysore, Madurai-Coimbatore, Ahmedabad-Surat and Kanpur. Districts in which large cities were located figured prominently as those containing a high rate of urbanization (over 60 per cent of population as urban. These highly urbanized districts contained the "millionaire" city or metropolitan areas and the cities (population over 100,000). Figure 7.1 gives the distribution patterns of urbanization. As per the 1991 census, only 217.18 million people (25.72 per cent) out of the total population of India is termed as urban. Districts with urbanization status less than 15 per cent were found in the agriculturally stagnant parts of the middle and lower Ganga plains, Telangana, non-irrigated sections of west Rajasthan, remote hilly tribal areas in the northeast, the flood-prone Mahanadi delta in Orissa, and eastern Madhya Pradesh.

The growth of urban population for 1961, 1971, 1981 and 1991 for the states is shown in Table 7.3. Within the states urban population for 1961 ranged between 5.2 per cent of the total population for Nagaland to 88.7 per cent for the highly urbanized union territory of Delhi. In 1991 Nagaland had moved up to 17.2 per cent, whereas the union territory of Delhi to 90 per cent. Table 7.4 gives the rate of growth in urban population during the decade of 1981-91. Nearly all the states registered healthy increases, from nearly

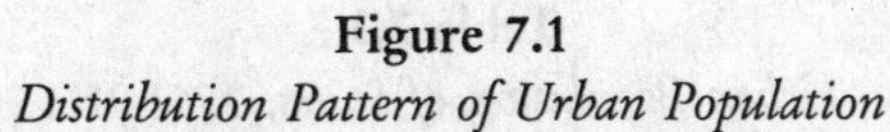

**Figure 7.1**
*Distribution Pattern of Urban Population*

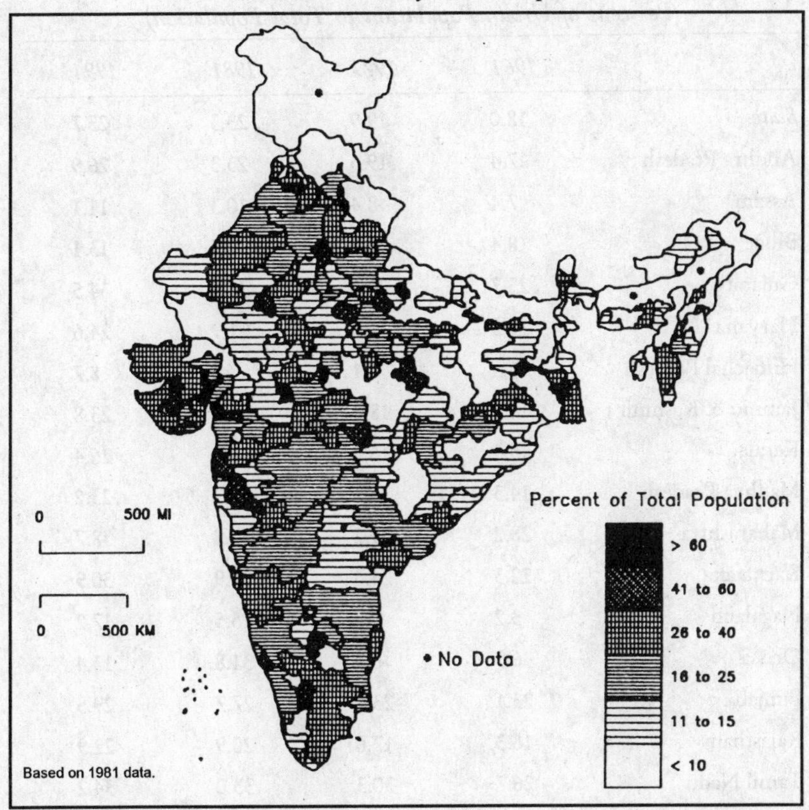


20 per cent for Tamil Nadu to 46 per cent for Jammu & Kashmir (the national urban population growth was 36.5 per cent).

Indian states are large in size and display a wide diversity of development levels. These statistics fail to portray the detailed regional patterns which only district-level statistics can present. But broad regional trends may, however, be noted. One, states with a low level of urbanization experienced higher growth rates. The backward states of Nagaland, Manipur, Orissa and Assam, each containing less than 10 per cent urban population, grew at faster than average rates. With the exception of Orissa, their high rates can be attributed to the emergence of new administrative towns and cities in predominantly rural territories, and a continuing influx of refugees from Bangladesh

**Table 7.3**
*Urban Population of States, 1961-91*
*(Per cent of Urban Population to Total Population)*

| | *1961* | *1971* | *1981* | *1991* |
|---|---|---|---|---|
| *States* | 18.0 | 19.9 | 23.3 | 25.7 |
| Andhra Pradesh | 17.4 | 19.3 | 23.3 | 26.9 |
| Assam | 7.4 | 8.4 | 10.3 | 11.1 |
| Bihar | 8.4 | 10.0 | 12.5 | 13.1 |
| Gujarat | 25.7 | 28.1 | 31.1 | 34.5 |
| Haryana | 17.2 | 17.8 | 21.9 | 24.6 |
| Himachal Pradesh | 6.3 | 7.1 | 7.6 | 8.7 |
| Jammu & Kashmir | 16.7 | 18.3 | 21.0 | 23.8 |
| Kerala | 15.1 | 16.3 | 18.8 | 26.4 |
| Madhya Pradesh | 14.3 | 16.3 | 20.3 | 23.2 |
| Maharashtra | 28.2 | 31.2 | 35.0 | 38.7 |
| Karnataka | 22.3 | 24.3 | 28.9 | 30.9 |
| Nagaland | 5.2 | 9.9 | 15.5 | 17.2 |
| Orissa | 6.3 | 8.3 | 31.8 | 13.4 |
| Punjab | 23.1 | 23.8 | 27.7 | 29.5 |
| Rajasthan | 16.3 | 17.6 | 20.9 | 22.9 |
| Tamil Nadu | 26.7 | 30.3 | 33.0 | 34.2 |
| Uttar Pradesh | 12.9 | 14.0 | 18.0 | 19.8 |
| West Bengal | 24.4 | 24.6 | 26.5 | 27.5 |
| *Union Territories* | | | | |
| Chandigarh | 82.8 | 90.6 | 93.6 | 95.0 |
| Delhi | 88.7 | 89.7 | 92.8 | 90.0 |
| Manipur | 8.7 | 13.2 | 26.4 | 27.5 |
| Meghalaya | 12.5 | 13.0 | 18.0 | 18.6 |
| Tripura | 9.0 | 7.8 | 11.0 | 15.3 |

All figures are rounded off to one decimal.
*Sources:* *Census of India*, 1971, Paper 1.
*Census of India*, 1981, Paper 1.
*Census of India*, 1991.

**Table 7.4**

*Urban Population Growth, 1981-91 (in per cent)*

| | | | |
|---|---|---|---|
| Andhra Pradesh | 43.2 | Maharashtra | 38.9 |
| Assam | 39.6 | Orissa | 36.2 |
| Bihar | 30.2 | Punjab | 28.9 |
| Gujarat | 34.4 | Rajasthan | 39.6 |
| Haryana | 43.4 | Tamil Nadu | 19.6 |
| Himachal Pradesh | 37.8 | Uttar Pradesh | 38.7 |
| Jammu & Kashmir | 45.9 | West Bengal | 29.5 |
| Karnataka | 29.6 | Other States | 60.8 |
| Kerala | 61.0 | Union Territories | 47.1 |
| Madhya Pradesh | 44.9 | All India | 36.5 |

*Sources: Statistical Outline of India*, 1996-97, Mumbai.

and West Bengal to the urban centers. High growth rates in Orissa are associated with the growth of mining and industrial activity. Above average urban growth rates were also recorded in Bihar, Madhya Pradesh and Jammu & Kashmir showing evidence of change resulting from new development plans, the building of transport links, exploitation of minerals and the establishment of heavy industry.

A second trend, which started in 1921, is the continuation of high growth rates for large cities. Figure 7.2 gives the location and growth of cities (over 100,000 population) between 1971 and 1981. As noted earlier, high growth rates for the cities are attributable mainly to heavy inmigration since 1951. In the decade 1951-61, however, there was a slackening in the growth rates as compared to the previous decades. This slowing in the acceleration of the growth of large cities suggests a slowing of inmigration to the cities—a situation perhaps resulting from rising unemployment in the large cities and a slight improvement of economic conditions in smaller cities and the countryside.

Thirdly, there has been an uneven spatial distribution of the degree of growth rates, resulting largely from the development of selected locations of heavy industry established during the Fourth, Fifth and Sixth Plans. The industrial centers of Dhanbad, Bhopal, Durgapur and Durg registered phenomenal gains, registering

**Figure 7.2**

*Growth of Cities with Population over 100,000, 1971-1981*

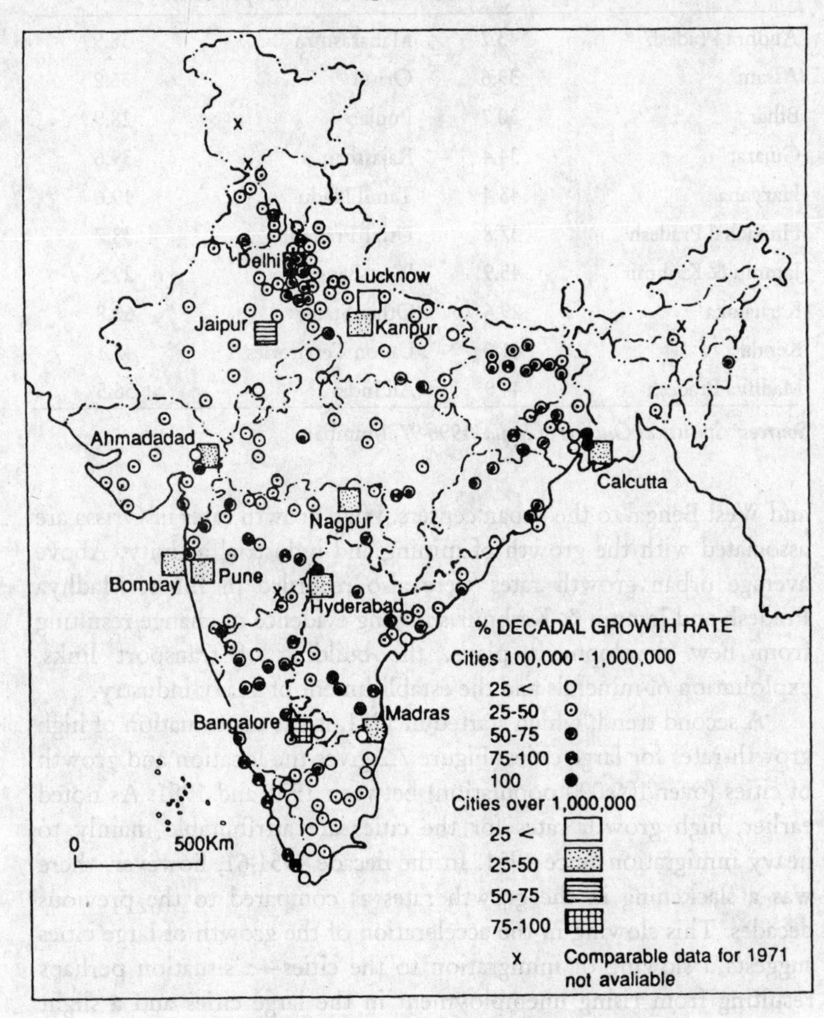


population increases over 75 per cent. Some administrative centers, such as Chandigarh, New Delhi and Bhopal, also grew over 75 per cent during 1971-81. Increased trade of the ports of Kerala was particularly noteworthy resulting in Quilon, Trivandrum, Kozhikode and Trichur growing between 75 and 100 per cent. The "millionaire" cities also recorded greater increases in population than the other cities. They registered between 24 per cent for Lucknow and 76 per cent for Bangalore. Delhi and Jaipur grew 57 and 58 per cent respectively during the same period.

## Internal Urban Forms

In internal forms, the urban settlement pattern often parallels that of rural settlements, as for example, the urban duplication of village residential segregation. However, these contain several additional features acquired during the British rule.

Indian towns and cities present a striking contrast to the American urban landscape. American cities are younger, typically exhibit planning in their gridiron layout and functional zonation, and have a basic morphological coherence. Well-planned cities in India, on the other hand, are few and of recent origin. The planned cities of Chandigarh, Bhubaneshwar, Gandhinagar, and the industrial townships of Bhilainagar, Durg and Rourkela belong to this category and were developed in the last thirty years. Among the notable exceptions to this general rule are the well laid-out, romantic pink city of Jaipur (built by Jai Singh, 1699-1743), and portions of old Delhi (Shahjahanabad), and the capital city of New Delhi built by the British between 1911 and 1931.

The rich, almost chaotic cultural, economic and architectural diversities of the Indian urban landscapes are products of a variety of influences, including different religions, languages and caste groups; and finally of the impact of British colonial rule. Through the medieval ages, Indian cities grew into amorphous assemblages of "period" pieces (historical remnants belonging to various periods) and juxtaposed areas of diverse communities, without any basic unit of layout and function (Brush, in Turner, 1962: 57-58). During the British occupation, Anglicized residential sections (known as "civil lines") and other military spaces were added to the old, indigenous parts of the cities, producing a dichotomous situation of contrasting patterns of the juxtaposition of the old, congested, unplanned parts of the city with the spacious and laid-out new sections.

### *The Indigenous City*

The older indigenous sections of the city (known in most cases as the "old city") are packed into a small space, usually a mile or less in diameter, but housing a majority of the city's total population. Streets are winding, irregular and narrow, made narrower by the protruding open shop fronts on the ground floor. The old city contains mostly one- or two-story brick structures and is not uncommonly surrounded by walls and an outer moat, broken by a few gates (e.g., in Delhi, Agra, Lucknow). The street pattern reflects the original defense consideration of containing residences, administrative offices and religious shrines within the wall. Though essentially residential in character, the old city shares its limited space with commercial and industrial uses. Thus, "a melange of convoluted streets, cul-de-sacs, alleys and by-ways gives access to residences and commercial uses and much small industry encroaching... on the public right of way" (Breese, 1966: 64). Outside the wall and the "old city" street pattern are more regular. In the southern India pattern of the indigenous city are less pronounced than in north India.

Population densities are very high in the old city. Some parts of Calcutta contain about 575,000 persons per square mile. Old Delhi has 450,000 persons per square mile and in parts of Mumbai there are over 350,000 persons to a square mile. Under heavy population pressures, the original land uses of residences have often disappeared. Dwelling units are often converted into shops or small factories, or sub-divided to house more families. Thus, the high density of residences is compounded by a mixture of land uses: residential, commercial and industrial. Generally, the lowest story contains commercial uses, whereas residences occupy higher stories. Service industries and, in some instances, small-scale industries are also interspersed in the residential areas. In general, residential areas are lacking in shopping facilities, open spaces, and such amenities as adequate water supply and sewage.

Within residential areas, various communities are segregated into areal groups (neighborhoods known as *mohallas*) by caste, religion, occupation and regional association, resulting in an essential lack of social cohesiveness. Brahmans and high castes are usually located in the best-built areas near the center of the old city. Muslims live in separate *mohallas* and so do labor castes. Menial outcastes or "untouchables" reside largely outside the walled portion of the city.

The hub of the retail merchandise activity, however, is the centrally located *chauk* (the main shopping area) or the *bazar*. It is crowded with countless small retail stores dealing in food, cloth, hardware, jewelry and other consumer goods. Bankers, moneylenders, health practitioners, dentists and public letter-writers all congregate here. Upper rooms of the buildings in the *chauk* are used as dwelling units. In larger cities, specialized services and goods tend to congregate into specific areas around the *chauk* and develop into shopping sub-centers or small *bazars* such as a cloth market, a grain *mandi*, a street of brassware shops, a *bazar* of goldsmiths or silversmiths or pottery-makers. In these small *bazars* craftsmen perform their work and display their wares in an open front shop. There are also the ubiquitous cart peddler, and sidewalk hawkers who do their work (ranging from corn-roasting to fixing eye-glasses) in the customer's presence, giving them a chance to supervise.

The density of streets and the functional mix in the old city are further complicated by a diversity of transportation in large urban areas. Pedestrian traffic is generally dense, although it fluctuates at different times of the day. Modes of transportation are many and varied—human, animal and vehicular. Between the sluggish donkey and the rapid commuter train (as in Mumbai) are: tonga, pedicab, motorcycle rickshaw, bicycle, ox-cart, taxi, private automobile, handcart, trucks of several kinds, bus, and street-car (as important modes of transportation of men and goods). The chaotic mixture of land uses produced by such diverse transportation facilities has been thus described by Breese: "The melange of facilities, sharing the right-of-way in generally uncontrolled fashion, is both the product and the creator of a high mix of land uses in Indian urban areas, especially in the old city sections" (1966: 57). The old and the new, the bullock-cart and the taxi jostle each other in chaotic complexity, and the *chauk* is always humming with life. Most smaller towns contain characteristics similar to that of the larger towns, but have a pronounced rural aspect. These are a little more than market places with incipient administrative functions.

### *Modern European Sections*

In contrast to the hodgepodge settlement patterns of the old city are the modern, well laid-out sections of the cities. These were built mainly during the twentieth century to house the British officers and civil administrators, the Christian missionary community and the growing elite community of Indian businessmen. Such sections are usually set apart from the "old city" and contain such settlements as civil lines, railway colonies, military cantonments and new residential colonies. These modern sections present an aspect of comparative spaciousness, comfort, greenery, peace and functional zonation. The tree-lined streets are broad and well-surfaced, and normally follow a gridiron pattern. Buildings are usually constructed of bricks and are frequently surrounded by large fences and landscaped lawns. House lots are distinctly marked and land use zonation is generally enforced.

The civil lines section typically is separated from the old city by a railroad and a transitional zone outside the old city walls. Originally designed during the British rule to house offices and residences of non-military (civil) branches of the government at the district or province (now state) level, civil lines persist in most of their original functions. Public offices such as the district magistrate's (administrator's) headquarters, the tax collector's office, law courts, police barracks, the jail, the government printing press, a public library, hospitals, and post and telegraph office are located here. In addition, hotels, cinema houses, colleges, clubs and stores, all tend to be congregated in the civil lines, thus creating a secondary "central business district" away from the old city *chouk*. Residences follow the Anglicized bungalow pattern of large houses containing spacious high-ceiling rooms and a big verandah or porch in the front, admirably suited to the hot climate. During the post-Independence period densities increased within the civil lines with a concomitant growth of middle-class "colonies" along the fringes. These new "colonies" are intermediate in density and planning between the old city and the civil lines.

The cantonments or the military stations are located farther apart in space only occasionally merging with the civil lines. After the Indian Rebellion of 1857 large military establishments were set up in all parts of the country with regional headquarters in urban cantonments in order to quell any future national uprising. Cantonments are spaciously laid out in geometric street formations containing barrack-blocks with rows of residences for soldiers, separate officer bungalows, hospitals, churches, officers' clubs, parade grounds, rifle ranges, ammunition depots and military supply warehouses. Cantonments are "the single most voracious land eaters in Indian urban areas" (Breese, 1966: 66). Usually, there is a small market area (*bazar*) mainly catering to the needs of the military personnel, although civilians from the nearby villages are also allowed to shop in the cantonment *bazar*. The entire cantonment complex is developed as a self-contained unit.

During and since the construction of the railroad network beginning in the middle of the nineteenth century, a large number of workers were engaged by the Indian railways. Large, sprawling colonies containing special subsidized quarters to house the railroad employees were constructed, usually along the railroad lines. These railway colonies exhibit a regularity of layout, a monotonous uniformity of structures and a graded hierarchy of housing amenities in direct relation to the wage scale and official status of resident workers. Unlike the Indian tradition of the caste- and religion-based settlement pattern of the neighborhoods (*mohallas*), the residential plans of the colonies transcend considerations of castes, religions and languages. In them a Punjabi Sikh may occupy quarters beside a Tamil Brahman or a Malayali Christian. The commercial and industrial sections of the railway colonies are less crowded than their counterparts in the old city.

Since Independence, the federal and state governments have constructed several colonies for employees working in its various civilian branches in New Delhi, Bhopal, Hyderabad, Chandigarh, and several other cities. Factory cities like Kanpur also have similar colonies. These colonies resemble the railway colonies in their gridiron street pattern and residential segregation of employees according to their wages and official status, although housing amenities are generally superior. Such government housing is frequently located at some distance from the place of work of the residents and the old city. Residents therefore must commute to work or shopping on bicycles or by city transit.

Despite the close juxtaposition of the old city and the newer European-style colonies, there is usually a transitional zone containing a railroad station, bus depot and cinema theater, and a residential spillover from the old city. This transitional zone of commercial-cum-residential functions catering to the needs of the old and new sections of the city reflects a fusion of the European and Indian lifestyles.

### *Bustees*

A special feature of most Indian cities, essentially in the last thirty years, has been the mushrooming growth of *bustees* or shanty towns. These are composed of cells of makeshift hovels which house thousands of squatters created by the large and continuing stream of inmigrants from the rural areas. Unable to find housing in the old city, migrants to the cities have spread out into *bustees* which are usually strung along the edges of the built-up areas, but are occasionally found well within the municipal areas of the city. The resident squatters (illegal occupants) generally choose vacant or underdeveloped areas of the city, such as parks, abandoned quarry sites, railroad plazas, railroad sidings or even the vacant portions of new colonies or the civil lines. Squatters' housing usually consists of galvanized iron-roofed tenements or one-roomed structures of mud walls. These appear in clusters of a few to several thousand units and each housing one to several families, thereby producing very high density areas with few public amenities. Shanty towns have added to the existing urban problems of congestions, poor transportation, inadequate public services, and an absence of proper functional zonation, and often impeding official efforts toward rational urban planning. Furthermore, as a result of the prevailing unhygienic conditions in the shanty towns, these have become the breeding grounds of diseases which may spread into neighboring areas. Municipal authorities constantly attempt, albeit futilely, to contain or eliminate the *bustees* by dismantling them, only to discover their reappearance soon after. Still another feature common to the larger cities is the occupation of sidewalk squatters who reside there at all times—rain or shine.

### *Urban Fringes*

With the expansion of the city, nearby villages are usually incorporated into its fold as "rural enclaves". These "enclaves" are generally deficient in such amenities as sewage disposal, water supply and electricity, and offer low-cost housing especially for those engaged in small-scale industrial operations such as furniture making and yarn weaving and dyeing.

Since Independence private residential colonies have also sprung up in spaces between the "rural enclaves" in a number of cities and in vacant areas close to the new city (civil lines, government colonies), although the number of residential units in these colonies still represent a fraction of those in government colonies. Private colonies usually contain small sub-urban neighborhoods designed for the requirements of middle-class and high-income groups. These private colonies range from ill-planned imitations of the government colonies to well-designed versions of foreign sub-urban development patterns. In general, such colonies are spaciously laid out and contain single-family, one- or two-story structures on wide streets with a modified gridiron pattern. However, they suffer from the great disadvantage of having inadequate public transit facilities to the rest of the city. Since high-class shopping centers are still lacking in these colonies, the residents usually commute to the old city *bazar* or civil lines for shopping. As a result, most sub-urban growth has sought locations along the few established lines of movement leading to the city core. Deficient siting and planning, and uncontrolled growth of these colonies, have created new problems of transport bottlenecks, and further taxed the provision of municipal services.

## The Role of Urban Centers

Cities in India, as elsewhere, have been the centers of cultural diffusion. Patronized by the princes and feudal lords during ancient and medieval times, these centers attracted artists and artisans from far and near. In modern times, in addition to their cultural role, these centers have become foci of manufacturing, trade and transport, thereby acting as magnets for varied employment opportunities. Thus, in the midst of a vast countryside still steeped in backwardness and apathy, urban centers propagate modern economic, social and political ideas. The cities are the places where social change originates and where the talent and organization necessary for the birth and execution of new ideas are available. Many cities function as administrative centers. They are also the loci of political, administrative and economic power. Industrial and commercial enterprises are generally located in them. The combination of all these forces means that they are likely to play an influential role in economic development.

Indian cities lead in the growth of literacy and female education and therefore help diffuse progressive modern ideas into the countryside. The census of 1991 reported 73.08 per cent of India's urban population and 44.69 per cent of rural population as literates. Most centers of specialized learning in fields such as medical sciences, engineering, vocational education, commerce and law are concentrated in the cities.

Modernizing influences generated by theaters, cinema houses and libraries are largely urban phenomena. Recent sociological studies report that several progressive tendencies are gradually taking root in Indian cities: the growing functional inoperability of caste distinctions, the breakdown of traditional *jajmani* system, a more positive response to family planning programs, and the development of political consciousness. In sum, cities are acting as great tradition breakers and playing an innovating and stimulating role in the country.

Nor is the role of cities wholly confined to their municipal limits or areas in their immediate proximity. As the spatial connections between cities and villages are developed by new roads and railroads, rural life is increasingly being transformed. The traditional rural economy, based on the *jajmani* principle, is gradually being replaced by monetized transactions and the services produced by traditional handicrafts are being displaced by machine-manufactured goods from the cities. As urbanization accelerates, the stationary subsistence rural economy will be inexorably drawn into a wider, expanding and capitalistic market nexus. On the social front, the power of the Brahmans, the landlords and the village elites, already slightly loosening, may decline further as a consequence of the diffusion of democratic ideas originating in the cities.

One consequence of the growing rural-urban links has been the cityward migration of rural population. Cities continue to absorb the courageous, and perhaps the more talented of the rural inmigrants. This represents a new loss of rural leadership, for inmigrants to cities often fail to return to the countryside. Despite greater urban unemployment, job opportunities are more numerous in cities and wages are also higher. Incomes of immigrants rise appreciably after they move to cities.

Cities, especially the larger ones, have become major recipients of massive investment funds through government planning and foreign aid programs. They thus tend to become the scene of most national development activity, new construction, industrial endeavor, commercial enterprise and educational opportunity. Since Independence, Indian cities have assumed a heightened role of economic power and responsibility.

Details on the locational characteristics, internal morphology, functional aspects, and other characteristics of major Indian cities and towns, have been given in the section relating to the regions of India (Chapters 14-17).

## Problems of Urbanization

If urbanization has been an instrument of economic, social and political progress, it has also been accompanied by serious socio-economic problems. Indian urban growth has mostly proceeded without proper planning. Physical expansion of the cities has not kept pace with the rapid increases in their population and economic activities. This has resulted in excessively high population densities, sub-standard living conditions and the growth of slums. The average densities of population in large metropolises are extremely high. Delhi averages 30,000 persons per square mile, Calcutta 40,000 and Chennai 30,000. High urban densities, especially in central parts of the cities, have resulted from a large and continuing stream of rural immigrants. The first large flow of migrants to the cities was in the "depression" of the late 1930s, as men moved in search of jobs. Later, during 1941-51, another 9 million persons moved to urban places in response to wartime industrialization and partitioning of the country in 1947. Immigration to the cities has now stabilized at about 2 million persons per year. The decade of 1981-91 thus registered a movement of 20 million persons. This inmigration has been strongly felt in the central districts of the city (the "old city"), where the inmigrants flock to their relatives and old friends before they look around for housing. Population densities beyond the "old city" decline sharply. Brush (1968: 369) refers to this situation of high density in the central parts of the cities as "urban implosion", which results from the concentration of people in the center of the city close to their work and shopping.

One of the grave problems facing the rapidly growing urban areas is housing. Chronic overcrowding has led to enormous housing deficits. An Indian National Sample Survey in 1959 indicated that 44 per cent of urban households (as compared to 34 per cent of rural families) occupied one room or less. In larger cities the figure of families occupying one room or less rose to 67 per cent (Turner, 1962: 280). Moreover, the current rate of housing construction is very slow. Indian cities require annually about 2.5 million new dwellings but less than 15 per cent of the requirement is being constructed. Major factors responsible for slow construction are: A shortage of building materials and capital, inadequate expansion of public utilities into sub-urban areas, poverty of urban immigrants, strong caste and family ties with ancestral dwellings and places of work in old sections of the city, and the lack of adequate transportation to sub-urban areas where most vacant land for new construction is located. The situation regarding the building needs of non-residential development is comparably depressing. Old buildings are obsolete or grossly inadequate for the spatial requirements of manufacturing, commerce, transport and service activities. In the absence of effective planning, this has led to a haphazard distribution of factories, warehouses, workshops, transport terminals, etc. This lack of zonation, along with the restricted availability of open spaces and inadequate public utilities, has made living conditions in many cities harsh and has jeopardized the smooth functioning of the economy. Without massive government aid there does not appear to be any immediate or long-range solution to the housing problem.

Problems of unemployment and destitution are no less serious. Urban unemployment, nation-wide, is estimated at 15 to 25 per cent of the labor force, and is even higher among the educated population. About one-half of all the educated unemployed in the country are concentrated in the four largest cities. Furthermore, urban incomes, although higher than those in rural areas, are nonetheless appallingly low. Nearly one-third of the urban population lived in poverty.

The task of Indian urban development is undoubtedly a difficult one and is linked with the broad question of economic health. It has been pointed out that "the prospects for India's urban future will remain clouded until employment oppörtunities increase, production in non-agricultural sectors of economy is accelerated, and a substantial rise in incomes makes available the savings and investment necessary for urban expansion... Urban development is merely a piece of the whole fabric of economic development" (John Brush in Zelinsky, 1970: 300-310).

How and in what regions should India's future economic development take place? Two opposing views with regard to the future locational strategy in Indian economic development have been expressed. Spearheaded by Lewis and Johnson, a strong plea has been made for a planned decentralization of India's industrial development and its relocation in medium-sized and small urban places (population 20,000-30,000) (Lewis, 1969: 197-199; Johnson, 1970: 157-177). Harris has advocated a contrary view favoring the expansion of large cities and centralizing future small-scale industries in them in order to maximize the returns from a smaller resource being invested at the ongoing centers of activities (Britton Harris in Zelinsky, 1970: 271-274). Both views, however, highlight the urgent need for urban and regional planning. In view of the acute and chronic problems of urban congestion, high unemployment and scarcity of available funds, there appears a need for development of small-scale industry in medium-sized cities within the broader framework of regional planning. Currently, there is a lack of such a focus in the location strategy of urban and regional economic development. Persuasive arguments have been advanced favoring replacement of the existing development concentrated in large metropolitan areas by a technologically progressive, geographically decentralized society organized along town-centered lines along with the development of villages. This plan envisages the location of new or expanded industrial enterprises in different size categories of settlements, such as villages, medium-sized cities and satellite towns proximal to large cities. According to the plan, the location of small agro-industries in villages and medium-sized cities will maximize the proximity of agriculture to concentrations of industrial and commercial activity. Dispersed industrialization would help in providing the kind of market environment that Indian agriculture needs, supplying the necessary network of development linkage between the polarized worlds of large cities and villages. In such a scheme, within these intermediate level cities, an integrated development of small local processing and fabricating enterprises, and the promotion of agro-industrial markets, has been proposed. It is further suggested that such a development of intermediate level cities would absorb a part of the migration stream which now crowds the larger Indian cities. A part of migration stream could be stemmed by being fruitfully utilized in new activities in the villages, while many of those who do, nevertheless, migrate could find work in new agro-industries of the medium-sized cities. The development of a graded hierarchy of settlements thus envisaged would be comprised of villages, towns, medium and large cities, which can become centers for production, trade and services offering employment opportunities for the young people who now are migrating to the metropolitan areas (Johnson, 1970: 167-171).

### *Notes*

Bala, R., *Trends in Urbanization in India, 1901-81*. Jaipur, 1986.

Berry, B.J.L. and Spodek, H., "Comparative Ecologies of Large Indian Cities", *Economic Geography*, Vol. 47 (suppl.), pp. 266-285.

Bose, A., *India's Urbanization, 1901-2001*. New Delhi, 1980.

Breese, G., *Urbanization in Newly Developing Countries*. Prentice Hall, Inc., 1966.

Brush, J.E., "The Spatial Patterns of Population in Indian Cities", *Geographical Review*. Vol. 587, 1968, pp. 362-391.

*Census of India, 1991, Series-1, of 1991*, New Delhi, 1991, *Provisional Population Totals Paper 2 of 1991*. New Delhi, 1991.

*Census of India, 1971, Paper 1 of 1971*, New Delhi, 1971.

*India: A Country Study*. Washington, D.C. 1985.

*India, 1997: A Reference Annual*. New Delhi, 1990.

*India: A Reference Annual*. New Delhi, 1973, 1977-78.

India: Town and Country Planning Organization, "Land Use Pattern of India's Cities and Towns," *Urban and Rural Planning Thought*, Vol. 11, 1965, pp. 188-190.

Johnson, E.A.J., *The Organization of Space in Developing Countries*. Cambridge, Mass., 1970.

King, A.D., *Colonial Urban Development*. London, 1976.

Lall, A. and Tirtha, R., "Spatial Analysis of Urbanization in India," *Tijdschrift voor Economische en Social Geograife*, Vol. 62,1971, pp. 234-248.

Lewis, J.P., *Quiet Crisis in India*. Washington, DC, 1969.

Mitra, Asok, *Delhi, Capital City*. New Delhi, 1970.

Mookherjee, D., *et. al.*, *Urbanization in a Developing Economy, Indian Perspectives and Patterns*. Berkeley, 1973.

Singh, R.L. (ed.), *India: A Regional Geography*. Varanasi, 1971.

Spate, O.H.K. and Ahmad, E., "Five Cities of the Gangetic Plain", *Geographical Review*. Vol. 40,1950, pp. 260-278.

Schwartzberg, J. (ed.), *A Historical Atlas of South Asia*. Chicago, 1978.

Turner, R., (ed.), *India's Urban Future*. Berkeley, 1962, especially chapter by John E. Brush.

Zelinsky, W. *et. al.*, *Geography and A Crowding World*. New York, 1970, particularly John E. Brush's article, "Some Dimensions of Urban Population Pressure in India", pp. 279-304.

# 8

# Political Geography

## Pre-Partition Patterns

On the eve of Independence, India was sub-divided into a fragmented patchwork of large and small political units. Administrative units, called provinces, were ruled directly by the British, while the remainder were the "native" states with their own rulers operating under the British suzerainty. This complex system resulted from the historical consolidation of the British interests in India, based in part on pre-British arrangements, but greatly changed by subsequent military and political developments.

The British provinces were created from the annexed territories acquired between 1757 (Battle of Plassey) and 1857 (Indian Revolt). Bulk of India's territory was acquired by force of arms. By 1815, the British had overcome French colonial competition in India and were firmly entrenched in the three key coastal locations of Bombay (now Mumbai), Madras (now Chennai) and Calcutta. From these three coastal points British occupation penetrated inland. British conquest was achieved by the amalgamation of territories acquired through wars into provinces for political and military convenience. Kingdoms or parts of kingdoms, annexed by conquest or treaties, were grouped arbitrarily into provinces under the British governors without consideration for traditional boundaries, cultural affinities or regional bonds.

Those remaining parts of the country which escaped direct

annexation largely occupied the interstices of the British provinces and were shared by over 600 native rulers. About half of these were in Kathiawad peninsula in western India and most of the remainder in adjoining Rajputana and central India. Each ruler enjoyed a degree of sovereignty under the mantle of the British power. At one extreme, some native states were as large and as populous as some of the important nations; for example, the states of Hyderabad and Jammu & Kashmir were each over 80,000 square miles in area, and the former had a population of 19 million people. At the other extreme, some states were very small, occupying areas of a only few square miles, containing a few hundred inhabitants, with few resources. Even the larger native states were unable to pose a united front to the British because they were flanked or isolated by the British provinces. By 1819, the British supremacy had become unchallengable. The British annexation of India had been completed by 1849. Only a few small insignificant coastal areas remained outside the British control in the hands of the French and the Portuguese, relics of the colonial struggles of the seventeenth and eighteenth centuries. India's state structure in 1857 and the French and Portuguese possessions persisted essentially unchanged until 1947 (Figure 8.1).

In 1974 when India attained Independence, the state system erected by the British underwent a major change through the creation of the Muslim state of Pakistan. Pakistan was formed in two separate blocks. While West Pakistan, carved out of northwestern India, contained ten native states, East Pakistan consisted solely of the eastern part of the province of Bengal and a segment of the adjacent Assam province. Some Indian observers feel that one of the basic reasons responsible for the creation of Pakistan was the British colonial policy in the twentieth century which contributed to the Hindu-Muslim schism. These scholars feel that this policy was directed at granting the Muslims greater political recognition as a means of countering growing nationalist urges of the Hindus. During the 1930s and 1940s, Muslims increasingly demanded a "homeland". By 1947, the Indian National Congress finally became reconciled to the demand of a "homeland" for the Muslims, and cooperated with the British government to make plans for partitioning of the country. The basic procedure followed on the granting of Independence was to create Pakistan from the portions of the British provinces having Muslim majorities in the population according to the census of 1941.

**Figure 8.1**
*Princely States, 1947*

(1) Hyderabad, (2) Mysore, (3) Travancore, (4) Jammu & Kashmir, (5) Gwalior, (6) Jaipur, (7) Baroda, (8) Jodhpur/Marwar, (9) Patiala, (10) Rewa, (11) Udaipur/Mewar, (12) Indore, (13) Cochin, (14) Bahawalapur, (15) Kolhapur, (16) Bikaner, (17) Mayurbhanj, (18) Alwar, (19) Bhopal, (20) Kotah, (21) Cooch Behar, (22) Junagadh, (23) Bastar, (24) Cutch, (25) Surguja, (26) Bhavnagar, (27) Rampur, (28) Manipur, (29) Nawanagar, (30) Pudukkottai. (The 30 most populated states are identified).

## *Territorial Consolidation, 1947-50*

Immediately after Independence, the governments of both India and Pakistan started tackling the difficult task of revising the complex administratively illogical state structure inherited from the British. The main problem was not the key British provinces but nearly 600 native states falling to India's share—a heterogeneous mass of territories. In addition to their remarkable variations in size, population and revenues, the native states had varying degrees of administrative cohesion and economic development. These territories, which were the accidents of history, did not coincide with the major economic, linguistic and cultural regions of the country. Furthermore, the British government had controlled external affairs, defense and communications of the native states. Such arrangements, which bestowed upon the British virtual control over the native states, may have been suitable for colonial administration, but in independent India, were clearly anachronistic.

At the end of the British rule in India, the native states were legally empowered to accede to either India or Pakistan. Most native states recognized their inadequacies as sovereign units and the wisdom of joining India or Pakistan. The new countries of India and Pakistan could ill-afford to allow the continued existence of independent kingdoms scattered throughout their territories—a situation which could seriously endanger the political stability and economic development of the new countries. Hence, the first job which the Indian government undertook was to establish its control over the native states and to create through consolidation a more logical system of states. By a process of territorial consolidation and reorganization, a union of all the native states within the overall administrative framework of the Union of India was achieved by 1950. The notable exceptions to this regrouping process were the large native states of Hyderabad and Jammu & Kashmir. The Muslim rule of Hyderabad (population 87 per cent Hindus in 1941) favored the idea of establishment of an independent state within the British Commonwealth. Under the Indian government's relentless pressure, followed by token military action, the ruler signed a treaty of accession in 1949.

The state of Jammu & Kashmir, another large area (over 84,000 square miles) posed a more formidable problem. The state was ruled by a Hindu, although 77 per cent of the population was of Muslims in 1947. The strategic importance of the state in proximity to the borders of China and the Soviet Union was heightened after partition because of its contiguity to both Pakistan and India. It was sought by Pakistan on the basis of its Muslim majority status (the Indo-Pakistani dispute over this state is discussed later in the chapter). Immediately following partition and Independence, while the ruler was still contemplating a possible course of action, tribesmen from Pakistani territory (with the possible connivance of the Pakistan government) attacked the state, forcing the ruler to make a desperate appeal to the Indian government for military help. This was quickly granted, but only after he decided to accede to the Indian Union, an action which made the state a part of India. Soon after, Pakistan rushed its own troops to support the invading tribesmen. India quickly moved its troops to halt the invaders. Thus, the Indo-Pakistan war over the state started.

Rather than flushing the invaders out of the entire state, the Indian government took the case of aggression from the Pakistani side to the United Nations which immediately appointed a commission to investigate the dispute. The United Nations Commission on India and Pakistan (UNCIP) proposed that a plebiscite be held to ascertain the wishes of the people of the state and also called on Pakistan and India to agree on a cease-fire line. A cease-fire line was delimited with areas of high altitude left undelimited, and accepted by both parties on January 1, 1949. The cease-fire line left India in possession of two-thirds of the state including the fertile valley of Kashmir lying east and south of the line, and area west and north of the line remained under Pakistan's control administered through the so-called "Azad" (Free) Kashmir government. Since then, the line has crystallized into a de facto boundary between the areas controlled by two countries.

By 1950, all the native states had acceded to India or Pakistan, and the first phase of the territorial readjustments of the former native states and the British provinces was over. In the readjustment process there were three types of arrangements which led to the country's territorial consolidation. Most of the small princely states, not viable as separate units, and located within, or adjacent to, the British governor's provinces (during British occupation) were merged with them to create large Part A states. Likewise, 275 medium-sized or small native states were integrated into five large administrative units (unions) with a separate legislature and a capital. Three of the largest states—Jammu & Kashmir, Hyderabad and Mysore—were allowed to retain their separate identities. These states and the unions of states were called Part B states. Finally, 61 medium-sized states were molded into new units as Part C states to be directly administered by the central government. Ajmer, Coorg, and the capital territory of Delhi were also accorded Part C status. Part C states were also allowed limited powers.

In addition to the British provinces and the princely states, there were a few French and Portuguese possessions in India, all located along or near the coasts. After Independence, India and France successfully negotiated, and in 1951, the French possessions of Chandernagore (near Calcutta) and Pondicherry (on the east coast) were handed over to India by France. Pondicherry has been retained as a union territory. The Portuguese possessions of Goa, Daman, Diu, Dadra and Nagar Haveli, all along the western coast, proved a real problem. Portugal remained unyielding in retaining these, whereas public opinion both in India as well as in the Portuguese possession, especially in Goa, strongly favored their merger with India. Unsuccessful in negotiations with Portugal, India forcibly annexed Goa, Daman and Diu in 1961 while Dadra and Nagar Haveli had earlier been taken over by India. Although India's occupation of Goa entailed little bloodshed, its annexation was widely condemned in the West. The former Portuguese territories have been retained as the union territories of Dadra and Nagar Haveli and Goa, Daman and Diu.

### *Reorganization Since 1950*

We have seen how the integration of the princely states into the Indian Union produced a patchwork of different categories of state structures. The political units devised by 1950, in many cases, lacked economic viability or a suitable administrative machinery. Even while carrying out the sweeping modification in the state system during 1947-50, the Indian government had announced its intention of devising a more rational reorganization at a future date. Such a reorganization would necessarily involve the creation of new units based on such considerations as distinctive regional, linguistic, cultural and economic characteristics. Against this background, and under mounting public sentiment favoring linguistic states and growing government concern, the Indian government created the first linguistically-based Andhra state, in the Telugu-speaking area of Madras in 1953. This quickly sparked renewed demands for political recognition by other linguistic and culture groups. This led to the establishment of a States' Reorganization Commission in 1954 which was entrusted with the task of making recommendations for a more rational sub-division of the country into states. The Commission's report was submitted in 1955, and was put into law as the State's Reorganization Act of 1956. Although the Commission considered several factors, language was the most important consideration in the creation of the new states. However carefully designed, no system of of Indian states could satisfy all the diverse pressure groups. After intense Marathi and Gujarati pressure, a fifteenth state of Gujarat incorporating the Gujarati-speaking areas of the Bombay state, was created in 1960. The new Bombay state became linguistically homogeneous and was named as Maharashtra. In 1956, it had already acquired the Marathi-speaking areas of adjoining Hyderabad and Madhya Pradesh (Vidarbha).

Numerous areas of political tension and many dissatisfied groups still remained. Two politically powerful groups were most vociferous in seeking territorial identification in the form of separate states: Sikhs in the Punjab and the Naga tribesmen in Assam along the Indo-Burma border. In addition, minor tribal groups were seething with discontent and agitating for their individual states. In Punjab, the Sikhs sought the creation of a Punjabi-speaking state. In reality, however, they used the linguistic issue to cloak their aspirations for a separate state in which their religion would be a majority. Initially, such demands were resisted by the Indian government out of a fear that further sub-division of the country would lead to. its undue fragmentation, especially along the strategically located international borders (e.g., the Punjab and Naga areas). However, the government yielded to the pressures of these insistent groups and created a new state of Nagaland in 1963. In 1966, out of the existing Punjab state, Punjabi-speaking areas were grouped into a new state of Punjab. Most of the remainder of the former bilingual Punjab state was named as Haryana and the balance went to Himachal Pradesh, whose status was raised from that of a union territory to that of a state in 1971.

Sweeping territorial rearrangements in the Assam hills resulted into the new state of Meghalaya which was carved out of the territory of the state of Assam in 1972 in order to satisfy the wishes of the people of the Garo, Khasi and Jaintia hills. These hill people did not wish to be dominated by the plains people of the Brahmaputra valley in Assam. In a concurrent process of reorganization the centrally administered union territories of Arunachal Pradesh (formerly North East Frontier Agency or NEFA bordering Tibet) and Mizoram (along the Burmese border) were created. In 1972, when Manipur, Meghalaya and Tripura were elevated to state status, there were 21 states and nine union territories. Following internal unrest, the Indian government accorded the status of an associate state to Sikkim in 1975, which had remained as a protectorate. The number of Indian states was raised to 22. Underlying Indian government's motivation in the creation of all these states was the hope that the creation of these political units would relieve tension in these border areas. The union territories of Mizoram and NEFA acquired the status of states in 1987 when the name of NEFA was changed to Arunachal Pradesh. Currently, the Indian Federation consists of 27 states and five union territories (Figure 8.2).

In contrast to the demands of smaller cultural groups to gain identification as separate states, the desirability of creating larger states has long been voiced. Even during discussions regarding the States' Reorganization Bill in 1956 in the parliament, a proposal for the creation of a United State of Bengal and Bihar was advanced but fell through for want of public support. Another proposal for the creation of a large southern state consisting of Dravidian India, to be known as *Dakshinapath* or *Dravidistan*, also failed to bear fruit. The proponents of larger states felt that factors of national security and economic progress were more important than cultural considerations. According to them, smaller cultural states would lead to the ultimate balkanization of the country, and be a potential hazard to the development of a viable spatial political organization. However, in several instances, the clamor for cultural "homelands" as separate states are localized in less developed areas and stems from alleged discrimination against or lack of recognition for the underdeveloped regions by political leaders from more developed parts of the state. The creation of new smaller states, therefore, can fulfill local politicians' aspirations for personal administrative role, and a greater participation in state politics.

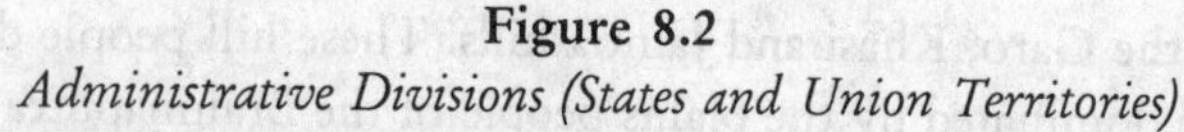

**Figure 8.2**
*Administrative Divisions (States and Union Territories)*

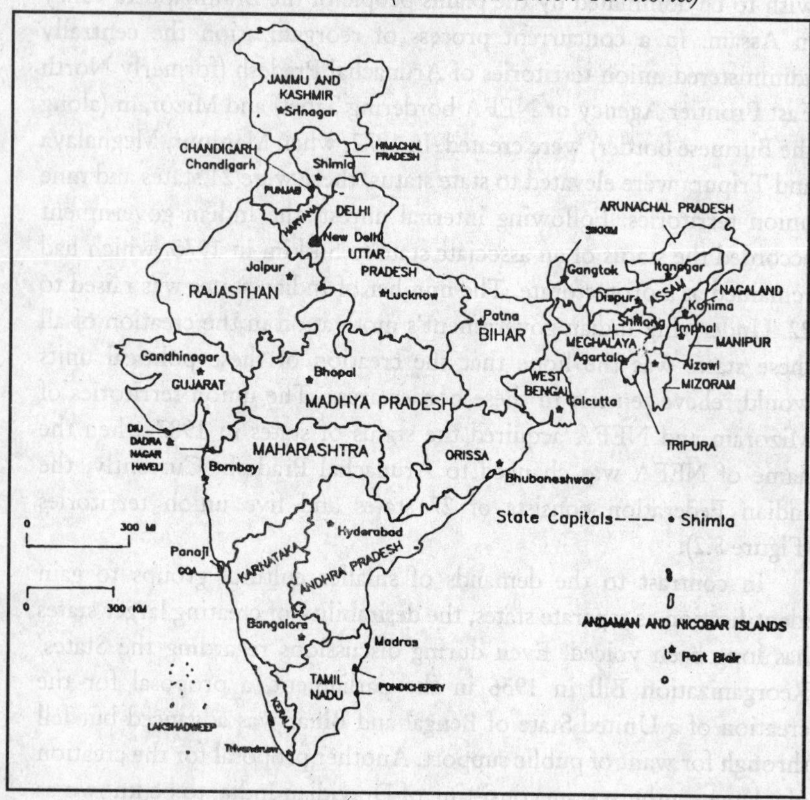


## Internal Administration

### *Administrative Territorial Hierarchy*

An experimental step toward the formation of larger political units was made in 1956 in the form of Zonal Councils. The State's Reorganization Act of 1956 divided India into five large zones, each containing a number of states and union territories functionally tied to an advisory body known as a Zonal Council. These were entrusted with purely advisory responsibilities of liaison between the states and the central government in matters of such inter-state disputes as border quarrels, water distribution and economic planning. Devoid of administrative, legal or legislative powers, their effectiveness is seriously limited. In view of the sensitive nature of the northeast

Indo-Tibet border, a special advisory body similar to the councils but invested with special functions was set up in 1973 for the states of Assam, Manipur, Meghalaya, Nagaland, Tripura, Mizoram and Arunachal Pradesh.

At the time of discussion on the states' reorganization in 1955-56, the Prime Minister pleaded unsuccessfully for the creation of larger states of the size of the Zonal Councils. Such a move envisaged five or six big administrative states for the country, each with a secretariat, legislature, one cadre of public services, one governor and a capital. It was argued that within such a large multilingual and multicultural state, various groups would not be apportioned among states, reducing the chances of clamor for cultural sub-division of a country. More importantly, it was argued that the creation of a few large states would be more conducive to the launching of coordinated economic development plans, and would result in substantial savings of public funds.

Unlike 27 states, seven union territories are directly administered by the union government in New Delhi through an administrator appointed by the President. Their legislative bodies possess reduced powers. The territory of Delhi is in fact a federal district. The Andaman and Nicobar Islands, as well as the islands of Lakshadweep (formerly Laccadive Isles) are the outlying island territories. Three are former French or Portuguese holdings: Goa, Daman (Damao) and Diu; Dadra and Nagar Haveli; and Pondicherry (ex-French) plus three other smaller territories. Other union territories lie on the northeast strategic borders with Tibet and Burma (Arunachal Pradesh, Mizoram). Figure 8.2 shows the major political units of India.

The states and the union territories are sub-divided into over 350 local administrative units, known as districts, below which are the sub-units of *tehsils*, *taluks* or sub-divisions, each comprising several (typically 100-300) villages.

### *Administrative Integration and National Security*

The territorial integrity and internal stability of India depends upon how well its administrative authority at the national level is balanced against the regional pulls of castes, languages, religions and other pressure groups. In a federal system, in which the states are accorded a large measure of internal administrative power, the overall administrative and social cohesion of the country has to be effectively protected by investing the federal government with special power to cope with the centrifugal tendencies that might be set in motion in the different states. In view of this, and despite the federal structure of this administration, India's Constitution provides for the government at the national level a large measure of centralized power. In fact, the rights of the states can be taken away from them temporarily by the federal administration by two-thirds majority of those voting in each of the two houses of the Union Parliament. States themselves can be created, abolished or divided by a majority vote. Since the complete revamping of the old state boundaries in 1956, several revisions of state boundaries have taken place, including the dismantling of old states and the creation of new ones.

The division of powers between the union and the states is clearly defined in the Constitution of India. Powers not specified in the Constitution (i.e., the residual powers) rest with the union. The federal (union or central) government has jurisdiction over national defense, foreign affairs, inter-state trade, regulation of mines, communications, railways, incorporation of banks and national insurance. The states have control over police, local government, public health, education, agriculture and irrigation. In addition to the union and state lists of powers there is also a concurrent list on which both the central and state governments can legislate. This list includes price controls and treatment of refugees, among other things. Although the states can exercise the power to tax, a large part of state revenues are derived from a revenue-sharing formula with the central government. The central government pays for many of the states' major economic development projects, hydroelectric dams, heavy electric machinery, locomotives or aircraft factories. Despite these overriding federal powers over the states, in actual practice, the states have exercised a good deal of political autonomy.

In practice, inter-state conflicts occasionally arise, e.g., disputes over boundaries, over the location of federal government development programs, or the sharing of river waters by several states. Such issues have often generated conflicts, even including acts of violence and riots carefully timed to draw central government's attention to the depth of regional feelings.

An important limitation on the powers of the states is the exercise of emergency powers by the union government at times when the security of India or any part of the territory thereof is threatened whether by war, internal disturbance, or external aggression, including domestic violence, failure of parliamentary government to function, or a breakdown of financial stability. In such cases "President's Rule" is imposed on the state, an act which must be approved by the Union Parliament within two months, and revolted every six months. The union government has exercised these emergency powers many times; nearly 10 times during 1950-67, 17 times during 1967-71 and 15 times during 1971-74. Over the last few years, President's Rule was declared in Punjab, Uttar Pradesh, Rajasthan, Jammu and Kashmir, Bihar and Goa.

The central government tries to maintain the overall unity of the country by such measures as the institution of all-India services, notably the Indian Administrative Service, Indian Police Service, and the various Economic Services. It has established integrated Indian Defense Services. The personnel of these services are obliged to serve anywhere in the country and help bring an inter-regional point of view wherever they serve. Furthermore, English has been retained as an associate language to promote national integration.

## External Relations

### *Sino-Indian Border Dispute*

India has a common, extensive and difficult border of over 2,400 miles with China, separating China from the Indian states of Jammu & Kashmir, Himachal Pradesh, Uttar Pradesh and Arunachal Pradesh. Difficult to demarcate on the ground because of rugged terrain and harsh environment, the boundary was delimited, though imprecisely, on the maps between India and Tibet (over which China has long claimed suzerainty). Although never legally accepting the boundary, China quietly ignored the the issue during the British rule in India. Not until 1959 China's claims to nearly 129,500 sq km of territory which the delimited boundary placed within India surfaced. China also laid claims to some areas of Bhutan resulting from the boundary dispute, although Bhutan was a protectorate of India.

The Sino-India border may conveniently be divided into three sectors: the western sector extends for 2,252 km from the eastern-most

**Figure 8.3**
*Northern Frontiers*

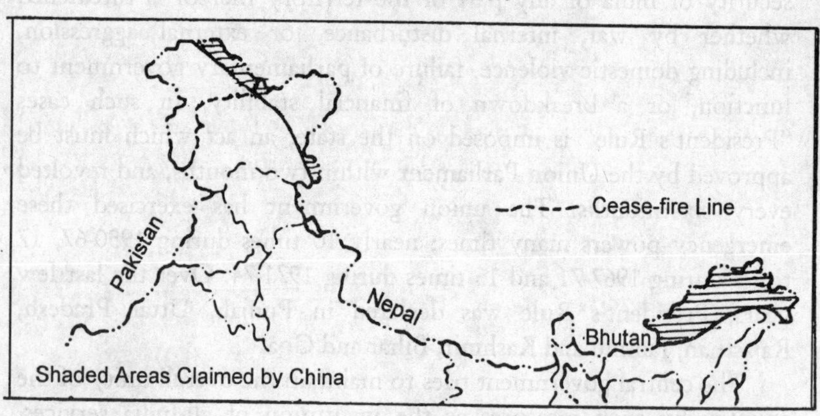


end of Ladakh in the state of Jammu & Kashmir; the middle sector lies for a few hundred miles between the Indian states of Himachal Pradesh and Uttar Pradesh and Tibet, known as the Sikkim-Tibet sector; and the eastern sector of about 1,287 km separates southeastern Tibet from the state of Arunachal Pradesh in India (Figure 8.3)

The western sector was never delimited on maps. The Chinese claims rest mainly on ethnic grounds, and on the assertion that the wastelands of the Aksai Chin in the disputed territory were always linked more with Tibet and Sinkiang. Ethnically this area is an extension of Tibet in language religion, and culture. But, the Chinese documentation regarding the actual occupation of the area by Tibet is inconclusive. The Indian case rests on the claims that the area has been historically administered by the state of Jammu & Kashmir since 1849, and that the Indo-Tibet Treaties of 1684 and 1842 confirmed the boundary between Tibet and Ladakh, although the precise boundary alignment was never undertaken (Van Eckelen, 1967: 165). China also claimed a part of the Hunza-Gilgit area in north Kashmir (ceded to it in 1963 by Pakistan), although the whole territory has been effectively under the British sovereignty since 1895.

The disputed territory in the middle sector is very small. The disputed area in the eastern sector lies to the south of the line, usually referred to as the McMahon Line, which in general follows the crest of the Himalayas between Bhutan and Burma for about 800 miles. India has stressed that the McMahon Line as the international boundary between Tibet and India as was agreed to between the governments of India, China and Tibet in 1914 and marked as such "on the map attached to the draft convention initialed by the plenipotentiaries of India, China and Tibet" (*The Sino-Indian Dispute*, 1963: 12-13). However, China has considered the McMahon Line illegal and unacceptable (*The Sino-Indian Boundary Question*, 1962:57), claiming that Tibet had no right to sign the 1914 Convention held in Simla which delineated the McMahon Line on the map. India challenges such a position, maintaining that Tibet was independent and in fact concluded several independent treaties which were considered valid by all parties, and were in operation for decades (*The Sino-Indian Dispute*, 1963: 12-13).

China denied the validity of the McMahon Line as an international boundary and laid claims to areas south of the line up to the foot of the Himalayan range in the Brahmaputra valley in the state of Assam. The disputed territory in Arunachal Pradesh of about 32,000 square miles has been administered by India since 1947. It has been inhabited by several tribal groups, such as the Miris, Abors, Daflas, Apa Tanis, Mishmis, Aikas and Monpas, each speaking its own distinctive tribal language of the Tibeto-Burmese family, and functioning more or less as an autonomous political unit under Indian jurisdiction. Racially, the tribes are closer to the Mongoloid Tibetans. The Monpas have generally adopted Lamaistic Buddhism, although some are animists. Contacts with the outside world have been few and varied from tribe to tribe, but more with the Brahmaputra valley than with Tibet in most cases. Until the nineteenth century, the British showed no interest in the tribes or their territory. It was only at the Simla Conference in 1914 that the status of the area was formalized.

Chinese governments never formally questioned the validity of the agreement between India and Tibet until 1959, although they never recognized it. From the Indian point of view there are two crucial elements in controversy; first, the fact that from 1914 onward Britain and, since 1947, India exercised jurisdiction, however weak it might be, over the area; and second that China did not dispute or formally protest against Indian control until 1959. Even in 1956 when the Chinese government's attention was drawn to certain Chinese maps showing the areas to be a part of China, Chou En-lai, the Chinese Prime Minister, promised to look into the alleged "cartographic errors" in the Chinese maps (*The Sino-Indian Dispute*, 1963: 16-17).

India's relations with China were marked by friendliness until 1959. Strangely, China's occupation of Tibet in 1950 which formed a part of the "outer ring" sphere of the British influence was designed to keep other major powers out. In fact, the leaders of independent India had disapproved of the British defense policy of keeping British influence in Tibet, which was based on the security system consisting of concentric "inner" and "outer" rings. During the British rule in India, the Himalayan countries of Nepal, Sikkim and Bhutan formed the "inner" ring in which complete border security was developed by a skillful alliance with the border states. Afghanistan and Tibet formed the "outer" ring in which diplomatic pressure kept other major powers out. After Independence, this defense policy for India's borders was abandoned. India's recognition of Chinese sovereignty over Tibet was accompanied by only a mild protest. In 1954 it relinquished its special privileges in Tibet which had been acquired from Britain, such as the right to keep military contingents in the town of Gyangste, in southern Tibet, and the control of Tibet's postal and electric services. These traditional rights were given up without trying to obtain Chinese endorsement of the McMahon Line (Van Eckelen, 1967: 193-97).

Chinese attempts to establish normal relations with India during the early 1950s when they were tightening administrative control over Tibet proved illusory. The Chinese moved slowly at first, consolidating their gains in Tibet, which they later extended to a part of Ladakh. By 1956 border incursions were growing in intensity. India, meanwhile, hoped that the insurmountable barrier of the Himalayas would leave China no choice but to follow a policy of peaceful coexistence. Indian apprehensions grew when in 1956 it was discovered that China had been building a road through Ladakh linking West Tibet with Sinkiang, and had quietly moved into Aksai Chin in eastern Ladakh—an area formerly under Indian control. Chinese military probes into NEFA were also viewed with alarm and dismay by the Indian administration. India formally protested in 1958, diplomatic notes were exchanged and both sides adopted increasingly hostile postures. Prospects for settlement through negotiations became clearly remote. Repeated small-scale armed clashes along the border between 1959 and 1962 escalated into a full-scale war in October of 1962 when China launched a major offensive in both NEFA and Ladakh. It was not quite clear how much territory China actually wanted to annex. Its demands kept shifting. As rapidly as the offensive was launched, China announced at the end of November that it would make a unilateral withdrawal from its advanced positions. The sudden withdrawal of the Chinese forces to the McMahon Line in the eastern sector puzzled most observers. Chinese forces had penetrated deep into the Himalayas, crossing the crest. Despite shipments of arms and ammunition from western countries, India could hardly fend off a swift bold push to the Brahmaputra lowlands in Assam. The Indian armed forces were outnumbered and outgunned. They were not well trained to fight a modern war in mountainous terrain. Surprisingly, Chinese forces pulled back without annexing the disputed territory in the eastern sector, a large portion of which had already fallen to them. In the western sector, however, they did not pull out of most of Aksai Chin.

The China war taught India many lessons. The myth that the Himalayas was an effective defense barrier was exploded. India's naive confidence in China's friendliness had dulled its perception regarding effective security measures in the Indo-China borderlands. The prompt and positive response of western countries in rushing military supplies to the war zone helped to improve the image of the West in Indian eyes. India realized that the posture of "non-alignment" was no substitute for defense preparedness.

The reason of expediency has been advanced for the sudden unilateral decision by China to withdraw its forces. Its advance troops would have been cut off from supply bases in Tibet in the winter of 1962 when the high passes in the Himalayas would have been closed by snow. The Chinese explanation was that they had no further territorial ambitions.

Since November 1962 an uneasy truce has been in force along the border. Unfortunately, neither side has called for a negotiated settlement of the dispute since the winter of 1962. India has tightened its security measures all along the border and only a few minor clashes have been reported. The Colombo powers, spearheaded by Ceylon (now Sri Lanka) tried to mediate a settlement, but failed to persuade China to vacate the annexed areas in Ladakh. The Soviet Union afforded some moral support to India without helping it with arms. Several countries condemned China as an aggressor. Cuba, Albania, and Portugal supported the Chinese.

One significant indirect effect of the war was the emergence of China as a superior military and political force in Asia—a position which was later bolstered when China became a nuclear power in 1973.

Clearly, it is in the interest of these two major neighboring powers to reach agreement in order to have peace in the area. In 1970 India's Foreign Minister spoke of his country's basic policy toward China in this manner: "Neither China nor India can change the geographical fact that our countries have a long common border. It is in the interest of both countries to settle the border question peacefully and normalize relations... When China is willing to take a concrete step in this direction, she will not find us lacking in response" (*India: A Reference Annual*, 1971-72: 524-25). More recently, the Chinese leaders have expressed similar views. The exchange of visits by the leaders of the two countries have only confirmed their intentions of achieving improvement in relations.

### *Border with the Himalayan Kingdoms*

Landlocked between India and Tibet, the Himalayan kingdoms of Nepal and Bhutan cover an area of a little over 194,250 sq km and share common boundaries totaling about 1,267 km with India. Historically, they were maintained within the "inner" ring of the British defense interest as "buffer" states. The dispute with China in the Himalayas has brought sharply into focus the strategic aspect of their location.

Nepal managed to survive as an independent country during the British rule over India, but consented to close attachments to India by treaties of friendship and protection from external aggression. Bhutan is protected from external invasion, although it became a fully sovereign nation and became a member of the United Nations in 1971. In Sikkim, which became a state within India in 1976, India exercised control over its external policy as well as its borders and internal administration. China has also laid claim to a part of Bhutan. Since the Sino-Indian hostilities of 1962, India has adopted defensive positions along the Sikkim-Tibet border designed to guard several strategic passes, and has reportedly stationed 25,000 men in areas which were the scenes of Chinese intrusions in the 1960s.

India formally protested in 1959 against China's claim over Bhutanese territory. China responded by denouncing Indian rights of protection of Bhutan, and asserted that any border dispute with Bhutan could be settled directly with Bhutan without Indian interference. The Indian government maintained its right to defend Bhutanese borders. It has been closely helping Bhutan in its defense efforts. There are engineering and other units of the Indian Army stationed along the Bhutan-Tibet border.

Like Bhutan, the independent kingdom of Nepal is sandwiched between Tibet and India. Neither China nor India has laid any territorial claims over it. China disputed Nepal's Tibetan border, but its claims were relinquished in 1961 following a treaty, and the entire northern border has now been demarcated.

Nepal has carefully pursued a policy of non-alignment, and has allowed both India and China to construct roads linking its capital city of Kathmandu with its two big neighbors. India considers Nepal to lie within its perimeter of defense interests and she is acutely aware of the dangerous potential of any possible future southward expansion by China made easier by the Tibet-Kathmandu road construction in 1968-70.

### *The Burma Border*

The 1662 km Burma border, extending from the limits of the Himalayan ranges southward roughly along the watershed between the Brahmaputra and Irrawady river systems, has only been demarcated recently. The border passes through the mountainous territory inhabited by several tribal groups, such as Nagas and Mizos. Distinct in culture, language and race from the plains people of the Brahmaputra valley, these tribesmen have shown considerable independence from the Indian and Burmese authorities and have occasionally used Burmese territory as a staging point for anti-Indian activities, often abetted by Communist guerrillas in Burma. In the pacification program of this border, the Indian government, as has been noted, yielded to political pressure and created the state of Nagaland in 1963 as a cultural-political "homeland" for the Nagas, and the centrally administered territory of Mizoram in 1972.

India and Burma have maintained good neighborly relations, despite Burmese deportation of much of its Indian population between 1948 and 1952. Mutually, they have tried to tighten border security against the insurgent guerrillas. Both countries reached an agreement in March 1967 regarding the demarcation of a portion of their common border.

### *Indo-Bangladesh Border*

The partitioning of the Indian subcontinent in 1947 into the two countries of India and Pakistan, created a common border of 1,943 km in East Pakistan (which since December 1971 has been separated from West Pakistan as an independent nation of Bangladesh). Never fully demarcated on ground, this border is criss-crossed by numerous streams. Only a few railroads crossed the international boundary of East Pakistan linking the Indian states of West Bengal and Assam. Rivers generally served as the main arteries of communications. East Pakistan, separated from West Pakistan by over 1,000 miles of Indian territory, was linked with its larger and politically more powerful western counterpart via the Bay of Bengal, Indian Ocean and Arabian Sea.

East Pakistan constituted a wedge of land in the east of the Indian subcontinent separating Assam, Tripura and Manipur from the rest of India until a new railroad constructed after Independence restored an effective communication link. Of crucial concern for India along this border was the possibility of the Bay of Bengal becoming hostile waters which could shelter nuclear vessels of Pakistan's allies in the event of a major Indo-Pakistan confrontation.

No major dispute arose along the border, although minor tension always prevailed. The continuously shifting streams, however, created territorial claims and counter-claims. Following the Indus Water Agreement between India and Pakistan in 1960, a dispute arose over the use of the Ganga river waters in East Pakistan. India had been planning to construct a barrage across the Ganga river at Farakka about 140 miles north of Calcutta, to rejuvenate West Bengal's river system. The Farakka plan was aimed at controlling floods, for producing more efficient water distribution between the northern and southern parts of the state of West Bengal, and helping to desilt Calcutta harbor. Pakistan feared that India's diversion of waters within the Indian state of West Bengal would seriously reduce water flow into East Pakistan, thus threatening its new irrigation projects and food sufficiency drive. India argued that the Farakka plan would not affect East Pakistan's water needs.

Negotiations failed to produce a settlement on the question of Farakka plan. India meanwhile started work on the barrage in the mid-1960s fearing that delay in construction would further jeopardize the shrinking water supply for Calcutta's metropolitan area as well as the rapidly worsening condition of Calcutta's harbor. During the 1980s a refugee influx of Bangladeshis (600,000 by 1989) into Tripura and Assam states created problems for India. India threatened to seal off its borders with Bangladesh, but never actually did it. These minor irritants apart, the emergence of an independent and generally friendly country of Bangladesh in place of East Pakistan has been viewed positively by India.

### *Jammu & Kashmir Dispute*

Unquestionably, the most critical problem between India and Pakistan has been the dispute over the territory of the former princely state of Kashmir (more accurately the state of Jammu & Kashmir), which has led to war between the two countries in 1948, 1965 and 1971, and more recently in 1999. Both countries have large political, economic and strategic stakes in Kashmir and for both the dispute has become a symbol of national prestige and international justice.

The state is not a single geographic, economic, cultural or linguistic unit, but a conglomeration of six distinctive regions which, except for Hunza-Gilgit and Nagar, were brought (under the administrative control) as a united political unit by Maharaja Gulab Singh, who entered into a subsidiary alliance with the British within the Indian Empire in 1849 (Figure 8.4). In every dispute both India and Pakistan have claimed the entire territory as if it were a homogeneous unit. Among the various regions the most significant is the historic Kashmir valley, a well-developed center of tourist attraction and politically the seat of central authority. Until quite recently it was accessible from India by a single road, which was seasonally snowbound until the introduction of snowplows in 1948. It is reached from the Pakistani side by a few roads. Overwhelmingly Muslim in population, this region is the most critical in the dispute.

**Figure 8.4**
*Jammu & Kashmir*

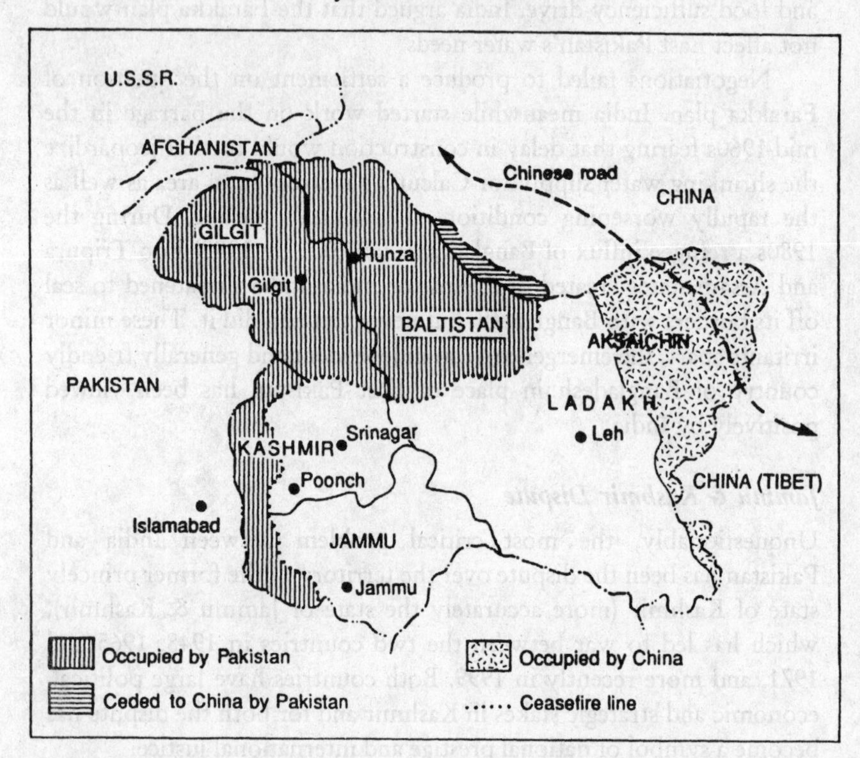


The minority of Kashmiri Brahmans have traditionally held positions of economic, social and political power, whereas the Muslim peasantry have remained absymally poor. The second region, Jammu, lies in the southern part and contains only one-seventh of the total territory. More than half the population of the Jammu region is Hindu (about 50 per cent in 1971). Its capital city of Jammu has been the state's winter capital, as well as the home of the state's ruling dynasty—the Dorga Rajputs. The third region, Gilgit, in the northern part of the state, is mountainous and almost entirely Muslim. It used to be reached from Srinagar, the state capital in the Kashmir valley, by crossing high mountain plateaus and glaciers, and is now only linked by a China-Pakistan road built in 1970s. The fourth region is that of Baltistan in the extreme northern part of the state, to the east of the

Gilgit-Hunza region, and like it, is not easily accessible and contains high mountains. It is reached by a road along the Indus river in Pakistan. It is overwhelmingly Muslim. The fifth region, Punch, lies to the north of the Jammu region and west of the Kashmir valley. Although, historically, a part of the Ladakh and is tied to it administratively, it differs from the rest of Ladakh in religion. It is mostly Muslim, adjoins the Pakistani border, and is easily accessible from Pakistan. Two important rivers, viz., Chenab and Jhelum, flow into the region from the Kashmir valley and through it into Pakistan where the waters are utilized for Pakistan's agriculture. The sixth region is that of Ladakh (also known as "Little Tibet") occupying over one-third of the state in its eastern portions. It is a vast, barren, high plateau resembling Tibet in culture, Lamaistic Buddhism, and Tibetan language and is Mongoloid in race.

We have noted earlier that the Kashmir state acceded to India in 1947. India's acceptance was subject to the determination of the wishes of the people by a plebiscite. Immediately upon its accession to India, Pakistan claimed that the state had violated the Maharaja's previous agreements with India and Pakistan, and that the accession was based on fraud and violence.

Pakistan's claim to Kashmir is largely based on religious and economic considerations. Kashmir state as a whole is a Muslim state, contiguous to and connected with Pakistan by several communication links. Furthermore, the four rivers, namely, Ravi, Chenab, Jhelum and Indus, vital to Pakistan's agricultural production, flow through Kashmir (more particularly the portion occupied by India). Pakistan has consistently maintained that a plebiscite of the entire state as recommended by the United Nations would result in a clear decision in its favor, and that India has been acting irresponsibly in its backing out of its original acceptance. India has held the position that Kashmir's accession had given India sovereignty over Kashmir. It had at first agreed to a plebiscite on condition that the Pakistani invading forces were withdrawn, a condition which was stipulated by the United Nations' resolution, but had never been realized. India, therefore, stresses the illegality of Pakistani support of the raiders and the Azad Kashmir government. Pakistan blames India for failing to withdraw its own forces from the Indian-controlled territory, and for supporting a regime prejudicial to the holding of a plebiscite.

Agreement on the demilitarization, considered essential to a fair plebiscite, was never achieved by either side. To India, Pakistan was an aggressor and should first withdraw its troops and its hold over the "Azad Kashmir" territory. Pakistan feared that if it withdrew its own forces, Kashmir would be left unprotected. The position of the two parties has been analyzed by Brown: "India has resolved not to permit a plebiscite, and Pakistan has not been able to force it. Whether or not India's stand on this issue is morally or legally justifiable, Pakistan's condonment of the invasion by the tribesmen, if not connivance in it, and later assumption of hostilities itself are seriously questionable... Each nation has pursued a policy in its own interest, and has had a less than a perfect case. The most potent consideration might possibly be that of need; Pakistan's was greater in 1947, but having reached an agreement with India on the Indus Basin's Waters in 1960, the extent of Pakistan's need seems to be diminished" (Brown, 1972: 200). (The Indus Basin Waters Dispute is discussed later in this chapter.)

Ever since accession, the Kashmir government in Srinagar has maintained a close relationship with India, and in 1952, negotiated a pact which has given it a special status within India. Since 1956 India has outright rejected the possibility of a plebiscite, blaming the United States for its military aid to Pakistan and thus acting in favor of Pakistani claims. India has feared and openly declared that American military aid to Pakistan would be used against India in the event of any resumption of Indo-Pakistani hostilities.

India has vigorously promoted several economic and development plans in Kashmir's Indian-controlled territory. In 1956 it arranged for the establishment of an elected constituent assembly for Kashmir which voted to make Kashmir a regular state within India. Since then, India has treated Kashmir as a constituent unit of the country, no longer open to a plebiscite.

Meanwhile, the United Nations has, over the years, tried several times to mediate in the dispute and has arranged Indo-Pakistan bilateral negotiations, but the two countries have not been able to resolve their differences.

Apart from the purely legalistic view of Kashmir's accession to India, two other major considerations have guided Indian policy toward the dispute. First, India has declared and considered itself to be a secular state. Surrender of Kashmir on religious grounds would be a denial of the nation's essential principles. Secondly, the strategic location of Kashmir, close to the Soviet Union and China, is believed to be of critical importance to India. China's aggressive policy in the late 1950s in its occupation of Tibet, and later a part of Ladakh and building of a road through Aksai Chin in Ladakh, as well as its threatening posture toward India have all lent urgency to the strategic aspect of Kashmir's location.

During and after the Sino-Indian border conflict of 1962, India and Pakistan negotiated under pressure from the United States and Britain to resolve their differences, and appeared ready to agree, in principle, to partition Kashmir. Talks fell through, however, for want of an agreement in the precise delimitation of the division. India appeared agreeable to accepting with minor adjustments, the cease-fire line as a basis for such a division, but Pakistan insisted on acquiring the Kashmir valley.

The Chinese occupation of a portion of Ladakh in 1955-56 further compounded the issue of Kashmir's partition. In 1963, a new dispute arose when China and Pakistan entered an agreement regarding the borders of Pakistani-held Kashmir—a move which represented a reversal of Pakistan's foreign policy. Before that time, Pakistan had always posed as a staunch western ally, a good member of CENTO and SEATO and had received large amounts of military aid from the United States, all ostensibly to meet communist aggression. In fact, prior to the Sino-Pakistan border pact on Kashmir, Pakistan had been loudly proclaiming the need for joint Indo-Pakistani resistance to the Chinese threat to the Indian subcontinent. China not only agreed upon the demarcation of the border, but also helped Pakistan to build a new road across its occupied area to Sinkiang. Pakistan, in return, ceded a large territory (2,000-2,700 square miles), a move denounced as illegal by India, as it claimed that the entire state (including the area controlled by Pakistan and ceded to China) belonged to India.

Meanwhile Indo-Chinese clashes along much of their common border erupted into a war in 1962. With the occupation by China of a part of Ladakh, India was forced to open a Kashmir front against China in addition to the deployment of forces against the Pakistani-held parts of Kashmir. In 1965 Indo-Pakistani fighting broke out in the Rann of Kutch (at the southern-most end of India's border with Pakistan), and later developed into a large-scale war along the entire India-West Pakistan border and the cease-fire line in Kashmir. With the efforts of the Soviet Union and the United Nations, an Indo-Pakistani agreement on cessation of hostilities was reached.

Then, in 1971, Indo-Pakistan conflict and the emergence of Bangladesh added another element of bitterness to relations between the two countries. Even fifty years after Kashmir's accession to India, the dispute remains unresolved. A more recent (1999) conflict between the two countries in Kargil sector has also adversely affected the efforts made by the leaders of both sides over the years to normalize relations. Since the prospect of holding of a plebiscite as a means of resolving the problem appears to become remote other means to solve the problem should be explored.

The main bottleneck in the solution to the Kashmir problem has been treatment of the state as one homogeneous unit. Legal, moral and economic grounds have been advanced by both parties to keeping the state as a unit. In reality, the cease-fire line, adjusted in 1972, has already crystallized into an international boundary and part of Ladakh has been occupied by China. Political analysis, including some in the United Nations, have favored the settlement on the basis of partitioning of the state because the state is only a conglomeration of regions. The valley of Kashmir could possibly be made an internally autonomous state whose territory could be jointly guaranteed by India and Pakistan under international trusteeship with both sides having access to it for trade. Currently, however, both parties are inflexible in their negative attitude toward a settlement negotiated on the basis of a partition formula.

### *The Indus Basin Waters Dispute*

Among the major problems that faced India during the early years of its Independence was the distribution of the waters of the Indus river basin between India and Pakistan. One unfortunate result of the partition of the Indian subcontinent in 1947 was Pakistan's critical dependence for its irrigated agriculture on river waters obtained from the river Indus and its four major tributaries (Sutlej, Ravi, Chenab, and Jhelum) which flowed through India or Indian-controlled Kashmir territory before entering Pakistan. Pakistan's irrigated area in the Indus river basin was nearly 20 million acres as compared to about 10 million acres that fell into India's territory. However, several of the headworks for the irrigation canals lay at sites in India. Thus, India could, if it wished, exercise control over the water of Pakistan's canals. Pakistan feared that India would divert this water to its own territory, thus seriously injuring Pakistan's agricultural base. Luckily for Pakistan, such fears were never realized. However, during times of hostilities, India did possess an edge over Pakistan through this threat.

Related to the problem of the Indus basin water allocation was the dispute over the construction of a large dam by India on the Sutlej river at Bhakra to provide irrigation facilities for India. Pakistan pointed out that India was trying to divert water to its advantage, thus depriving Pakistan of its share of water. India explained that the Bhakra scheme was an old scheme planned almost 30 years prior to partition.

Immediately following partition the water allocation problem of the Indus waters surfaced, and the negotiations between the two countries began in 1952. After eight years of hard bargaining and mediation by the International Bank for Reconstruction and Development (IBRD), an agreement was reached in 1960. According to the Indus Waters Treaty, waters of the three eastern rivers, viz., Ravi, Beas and Sutlej, were allocated to India to use to the maximum of her needs, while the western three rivers, viz., Indus, Jhelum and Chenab, were awarded to Pakistan. Since it was along the three eastern rivers (which were allotted to India) that most of Pakistan's irrigated territory lay, the treaty called for the construction of costly link canals from the Chenab, Jhelum and Indus rivers (awarded to Pakistan) to the Ravi and Sutlej rivers in order to keep the eastern rivers supplied with water. These extensive engineering projects were to be financed by loans from the IBRD and by contributions from India and Pakistan.

The two countries have displayed remarkable prudence in adhering to the terms of the treaty, even during the hostilities in 1965 and 1971. Irrigation development schemes have recently been expanded, and now encompass over 35 million acres. It might be observed, however, that instead of water allocation, an integrated water utilization scheme for both countries would be more beneficial in terms of economy and efficiency. Such a scheme, however, depends on the normalization of political relations between the two countries (Michel, 1967: 521-25).

## The Indian Ocean

India's long coastline of about 6,250 km projects into the Indian Ocean. Over 85 per cent of India's external trade is maritime, carried through the coastal ports. Imports of raw materials used for the country's industrialization and economic development arrive almost entirely by sea.

The Indian Ocean is strategically located with reference to India. The Indian subcontinent seemingly forms a keystone in the arch of the coasts that border it. It is also girdled by land on the other two sides as well with the continent of Africa forming its western, and Southeast Asia and Australia its eastern walls, leaving its southern side open. This "landlocked" nature of the Indian Ocean has given India a commanding position. From the eastern coast of Africa and the shores of the Persian Gulf to the Strait of Malacca no other country rivals India's dominant location in the Indian Ocean.

For over 1800 years Indians have used the Indian Ocean for trade, defense, colonization and diffusion of their culture particularly in Southeast Asia. As early as the fourth century B.C., the Mauryan kings had established ports on the coast of the Bay of Bengal. The large naval kingdoms of the Cholas and Chalukyas were set up in south India. Sri Vijaya empire, set up by the Indian rulers in Southeast Asia from eighth to eleventh centuries, maintained strong cultural and commercial ties with south India through the Indian Ocean. The European thrust into the area started in the sixteenth century, after Vasco da Gama's landing on the west coast of India in 1498. Major European powers were eventually drawn into a long and bloody struggle for power in the Indian Ocean. Eventually, the British gained supremacy over the Indian Ocean and the Indian subcontinent.

According to an Indian historian and diplomat, K.M. Panikkar, the history of British control in India illustrates the basic geo-political principle that the power which rules the sea eventually rules the adjoining land. Panikkar (1945) observes that the pre-British invasions and the land-directed conquests of India led to the founding of political dynasties, which in a short period were Indianized. Only the British rulers could remain unassimilated as they could draw their strength from England through the naval supply lines.

India's current defense posture necessarily includes adjustment in the political and strategic task of filling up the vacated role of the

British in the Indian Ocean. The United States and the Soviet Union have also shown interest in the area. The indivisibility of the seas makes it possible for them to exercise their power even though they may be far removed from the Indian Ocean. India wishes that outside interests remain limited to trade. This is only possible if she maintains an adequate defense capability, and can play the role of a dominant power in the Indian Ocean.

Since 1970 the Soviet Union (now Russia) has been steadily building trade and economic connections with India and the countries bordering the Indian Ocean. It is constructing naval radio stations and ammunition depots at the mouth of the Red Sea. It has obtained access to port facilities in Somalia, Mauritius and Singapore. The Soviet Union is also trying to gain a foothold in the oil-producing states of the Middle East. The Russian push in the Indian Ocean is reportedly aimed at containing Chinese influence in the African and Arab countries, and in Southeast Asia. Since Indian and Australian naval capabilities are not significant, the western countries are deeply concerned over expanding Soviet presence in the Indian Ocean by the establishment of bases in Diego Garcia—a move opposed by the Indian government. India's current preoccupations with its domestic economic development programs, and its modest financial resources make it beyond its means to develop a strong naval capability to deter encroachment by other major powers on India's role in the Indian Ocean.

India was able to establish its supremacy in the Indian Ocean in 1988 when Sri Lankan mercenaries invaded Maldives, and Indian forces had to come to its rescue. Maldives has consistently turned down lucrative offers by big powers to provide facilities for establishing military bases on its soil in order to make the Indian Ocean a zone of peace, much to India's wishes.

### India and the World

Earlier in this chapter, problems related to India's international borders were discussed. India's international trade and associated aspects of its foreign relations are examined in Chapter 12. Only a brief review of some other aspects of India's foreign relationships is presented in this sub-section.

Immediately following Independence, India was too engrossed

with its numerous domestic problems to be particularly concerned about the outside world. Basically, its foreign policy was that of non-involvement in the cold war between the western and the communist countries. Soon, however, its size, large population, resource base, location in the Indian Ocean, and newly attained free status thrust it into prominence among the developing countries. It possessed a great cultural heritage, a long-history of struggle for independence and a responsive leadership; and the newly independent countries thus looked to it for guidance. India's policy of non-involvement commanded respect. It was in 1962, during the Sino-Indian war, that its policy of non-alignment suffered a setback when India had to reassess the basic premise of its foreign relations, and fixed new priorities regarding its defensive posture. The Indo-Pakistani war of 1965 pushed it further into a defensive build-up. Since 1963, India has sought successfully to strengthen relations with its immediate neighbors: the Himalayan kingdoms, Burma and Sri Lanka. It is only with Pakistan and China that border issues remain unresolved.

In addition to the Indo-Pakistan conflict over Kashmir, problems rooted in the partitioning of the subcontinent itself existed in some other areas. The first problem was that of the rehabilitation of millions of displaced persons who had crossed international borders, and the related issue of the disbursement of the evacuee properties abandoned on both sides. Relations between the two neighbors were strained on the matters of the evacuee property compensation by both governments. The question of the settlement of the cash balances and assets the new administration of Pakistan left in India posed another threat to Indo-Pakistani relations. In 1960, after years of negotiations, these issues were resolved. In addition to the war over Kashmir, Indian and Pakistani forces clashed on the mudflats of the Rann of Kutch over minor boundary disputes in 1965. The two sides agreed to demarcate the boundary lines suggested by an international tribunal, a work completed in 1969. Indo-Pakistani relations deteriorated again as a result of the conflict in Bangladesh in 1971 and more recently (1999) in Kargil sector of Kashmir valley. Efforts to normalize relations are now being attempted. The Indo-Pakistan-Bangladesh agreement of 1974 regarding the issues of the recognition of Bangladesh and repatriation of all the Pakistani prisoners of war contributed toward the normalization of relations between these countries and toward stability in the subcontinent.

The international boundary partitioned the Indian subcontinent with inadequate regard to lines of communication, canal waters and the complementary nature of the economies of the countries. India and Pakistan have so far lived under a shadow of mistrust and hostility and, therefore, have not been able to establish the desirable goal of coordinated regional, economic and defense planning. With the recent prospect of normal relations such a scheme of coordinated planning in the subcontinent can be achieved.

India has been concerned about its friends along the Himalayan border and has stepped up economic aid to Nepal and Bhutan. Nepal has consistently pursued a policy of non-commitment to either India or China, and has accepted economic aid from both. India built a highway linking Nepal's capital at Kathmandu with north India, and China has similarly linked it with the Lhasa-Sinkiang road across difficult terrain and high altitudes. India does not particularly appreciate Nepal's policy of non-commitment, and would favor a more pro-Indian policy. Nepal's entry to the outside world is essentially through Indian ports making it dependent on India. Relations between the two countries soured in 1989 when Nepal sought to diversify its trade, and established closer ties with China. India thought it to be an act of contravention of the old trade pact between the two countries, and closed 13 of the 15 cross-points on the border. Later, a new trade agreement in 1990 restored the old congenial relations when the cross-points were reopened. Bhutan, for long an Indian protectorate, became a sovereign state, but retains a measure of Indian protection, and friendly relations. Sikkim's merger into India in 1975 stemmed mainly from tightening India's defense along Tibet border.

Indo-Burmese relations have generally been friendly. The principle irritant in the relations has been the status of about 0.5 million Indians (in 1960) in Burma (now Myanmar). During the 1950s and 1960s Burma passed strict land appropriation laws, nationalized its trading companies, and subjected them to discriminatory treatment, and causing large numbers to move to India. As of 1969 less than 100,000 Indians were left in Burma. Relations between India and Sri Lanka (formerly Ceylon) have been somewhat strained over the issue of about a million persons of Indian origin who were working mainly on tea plantations. They had not applied for Sri Lankan citizenship. Sri Lanka refused to grant citizenship to these "stateless" persons. In 1964, an agreement over this issue was reached which called for repatriation to India of a portion of the involved persons. Furthermore, Sri Lanka's population includes a sizable Tamil-Hindu minority (24 per cent of the total population) whose relations with native Sri Lanka Buddhists (68 per cent of the population) have not been harmonious. Since the 1964 agreement, however, relations between India and Sri Lanka considerably improved.

In the late 1980s relations between Sri Lanka and India deteriorated over the presence of Indian peace-keeping force in Sri Lanka deployed since the 1987 accord in the Tamil-dominated northeast to disarm Tamil guerrillas fighting for an independent homeland. Sri Lankan government had offered Tamils limited autonomy in exchange for arms surrender. The guerrillas, however, rejected the agreement and turned their guns on the Indians and sustained heavy losses on them. Relations, however, began to improve as India withdrew its troops in 1990.

India's relations with the West, especially with the United States, stem from its basic posture of non-alignment. It has looked upon the western military alliances, like NATO, CENTO and SEATO as creators of tensions in an already troubled world because these alliances bring more arms and armaments and lead to a chain reaction of communistic defensive postures. To India, containment of communism, an implicit objective of the western military alliances, is of secondary importance. India has argued that peace can best be achieved by non-entanglement in such military blocs. India has suggested that world peace will not be promoted by generating ideological confrontation between communism and democracy, but by peaceful coexistence.

The West, on the other hand, has not been able to reconcile India's posture of non-alignment with its attitudes toward Pakistan and its annexation of Goa. Indo-American relations deteriorated markedly after 1954 when the United States started providing military assistance to Pakistan. The U.S. assurance that its arms could not be used against India did not convince Indians, who feared that the U.S. arms would eventually be used against them in the event of

Indo-Pakistan hostilities. India's fears were realized during the Indo-Pakistani wars of 1965 and 1971. To counteract military assistance to Pakistan, India purchased arms and technical help from the Soviet Union to build armaments in Indian factories. Many persons in the United States accused India of having gone over to the communist side. Indo-American relations, which had appreciably improved in 1962 when the West rushed military supplies to check the Chinese advance into Indian territory along her Chinese borders, deteriorated again in 1972 on account of the pro-Pakistani stance of the U.S. government during the Bangladesh crisis. Despite these fluctuations in Indo-American relations, there has always flowed a strong undercurrent of mutual goodwill. The United States has remained one of the leading trading partners of India. India's continuing dependence on the United States for economic aid makes it imperative for India to gain American goodwill. Its geographic location, size, population, resource base and democratic form of government have given India an importance which the United States cannot easily ignore. As a political observer points out, "given the coincidence of political values between the U.S. and a democratic India, and the absence of major conflicts of interests between the two countries, a regional security structure on the eastern flank of the volatile Gulf, and the Middle East that reflects India's power and strength should be appealing in Washington. Such a regional arrangement... could lend great stability to the Indian subcontinent" (Mohammed, 1991: 182).

Since Independence, India has maintained cordial relations with Britain and has remained an influential member of the Commonwealth of Nations. Britain has consistently adopted a friendly attitude toward Indian foreign policies, and has cooperated in resolving a few Indo-Pakistani disputes (e.g., the Rann of Kutch dispute). Britain has pulled its land, sea and air forces out of the Indian Ocean. The British stand on the Kashmir issue has been one of the few irritants in Anglo-Indian relations.

Indian relations with the Soviet Union have been friendly especially since the downfall of the Stalin regime. Close economic and diplomatic ties, and a degree of military cooperation have been maintained. The Soviet Union's open support of India over the Goa and Kashmir issues has been widely appreciated in India. The Soviet

Union has aided India with its economic development programs. After 1966, however, a more even-handed policy aimed at Indo-Pakistani power parity in the subcontinent has been pursued by the Soviet government. A Soviet decision in 1968 to aid Pakistan militarily had generally been interpreted in India as the formation of a Soviet lever against Pakistan's growing dependence on China for arms and friendship.

India's relations with other communist countries, with the exceptions of China and Albania, have been marked by economic and diplomatic cooperation. Trade between the East European countries and India has been growing.

India has tried to play the role of a spokesman for several African and Asian nations and has often inspired the cause of their national freedom movements. It enjoys good relations with those countries with the exception of South Africa and Israel. Its cool relations with Israel basically resulted from her conciliatory attitude toward the Arabs. India has adopted a pro-Arab stance in the Arab-Israeli conflict. It imports large amounts of oil from the Middle East. Its relations with Turkey are diplomatically correct and with Iran friendly. During the 1965 Indo-Pakistani war, Turkey and Iran provided military and moral support to Pakistan. Indo-Afghanistan relations have been particularly friendly, especially in view of India's firm moral support to Afghanistan on the Pakhtunistan issue. India is a leading participant in the SAARC (South Asian Association of Regional Cooperation).

Many countries in Africa and islands in the Indian Ocean, Guyana and the West Indies, contain large population components of Indian origin. In Kenya, Tanzania, Mauritius and Fiji (in the Pacific Ocean), Indians have achieved success in banking, commercial and trade enterprises. In 1971 many Indians were dispossessed of their residence status and properties and were expelled from Uganda.

India's relations with Japan have been more important economically than politically. India rightly considers Japan to be the most developed country in Asia and its industrial strength to be a major factor in Asian power politics. Both countries are important trading partners.

### *References*


Brown, W. Norman, *The United States and India, Pakistan and Bangladesh.* Cambridge, Mass., 1972.

Frankel, E.G., *India's Political Economy, 1947-77.* Princeton, 1979.

Government of India, *India, 1997: A Reference Annual.* New Delhi, 1990.

Hardgrove, R.L. Jr., *India Under Pressure: Prospects for Political Stability,* Boulder, 1984.

Harrison, S.S., *India, the Most Dangerous Decades.* Princeton, 1966.

*India: A Reference Annual, 1971-72.* New Delhi, 1972.

*India: A Reference Annual, 1976.* New Delhi, 1977.

Kothari, Rajni, *Politics in India.* Boston, 1970.

Lamb, Alastair, *The China-India Border.* London, 1964.

Lamb, A., *The Kashmir Problem.* London, 1966.

Maxwell, N., *India's China War.* New York, 1970.

Menon, V.P., *The Story of the Integration of Indian States.* New York, 1956.

Michel, A.A., *The Indus Rivers: A Study of the Effects of Partition.* New Haven, 1967.

Mohammed, Ayoob, "Dateline India: The Deepening Crisis", *Foreign Policy,* Vol. 85, Winter, 1991-92, pp. 166-84.

Morris-Jones, W.H., *The Government and Politics of India.* New York, 1967.

Panikkar, K.M., *India and the Indian Ocean.* London, 1945.

Schwartzberg, J.E. (ed.), *A Historical Atlas of South Asia.* Chicago, 1978.

*The Sino-Indian Boundary Question.* Peking, 1962.

*The Sino-Indian Dispute.* Government of India, Delhi, 1963.

Spate, O.H.K., "The Partition of India and the Prospects of Pakistan", *The Geographical Review,* Vol. 38, 1948.

Tinker, Hugh, *India and Pakistan: A Political Analysis.* New York, 1962.

Van Eckelen, W.F., *Indian Foreign Policy and the Border Dispute.* The Hague, 1967.

# Resources and Economic Development

# 9

# Agriculture and Rural Development

For millennia agriculture has played a central role in Indian economy and will undoubtedly continue to do so in the near future. Close to 70 per cent of the population derives its livelihood from agriculture and related occupations. Likewise, about 35 per cent of the country's national income is derived from the agricultural sector. Major problems of the country such as food sufficiency, rural unemployment and economic, political and social discontent are directly related to its agricultural systems.

In statistical terms, India ranks high among the producers of agricultural goods. It stands third in the world in wheat production, after the Soviet Union and the United States, and second in rice and millet production after China. Producing between one-quarter and one-third of all the world's tea, jute, peanuts and hemp-fiber, it leads all nations in their productions. It produces exportable amounts of several other cash crops in which its world status is high: third in the production of sugarcane, tobacco and oilseeds and fourth in cotton. It has the largest bovine (290 million or one-quarter) population of the world. These impressive figures must be matched against a large population rising at nearly 2 per cent a year, low agricultural yields and poor per capita production. Although subsistence levels in foodgrains have been recently reached and prospects for future foodgrain increases have brightened up as a result of diffusions of agricultural modernization, the race between food and population remains a close one.

Despite some breakthroughs, the basic structures of the agrarian economy remain traditional. Established centuries ago, these structures of a self-contained rural economy, founded in caste-derived occupational land tenures made complex by absentee and parasitic landlords, have been slow to respond to modernization. This agrarian economy has also directly affected other segments of economy such as food self-sufficiency, industrial development, unemployment, foreign trade, social organization and administrative stability in the post-colonial era.

Agriculture is a multidimensional problem, its various components are so inextricably interconnected as to render analysis and classification of problems difficult. One example of the interconnectedness of the problems is the land tenure system. Landholdings are tiny, fragmented, and not particularly amenable to modernization. The problem is at once social, economic and political. Inheritance laws, the colonial *zamindari* system, caste-based agricultural practices, all are bound up with the problem. Legislative efforts to ameliorate the situation are tied up with the social and political systems. Compounding the situation has been the mounting demographic pressure, which has wiped away gains in agricultural productions during the last three decades. The process of agricultural rehabilitation, therefore, would consist of effecting improvements on several fronts, in techniques, in social institutions and in reducing demographic pressures. Some major agricultural problems are analyzed below as units, although each is interconnected with the other.

### *Land Tenure System*

Statistics regarding land distribution among rural families are few and largely based on estimates. Table 9.1 indicates the general pattern of the agricultural landholdings. At the bottom of the ladder are the 59 per cent of households who own less than one hectare of land each. It is estimated that within this category 10 per cent of the households are totally landless. By comparison, one-half million families, or the richest less than 2 per cent of the households, own 17.5 per cent of the land used for agricultural production. While this highly unequal distribution of agricultural land suggests a serious land problem, even more serious is the unequal farm wealth or income. In reality, the smaller farms are not only uneconomic but also indicates a serious problem of agricultural inefficiency. Figure 9.1 displays the distribution of spatial variations in landholdings in India.

**Table 9.1**

*Agricultural Landholdings*

| *Size of Holdings (ha)* | *Percentage Distribution* | | | | | |
|---|---|---|---|---|---|---|
| | *No. of Holdings* | | | *Area of Holdings* | | |
| | *1990-91* | *1985-86* | *1970-71* | *1990-91* | *1985-86* | *1970-71* |
| Below 1 | 59.0 | 58.0 | 51.0 | 14.9 | 13.1 | 9.0 |
| 1-2 | 19.0 | 18.2 | 18.9 | 17.3 | 15.5 | 11.9 |
| 2-4 | 13.2 | 13.6 | 15.0 | 23.2 | 22.2 | 18.5 |
| 4-10 | 7.2 | 8.2 | 11.2 | 27.2 | 28.7 | 29.7 |
| 10+ | 1.6 | 2.0 | 3.9 | 17.4 | 20.5 | 30.9 |
| Total | 100.0 | 100.0 | 100.0 | 100.0 | 100.0 | 100.0 |

*Source:* *Statistical Outline of India*, 1996-97, Bombay, 1997, p. 59.

A major problem of landholdings has been their fragmentation. One household may own tiny bits of scattered plots. The origin of fragmentation relates to Hindu and Muslim inheritance laws and customs under which all sons would receive equal shares in good and bad sections of the land based on soil conditions, type of crops, topography, drainage and fertility at the time of inheritance. Further fragmentation resulted from partial confiscation of a lot by revenue intermediary in case of delinquency of payments. For poor cultivators, a part of the land may have to go to the landlord as payment of the debt incurred. Often villages of the size of 5,000 hectares would be divided into 50,000 fields. Since the enactment of legislation on consolidation of holdings, millions of tenancy suits lie pending in the already overburdened civil courts. Fragmentation clearly creates uneconomic holdings. Apart from the duplication of effort in irrigation and cultivation of scattered fields, a farmer must lose much land in the earth-ridges to separate his fields from others. For the same reasons fragmented fields do not lend themselves to mechanized farming or the application of chemical fertilizers.

Another problem is that of landless or tenant cultivators. The rise

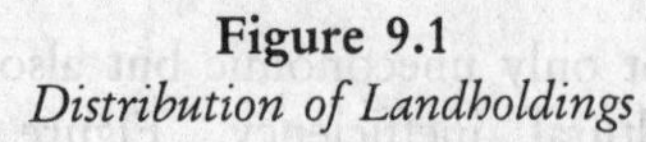

**Figure 9.1**
*Distribution of Landholdings*

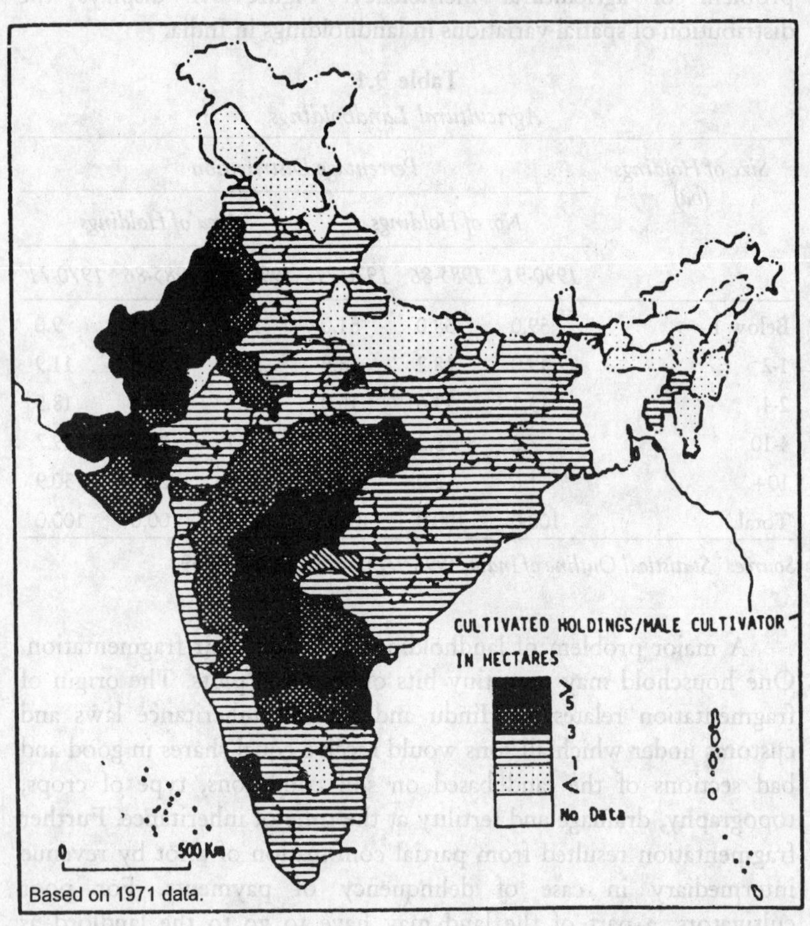


of landless cultivators, though common in most countries, assumed tragic proportions in India during the British rule. The dominant prevailing system of land tenure during the Mughal rule was the *ryotwari* or peasant-proprietor (small holding) system. The peasant was the owner of the land and possessed the decision-making and inheritance rights. The government collected revenues from the peasants directly through the village governments. During the late Mughal period revenue collection was increasingly assigned to the intermediaries or *zamindars*, who, during the British rule, virtually became the landowners. Often repressive in collection practices, the *zamindars* were given rights to confiscate lands of delinquent revenue payers. At the time of Independence there were two main systems of land tenures, namely, the *ryotwari* and the *zamindari*.

A characteristic feature of the *zamindari* system was that the actual cultivators had no contact with the government, and were at the mercy of the intermediaries. The *zamindari* system became a fundamental issue in the land reforms after Independence. In the wake of such slogans as "land to the tiller", many laws of land redistribution, land consolidation, land ceilings and the abolition of the *zamindars* and absentee landlords were enacted during the 1950s and 1960s. Progress undoubtedly has been achieved in a number of states in legislation enactment, but implementation remains difficult. Before Independence 40 per cent of the cropped area was under the *zamindari* system which by 1972 had been largely eliminated. After 1947, nearly 20 million farmers were given landowning rights and nearly 6 million hectares of land allotted to the landless farmers. By 1972 scattered landholdings of 32.6 million hectares had been consolidated. By 1985 nearly 52 million hectares, representing only one-third of total cropped area of the country, had been brought under consolidated holdings. However, only two states, viz., Punjab and Haryana, could claim 100 per cent consolidation in landholdings. Land ceilings ranging between 4 and 30 hectares have been fixed by the states. Despite these legislative measures, nearly one-half of the households still possess landholdings of 20 hectares or more.

A related problem was that of the absentee landlordism. Over the centuries, particularly during the British rule, a large cadre of absentee landlords or *zamindars* had arisen. Although owning most of the cultivable land, the *zamindars* resided in the urban areas, leaving the chores of cultivation to the tenant farmers (sharecroppers), some of whom would in turn occasionally hire laborers to do the work. A system of sharecropping was soon established. Periods of economic depression, unforeseen weather and constant borrowing of money from the *zamindars* often ruined the sharecroppers, but enriched the *zamindars*. Improvements in agricultural practices by the sharecropper under such conditions were impractical. The absentee landlord was content with whatever returns he could get, and was disinclined to make investments for agricultural improvements, whereas the tenant-farmer did not possess the capital necessary to do so. He was content with whatever share of the crop he could get without any additional investment, in the true fatalist tradition.

*Peasant Indebtedness*

Although peasant indebtedness is universal among subsistence farmers, but its impact is perhaps nowhere as crushing as in Indian agriculture. Estimates of the extent of Indian peasant indebtedness vary, but all point to an incredibly high level. Following a period of depression in the 1930s, rural debt in British India (excepting the native states) was estimated to be over Rs 12,000 million, which probably equaled the average annual rural income. This, in all likelihood, has been on the increase since then. An estimate of the All-India Rural Credit Survey of 1954-56 indicated that nearly 70 per cent of all the cultivating families were in debt. New statistics confirm the continuance of large-scale and deep-rooted rural poverty in the country. The average annual per capita income of an Indian farmer (cultivators, landlords and moneylenders included) is about Rs 200, whereas the average indebtedness of the rural families in debt (70 per cent of the total) amounts to Rs 500 (in the early 1980s).

The lot of the landless cultivators is even more pitiful than indicated by the average figures. To an average farmer indebtedness is a way of life. Factors responsible for indebtedness are many. To name a few: uneconomic, fragmented landholdings, vagaries of monsoons, capital deficiency to tide over sickness, drought, floods, cattle diseases, lack of storage facilities forcing sale of the harvest even at unprofitable times, and above all, the incurring of customary extravagant expenses on certain social occasions, such as the birth of a son or the marriage of a daughter.

Rural credit services are notoriously inept in organization and inefficient. Although government-run cooperative societies offering credit facilities to farmers in 1996 numbering 1.5 million, these served only 35 million agricultural families and provided to one-third of the agricultural credit requirements of the country. Specialized cooperatives for sugarcane, fishing, marketing, agricultural management, and the processing of agricultural output were also established during the period. Undoubtedly, the facilities have been improving, but mainly for the upper class farmers who are more creditworthy. The bulk of farmers lacks vision, initiative and resourcefulness to make use of the facilities. They still approach the local moneylender, usually the big landlord, who charges exorbitant rates for commercial loans. Interest of 20 to 36 per cent a year on loans is not uncommon.

### *Problems of Mechanization*

In large measure, the Indian peasant's agricultural inefficiency has resulted from the work of nature and society (e.g., lack of irrigation facilities, poor soils, a caste-based system of exploitation and deficiency of capital). His conservatism was not merely an expression of his resistance to new ideas or techniques. It stemmed from traditional practices, which were intimately adjusted to the environment. His tools and practices were simple and the land seemed to yield enough.

The need for diffusion of innovations to pull the peasant out of tradition and poverty has been widely recognized, but the farmer's resistance to innovations still persists. The traditional perceptions must be altered. Evaluation of innovations before initiation has also to be clearly undertaken. The use of an ordinary light plough may serve as an example. The criticism that the light plough merely scratches the surface of the soil is not entirely tenable for conditions under which an average Indian farmer has been working. The traditional "surface scratcher" light plough is all that his bullocks can draw and that he can carry on his shoulders to and from his scattered fields. Furthermore, it has been demonstrated that deep ploughing is useful only for certain crops and in specific soil conditions. The key to good ploughing is the efficient use of soil moisture and irrigation.

The introduction of innovations in techniques (machines, irrigation, better crop varieties and seeds) must, therefore, be based on and coincide with the given circumstances of farmers' capital resources (credit facilities), tenancy laws regarding viable and consolidated landholdings and infrastructural bases. Simple, improved tools such as seed drills, threshing and winnowing appliances, water lifts or electric wells are cost-effective devices that can be more useful than huge machines to an average Indian farmer. Mechanization, in conformity with Indian conditions, however, has to be introduced. Simple farm implements are no doubt, useful and in the long run cost-effective. For example, a small tractor can perform a multiplicity of farm jobs from clearing, cultivating, harvesting and the transporting of produce and implements, although its initial expense for an average farm could be large. A large harvester-combine, on the other hand, may not be suited for small Indian farms. By 1994, there were about 1,100,000 tractors and 53,000 power tillers in operation.

Associated with mechanization is the problem of unemployment among agricultural labor which would undoubtedly be aggravated if mechanization programs were carried out. The problem of agricultural unemployment has indeed taxed the minds of agricultural analysts. An answer to it is the proposal that the massive agricultural labor force that would be released from the farms consequent upon mechanization, could be usefully channeled into newly developed agro-based rural industries.

The use of mechanical implements has been on the rise in the country since 1961. An efficient iron plough has replaced the traditional wooden one in several parts of the country. Its use has become significant in western Uttar Pradesh and Tamil Nadu, where 100 iron ploughs were used for each 1,000 hectares of cultivated land, as compared to 26 ploughs/1,000 hectares for the nation in 1966—a dramatic rise since 1961. By 1966 tractors had become widely used in Punjab and Haryana (4 to 5/1,000 ha of cultivated area); elsewhere the use was negligible. In 1995, about 2 million tractors were in use in the country. The sale of tractors and power-tillers touched an all-time high during 1998-99, viz., about 262,322 tractors and 14,488 power-tillers. The use of specialized implements, namely, seed-drills, threshers and sprayers is now common in Gujarat, Karnataka and Andhra Pradesh. Power-driven water-lifts have become widespread in Punjab, Haryana, Gujarat and parts of Tamil Nadu (15 electric or diesel wells per 1,000 hectares of cultivated land) in the 1980s. Great regional variations clearly exist in the diffusion of these innovations. The most noticeable shift from the traditional farm operations to mechanization in Punjab and Haryana has paralleled the initial diffusion of high-yielding varieties (HYVs) of wheat and millets in those states.

### *Infrastructural Problems*

Rural communication in India is generally inadequate, although the country's road network is one of the largest in the world. Poor unsurfaced roads serve most rural interiors, straining the means of transport of goods by bullocks. This has been one factor which has hampered development of industries based on agricultural produce, such as canning and dairying.

The Community Development Projects, during the early phase of planning (1951-66), laid considerable emphasis on the development of communications in rural areas. More than $1,200 millions have been spent on improvement and expansion of rural link roads. Road track connected with the rural interiors expanded from 243,000 km to 721,000 km between 1951 and 1966, and in 1992 the track consisted of over a million kilometers of surfaced roads.

A major task of rural development is the provision of marketing facilities for the middle and small farmers. Only in a few states such as Punjab, Maharashtra, Andhra Pradesh and Gujarat does a fairly well-developed network of officially inspected markets exist. Elsewhere, conditions for the storage, credit and transportation of commodities are poor. Even in areas of regulated markets, the small farmer rarely utilizes the existing facilities. Moreover, he is still at the mercy of unscrupulous traders in obtaining a fair price for his products despite the national pricing policy. A small farmer is easily exploited by secret brokerage, false weights and payment of inflated commissions. The key to agricultural progress lies, in a large measure, in the creation of an adequate infrastructural base.

### *Agricultural Productivity*

As the Table 9.2 showing major crops of selected countries indicates, Indian agricultural yields are among the lowest in the world, although a steady rise in yields since 1950 was recorded.

Scientific fertilization is a recent phenomenon and practiced by few. An average farmer, deep in debt, does not have the resources to apply nitrogen, potash or potassium to the crops. Investment in chemical fertilization could lead to his economic ruination. Burnt stubble, branches, leaf mold or animal manure application are his chief means of fertilization. Even these are inadequately applied. Social customs frown upon the utilization of human excreta for the fields. Occasionally cow dung is used but 60 per cent of it is burnt as a fuel or lost for mixture in plaster coating to floors or walls in rural areas. Recent breakthroughs in yields by the introduction of HYVs of crops have demonstrated the need for chemical fertilization. The Five

Year Plans, therefore, rightly emphasized expansion in the output of chemical fertilizers.

**Table 9.2**

*Average Yield of Selected Crops (kg/ha)*

| | *India* | | | *USA* | *UK* | *Japan* | *USSR* |
|---|---|---|---|---|---|---|---|
| | *1950-51* | *1970-71* | *1994-95* | ---------- | *1986* | ---------- | |
| Rice | 668 | 1,134 | 1,921 | 6,334 | * | 6,322 | 4,242 |
| Wheat | 663 | 1,299 | 2,553 | 2,317 | 6,987 | 3,568 | 1,894 |
| Jowar (Sorghum) | 353 | 470 | 783 | 4,249 | * | * | 1,003 |
| Sugarcane (Gur) | 3,342 | 4,966 | 6,800 | 8,000** | * | *** | *** |
| Cotton | 88 | 108 | 260 | 1,624 | * | * | 1,700** |
| Jute | 1,043 | 1,177 | 1,983 | *** | * | * | * |
| Groundnuts | 775 | 832 | 1,042 | 2,699 | * | 1,918 | 2,143 |

* Crop not important.
** Estimate.
*** Data not available.

*Sources:* 1. *F.A.O. Production Yearbook*, 1988, Rome, 1988.
2. *Government of India, Seventh Five Year Plan*, Vol II, New Delhi, 1985.
3. *Statistical Outline of India*, 1996-97, Mumbai, 1997.

The use of chemical fertilization has steadily gained ground since the early 1960s. The most rapid growth in the use was in the Haryana-Punjab area, which was responsible for 32 per cent of national nitrogenous fertilizer consumption during the Fifth and Sixth Plans. Other areas where fertilizer consumption markedly improved ware in Tamil Nadu and parts of Andhra Pradesh. Other areas have been slow to accept fertilizers, partly because these areas were poor in irrigation, water management schemes and rural electrification to sustain the consumption of fertilizers.

Potentials for diffusion of scientific fertilizer utilization appear bright if other capabilities are improved. About three-fourths of the area given to wheat and rice cultivation and 50 per cent devoted to other foodgrains were still not utilizing chemical fertilizers in 1986. The success of fertilization varies by regions. Universal diffusion is neither possible nor required, but considerable potential for diffusion exists for the watered areas of the Ganga plain and Maharashtra.

## The Use of Land

Records of India's land classification have been faithfully maintained since the beginning of the twentieth century, although their quality has suffered for lack of statistical verification, or improvement in terminology and classification systems. The reporting agencies, mostly at the village level, have not been above reproach. Terminology is often ambiguous. For example, the category of "culturable wasteland" has never been clearly defined. Although it suggests potentially usable land for cultivation, in reality it contains only a fraction of land presently fit for cultivation. "Area not available for cultivation" includes permanent pastures, grazing land and land under miscellaneous tree crops. "Current fallow" represents area lying fallow for less than one year. "Other fallows" are cultivable lands which have not been ploughed for between one and five years. Areas not cultivated for over five years lapse into the "culturable waste" category. Problems are compounded by use of such terms as "adjusted", "unadjusted", "provisional", "final estimates", and "not reported". Table 9.3 gives an idea of the land uses in India for 1950-51, 1970-71, 1981-82, and 1992-93.

**Table 9.3**

*Land Utilization (in million hectares)*

| | *1950-51* | *1960-61* | *1970-71* | *1980-81* | *1993-94* |
|---|---|---|---|---|---|
| Total reported area (according to village papers) | 284.3 | 298.4 | 303.7 | 304.1 | 304.8 |
| Forests | 40.4 | 54.0 | 63.9 | 67.4 | 68.4 |
| Net area sown | 118.7 | 133.1 | 140.7 | 140.2 | 142.0 |
| Area sown more than once | 13.1 | 19.6 | 25.0 | 33.0 | 44.3 |
| Total cropped area | 131.9 | 152.8 | 165.8 | 173.3 | 186.4 |
| Area not available for cultivation | 47.5 | 50.7 | 44.6 | 39.6 | 41.0 |
| Current fallows | 10.6 | 11.6 | 10.6 | 14.8 | 14.3 |
| Fellow land other than current fallow | 17.4 | 11.1 | 8.7 | 9.8 | 9.7 |

*Sources:* 1. Statistical Abstract of India (1992).
2. Statistical Abstract of Haryana (1996-97).

Topographically, two-thirds of India's surface area is usable. The rest is either in high and rugged terrain, is climatically unsuitable for cultivation, or is barren. About 177 million hectares of land are given to cultivation, representing nearly one-half the surface area of the country—a figure placing India among the leading countries on the basis of percentage of its land devoted to cultivation. On a per capita basis, however, it amounts to only 0.20 hectare. About one-fifth of the cultivated area is double-cropped (cultivated more than once), while the same amount of the cultivated area and most of the double-cropped area is under irrigation. Surprisingly, a mere 4 per cent of the land is in permanent pasture and grazing, in a country which contains the largest bovine population in the world. Could it be that a part of the "culturable waste" belongs to the "grazing" category?

A steady but modest extension of cropland amounting to 30 million hectares took place during the three decades between 1950-51 and 1981-82. Area given to forest also increased, while fallow lands decreased in area mainly due to classificatory changes. A more notable increase was in the area under irrigation and in double-cropped land. Wastelands, however, did not shrink appreciably.

In 1981-82, about 143 million hectares of land was under cultivation, and only 26 million hectares of which yielded two or more harvests. Cropping intensity, therefore, is low. In geographical distribution, both the amount of cropland and cropping intensity showed great regional variations. In most parts of the Ganga plains, eastern Gujarat and the lava regions of Maharashtra, nearly 85 per cent of the land was under crops. Rajasthan, Meghalaya, Manipur, Sikkim, Jammu & Kashmir, Nagaland and eastern Madhya Pradesh had the smallest proportion of land (less than 20 per cent) given to cultivation. The double-cropped area, however, did not positively correlate with regions containing high proportion of cultivated land. In fact, double-cropping was more significant in a few hilly areas, such as Himachal Pradesh (where, in places, the double-cropped area included up to 60 per cent of sown land). In these areas of poor economy farmers generally depend on a second cash crop. In addition to growing cereal crops (wheat, maize and rice) in these areas of adequate rainfall, a second crop of potatoes and vegetables is grown as

**Figure 9.2**
*Cropping Intensity*

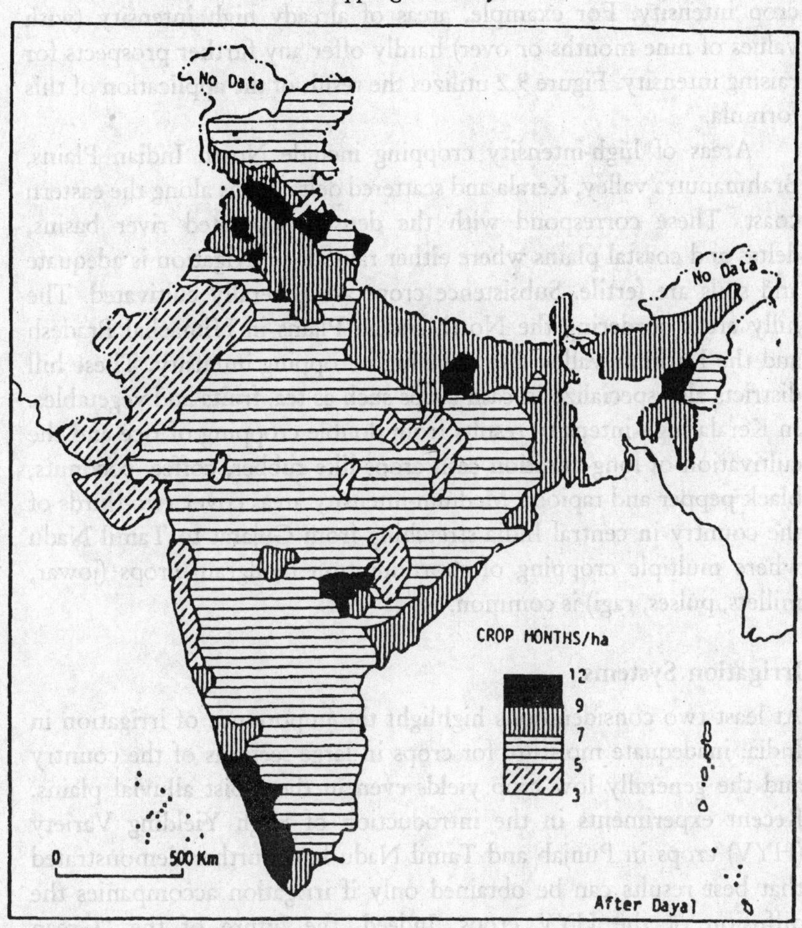


cash crop on small hillside (or valley) holdings. Elsewhere, the double-cropped area in general amounted to less than 20 per cent of the cropland.

The amount of cropland and the double-cropped area reflects the intensity of cropping and the relationship between the two is particularly significant. A refinement which incorporates the duration of crops in the field as a measure to indicate cropping intensity has been proposed by Dayal (1979: 289-95) who assumes that the length of crop duration accounts for the inputs of labor, land and irrigation water. His index gives crop months per hectare of net area sown. A map based on such an index helps identify areas capable of enhanced crop intensity. For example, areas of already high intensity (with values of nine months or over) hardly offer any further prospects for raising intensity. Figure 9.2 utilizes the result of the application of this formula.

Areas of high-intensity cropping include North Indian Plains, Brahmaputra valley, Kerala and scattered delta plains along the eastern coast. These correspond with the densely-populated river basins, deltas and coastal plains where either rainfall or irrigation is adequate and soils are fertile. Subsistence crops are generally cultivated. The hilly areas bordering the North Indian Plains in Himachal Pradesh and the Kashmir valley also have high cropping intensity. These hill districts also specialize in cash crops such as tea, fruits and vegetables. In Kerala high intensity results from double cropping of rice and the cultivation of long-duration cash crops like rubber, coffee, coconuts, black pepper and tapioca. Medium-intensity areas cover two-thirds of the country in central India stretching from Gujarat to Tamil Nadu where multiple cropping of short-duration foodgrain crops (jowar, millets, pulses, ragi) is common.

## Irrigation Systems

At least two considerations highlight the importance of irrigation in India: inadequate moisture for crops in large sections of the country and the generally low crop yields even in the moist alluvial plains. Recent experiments in the introduction of High Yielding Variety (HYV) crops in Punjab and Tamil Nadu have further demonstrated that best results can be obtained only if irrigation accompanies the diffusion of the HYV crops. Indeed, the future of the "Green Revolution" rests in a large measure upon the expansion and intensification of irrigation.

In 1972-73 roughly one-fourth (24.1 per cent) of the total cultivated area was irrigated. Canal systems accounted for 35 per cent, wells 30 per cent, and tanks 1 per cent of the irrigated area. The Fifth Plan estimated that by the end of 1973-74 about 43 million hectares of cropland were irrigated and that the potential existed for future extension to twice the current level of irrigation by 1979. In 1992-93 close to 36 per cent of the cropped area was under irrigation.

Within the country there are wide spatial variations in the amount of irrigation provided to farmlands. It ranges from 8 per cent in Tripura (high rainfall area) to 16.5 per cent in dry Rajasthan to as much as 76 per cent of the cultivated land in Punjab where irrigation facilities are the most extensive. Generally speaking, irrigation facilities are most widespread in the North Indian Plains, serving one-quarter to one-half of the farmland. It was here that the most extensive network of irrigation was initially laid out during the British rule. In hilly sections of the country (Himachal, Pradesh, Jammu & Kashmir) where cultivated area is restricted, comparatively large areas (9 to 15 per cent) are under irrigation. Inter-state variations result from differences in physical features but also from the developmental processes during the nineteenth and twentieth centuries. An extensive irrigation network was developed under British rule, and has been extended since Independence. Irrigation became a key element in Indian planning. By the end of the Fourth Plan (1969-74) 45 million hectares or 25 per cent of the cropped area was under irrigation. In 1991-92, the figure had risen to 64.6 million hectares or 35 per cent of the cropped area.

As a natural consequence of rapid population growth and economic development programs during the 1960s and 1970s, the demand for water, both for extension of irrigation as well as for power generations has increased sharply. Population pressures made foodgrain imports necessary, and increasing food production through extension of irrigation facilities became a matter of urgency. Many new canals and other facilities were constructed. By the end of the Fifth Plan (1975-81) irrigated area had increased to 51.6 million hectares from 22.5 million hectares in 1947. Irrigation expansion under the Fifth Plan focused mainly on the areas where the HYV crops were introduced.

Figure 9.3 exhibits major areas of irrigation in the country. The distribution reflects the federal nature of Indian planning which enables the states to seek financial support from the central government. Although India's Planning Commission, within the federal framework, articulated a policy of achieving so-called "balanced regional development" for the purpose of locating new schemes of irrigation, most schemes became objects of political pressures as each state advocated location of projects within its

**Figure 9.3**
*Irrigation*

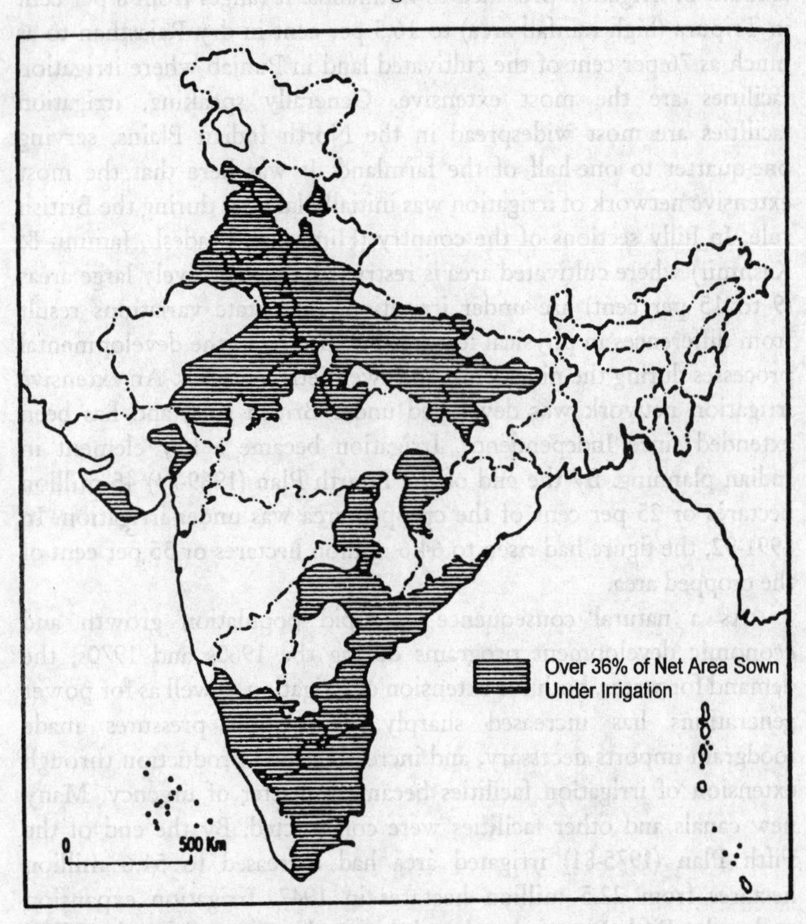


boundaries. Revision of political boundaries in 1956 and adjustment since then often complicated the picture in some instances. For example, the Tungabhadra project, originally planned before 1947, was held up for several years on different occasions. With its principal dam in Karnataka, its service area fell mainly inside Andhra Pradesh after the 1956 revision of state boundaries.

Occasionally, two states jointly sponsor a scheme. Bihar and Uttar Pradesh thus utilize the Gandak project and even share it with Nepal under the 1956 Indo-Nepal agreement. A major part of it is now complete. Noteworthy among the inter-state projects is the DVC (Damodar Valley Corporation) which was conceived for the integrated development of a river basin. It envisaged comprehensive planning of the Damodar basin in West Bengal and southern Bihar in irrigation, flood control and power generation on the TVA pattern. Commissioned in 1948, by 1973-74 it provided irrigation to 350,000 hectares and had reached roughly 90 per cent of its target. Another scheme, Bhakra Nangal, is a unique multipurpose project, aimed at providing irrigation to 1.5 million hectares in the states of Haryana, Punjab and Rajasthan. Special beneficiaries will be the semi-arid lands of Haryana and Rajasthan. An integrated system of link canals designed to raise agricultural productivity has already been responsible for the diffusion of HYV crops in these areas.

Most major *irrigation canals* are aligned along the axes of interfluxes so as to provide irrigation to a large territory. The canals in north India are fed by the perennial rivers and are seldom completely dry as their flow is regulated at masonry headworks. Those of the Deccan rivers lie in the deltas of the eastern coast during monsoons and are only seasonal (Figure 9.4). Canals account for over 35 per cent of the area under irrigation. The net area under canal irrigation is nearly 17 million hectares.

Among the disadvantages of the perennial system is the collection of fertile silt at headworks, leaving little for the canal to carry to the irrigated land. Anther problem experienced has been over-irrigation in areas of poor drainage and excessive flow causing the accumulation of alkali pans and salts (known as *thur*) in the cultivated lands, due to rapid evaporation and the rise of salts to the surface in dry areas. Problems of waterlogging and salt formation have become serious in parts of Haryana, Punjab and Rajasthan. Recent example of waterlogging is that of Indira Gandhi Canal Command Area in Rajasthan which has affected several thousand acres of productive land and pastures in the districts of Ganganagar, Bikaner and Jaisalmer. Waterlogging has also been reported in the Gandak and Kosi project areas in Bihar.

In the Deccan problems of water storage and flow arise because rivers are non-perennial and canals become dry for a considerable time. Water may have to be pumped from the natural storage points in the river. Conversion of inundation to perennial canals has been recently attempted. It entails a costly procedure of enlarging river

**Figure 9.4**
*Canal Irrigation*

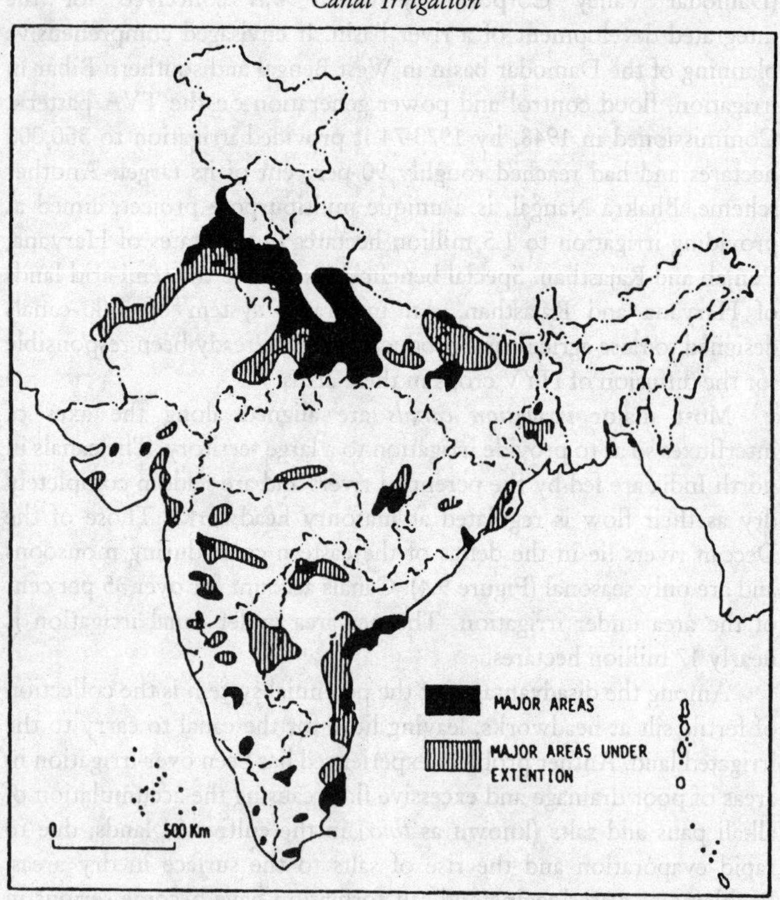


storage areas to a level required for adequate water distribution.

Unique among the recent projects is the proposed construction of a National Grid System linking the Ganga and Kaveri river basins and carrying the surplus water of the Ganga 2,200 miles southwards into the Kaveri by hydraulic techniques.

Irrigation by *wells* (traditional and, more recently, power-driven) amounted in 1992-93 to 51.2 per cent of the irrigated area and occupies an important role in agriculture. The use of power-driven wells has been on the increase since the 1960s and 1970s particularly in Punjab, Haryana and western Uttar Pradesh, but traditional well irrigation

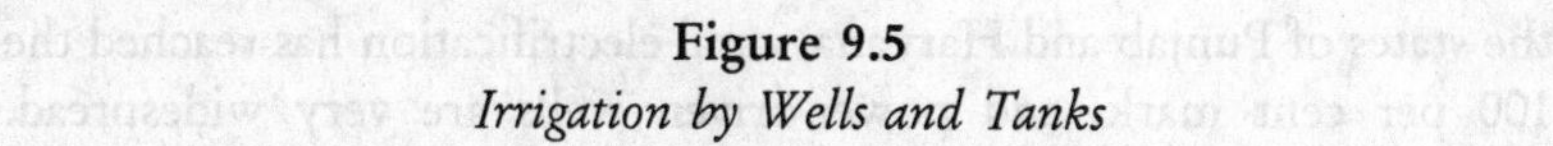

**Figure 9.5**

*Irrigation by Wells and Tanks*

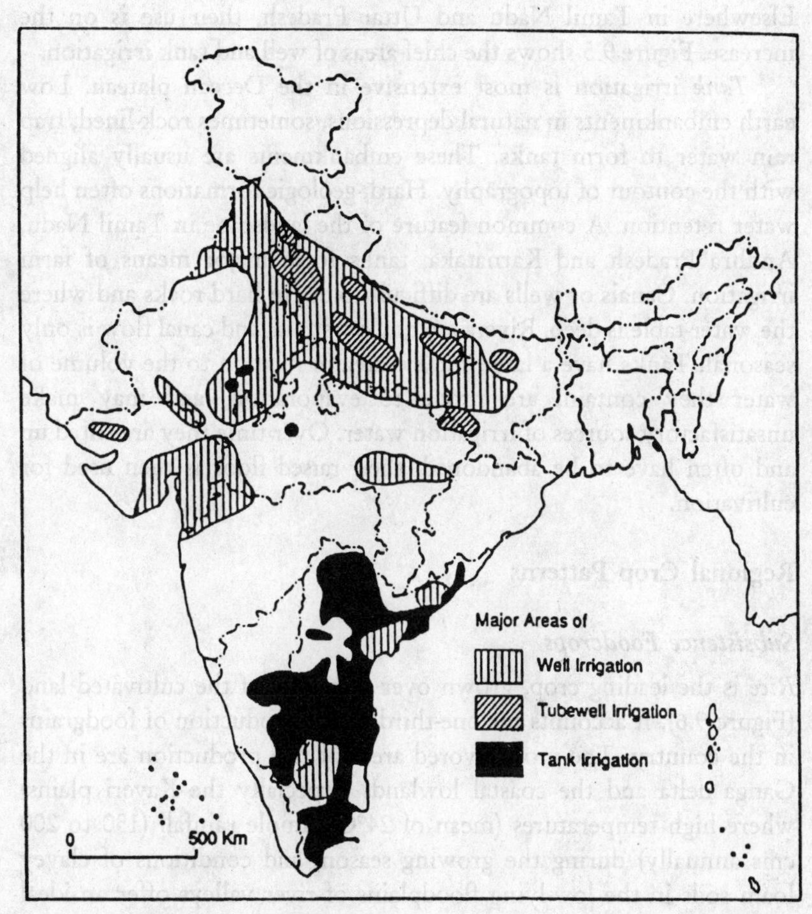


operated by human and animal muscle remains quite important as the backbone irrigation in north India, an area suited admirably to the construction of wells because of its high water-table, absence of hard sub-surface, and consequent ease of digging. The traditional well, whether masonry (brick) or non-masonry, despite its ease of construction, is cost-effective only for a small farmer, who has to irrigate small patches for growing vegetables, rice, seedlings, sugarcane or pulses as cash crops. Larger holdings demand a more efficient system. Big farmers utilize canal irrigation or power-driven wells. In the states of Punjab and Haryana, rural electrification has reached the 100 per cent mark, and power-driven wells are very widespread. Elsewhere in Tamil Nadu and Uttar Pradesh, their use is on the increase. Figure 9.5 shows the chief areas of well and tank irrigation.

*Tank* irrigation is most extensive in the Deccan plateau. Low earth embankments in natural depressions, sometimes rock-lined, trap rain water to form tanks. These embankments are usually aligned with the contour of topography. Hard, geologic formations often help water retention. A common feature of the landscape in Tamil Nadu, Andhra Pradesh and Karnataka, tanks form major means of farm irrigation. Canals or wells are difficult to dig in hard rocks and where the water-table is deep. Rivers are non-perennial and canal flow is only seasonal. Tanks have a large surface area in relation to the volume of water they contain, are prone to evaporation, and may make unsatisfactory sources of irrigation water. Over time they are silted up and often have to be abandoned; their raised floor is then used for cultivation.

## Regional Crop Patterns

### *Subsistence Foodcrops*

*Rice* is the leading crop, grown over a quarter of the cultivated land (Figure 9.6). It accounts for one-third of the production of foodgrains in the country. The most favored areas of rice production are in the Ganga delta and the coastal lowlands (especially the Kaveri plains) where high temperatures (mean of 24°C), ample rainfall (150 to 200 cms annually) during the growing season, and conditions of clayey loam soils in the low-lying floodplains of river valleys offer an ideal environment for growth. In these areas and in Tripura, Assam, Manipur and Mizoram more than 60 per cent of the cropland is devoted to rice cultivation.

Rice is virtually monocultural in a few areas. It ranks as the premier crop in the middle Ganga basin (Bihar, eastern Uttar Pradesh), Assam valley, Meghalaya, Maharashtra coastal plains, and Kashmir valley where 15 to 45 per cent of the cropland is under rice cultivation (Figure 9.6). Essentially a "wet crop", it is grown only under irrigation in western Uttar Pradesh, Punjab, Haryana and central India. Tolerant of heavy alkaline soils, its cultivation has

**Figure 9.6**
*Rice*

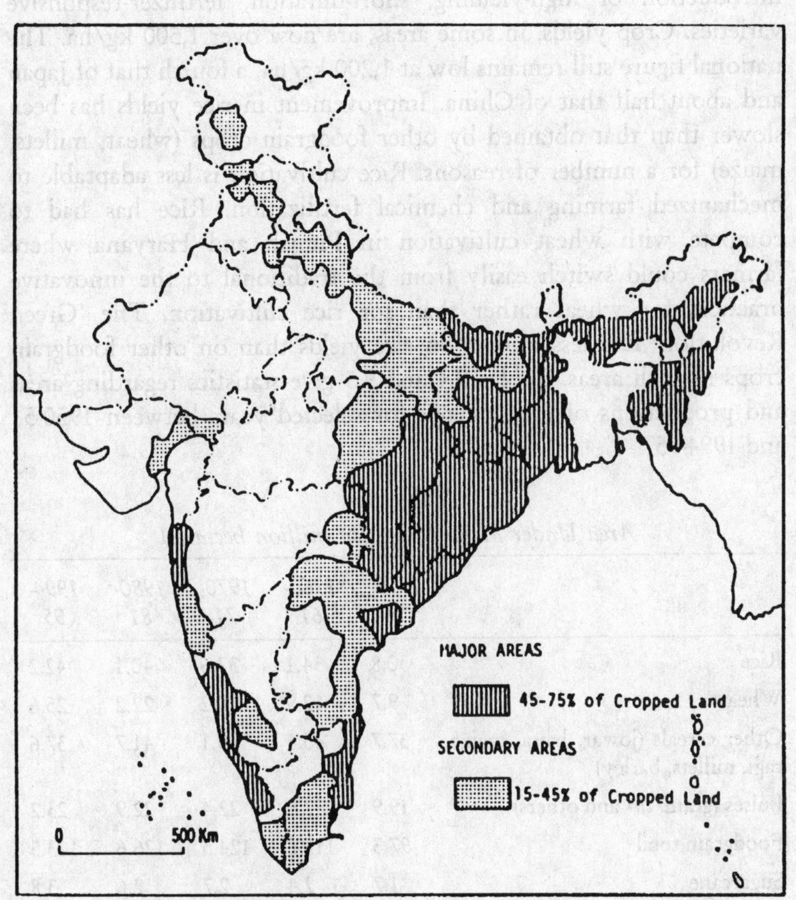


recently expanded in the waterlogged, alkaline areas of Haryana Punjab and Uttar Pradesh. The area given to rice cultivation has steadily increased between 1950-51 and 1994-95, growing from 31 million hectares to 42.2 million hectares. Likewise, the production rose from 20 to 30 million tons per year during the same period, a consequence modest gains in the extent of cultivated area but substantial gains in yields. Increased irrigation facilities in drier areas, reclamation of waterlogged soils, and introduction of new high-yielding strains crops (particularly in Haryana and Tamil Nadu) made this possible. Yields have benefited particularly from the introduction of high-yielding, short-duration, fertilizer-responsive varieties. Crop yields, in some areas, are now over 1,500 kg/ha. The national figure still remains low at 1,200 kg/ha, a fourth that of Japan and about half that of China. Improvement in rice yields has been slower than that obtained by other foodgrain crops (wheat, millets, maize) for a number of reasons. Rice cultivation is less adaptable to mechanized farming and chemical fertilization. Rice has had to compete with wheat cultivation in Punjab and Haryana where farmers could switch easily from the traditional to the innovative practices for wheat rather than for rice cultivation. The 'Green Revolution' had less impact on rice yields than on other foodgrain crops in such areas. Tables 9.4 and 9.5 give statistics regarding areas and productions of major crops for selected years between 1950-51 and 1994-95.

**Table 9.4**

*Area Under Major Crops (in million hectares)*

| | *1950-51* | *1960-61* | *1970-71* | *1980-81* | *1994-95* |
|---|---|---|---|---|---|
| Rice | 30.8 | 34.1 | 37.6 | 40.1 | 42.2 |
| Wheat | 9.7 | 12.9 | 18.2 | 22.2 | 25.6 |
| Other cereals (jowar, bajra, maize, ragi, millets, barley) | 37.7 | 45.1 | 45.1 | 41.7 | 37.6 |
| Pulses (gram, *tur* and others) | 19.9 | 22.5 | 22.4 | 22.9 | 23.2 |
| Foodgrain total | 97.3 | 115.6 | 124.3 | 126.6 | 123.5 |
| Sugarcane | 1.7 | 2.4 | 2.7 | 2.6 | 3.8 |
| Oilseeds (linseed, rapeseed, sesamum, mustard, castorseed, peanuts) | 10.7 | 13.8 | 16.6 | 17.5 | 25.3 |
| Cotton | 5.9 | 7.6 | 7.6 | 7.5 | 7.9 |
| Jute (mesta) | 0.6 | 0.6 | 0.7 | 0.9 | 0.7 |

*Sources:* 1. *Statistical Outline of India*, 1997, Mumbai, p.51.
2. *India: A Reference Annual*, 1990.

*Paddy* (wet rice) is a *kharif* crop sown after the rains have set in from June to August, and harvested between November and January. In West Bengal two crops are the general rule: *aus* (June to September)

**Table 9.5**

*Production of Major Crops (in million metric tons)*

| | *1950-51* | *1960-61* | *1970-71* | *1980-81* | *1994-95* |
|---|---|---|---|---|---|
| Rice | 20.6 | 34.5 | 42.2 | 53.6 | 81.2 |
| Wheat | 6.5 | 11.0 | 23.8 | 36.3 | 65.5 |
| Other cereals | 15.4 | 23.7 | 30.5 | 29.0 | 30.3 |
| Pulses | 8.4 | 12.7 | 11.8 | 10.6 | 14.1 |
| Foodgrains total | 50.8 | 82.0 | 104.4 | 102.9 | 191.1 |
| Sugarcane | 5.7 | 11.4 | 12.6 | 15.4 | 25.8 |
| Oilseeds | 5.2 | 7.0 | 9.3 | 12.1 | 21.4 |
| Cotton | 2.9 | 5.6 | 4.7 | 7.0 | 12.1 |
| Jute (mesta) (in 1000 bales) | 3.3 | 4.1 | 4.9 | 6.7 | 8.3 |

*Sources:* 1. *Statistical Outline of India*, 1997, Mumbai, p.50.
2. *India: A Reference Annual*, 1990.

and *aman* (November to January), normally accounting for most of the cultivation, with an occasional third crop of *boro* (February to March). In such areas rice tends to be monocultural. Requiring painstaking labor, rice is intensely raised in tiny, flooded, heavily-manured beds, transplanted by hand, and harvested manually.

Climatically, topographically, and in the provision of plentiful farm labor, the densely populated lower Ganga basin and the coastal plains are ideally suited for rice cultivation.

In terms of caloric value per hectare rice crop yields are more productive than other foodgrain crops. Not surprisingly, rice is the major crop in densely populated areas, provided that the environmental conditions are favorable. In Punjab, western Uttar Pradesh and central India, wheat and other grain crops are the staple food of the common man; rice is used for festive occasions.

*Wheat* is characteristically different from rice in its climatic needs. It grows best in dry, irrigated areas during the cool season (*rabi* crop). Particularly adapted to the dry, irrigated lands of Punjab, Haryana, and western Uttar Pradesh and the Malwa plateau, it is grown as a premier crop and is the main staple food of the people of these areas. The principal areas of production are Sultej basin, Ganga plain and Malwa plateau where nearly one-third of the cultivated land is devoted

**Figure 9.7**
*Wheat*

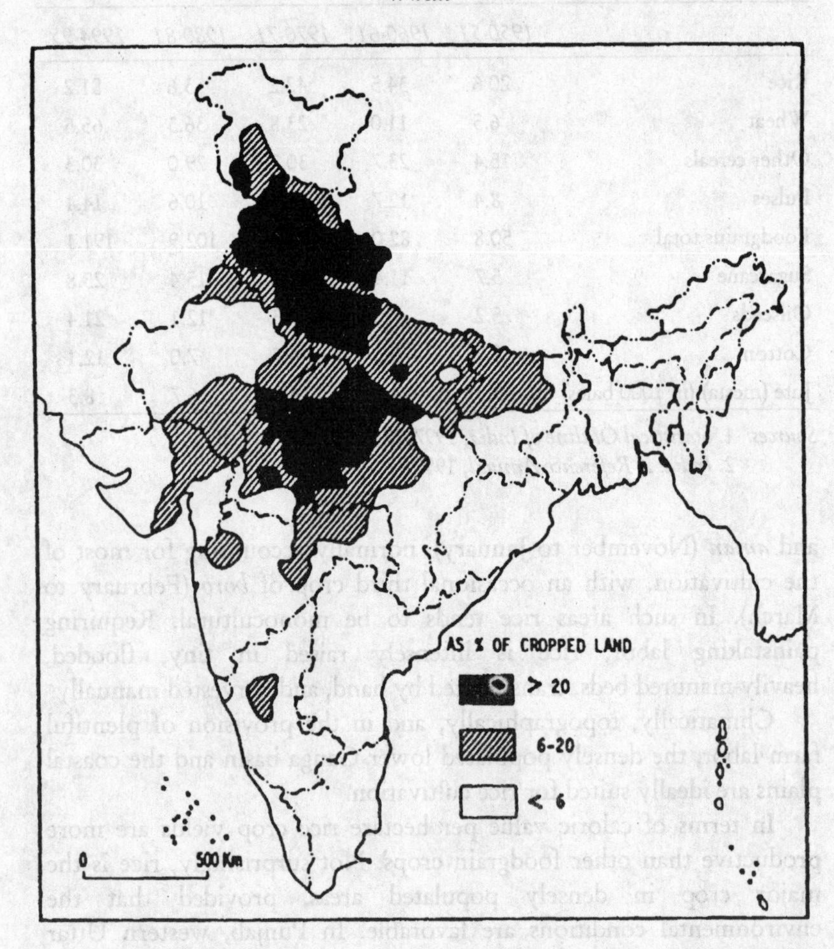


to wheat cultivation (Figure 9.7). Its proportion of cropland declines eastwards in the middle and lower Ganga plains where rice gains primacy, and southward from the Malwa plateau where millets, groundnuts and oilseeds predominate. In the peninsula, wheat cultivation extends from Gujarat to as far south as Karnataka on black lava soils but on merely 4 to 5 per cent of the cropland. Generally, the crop grows best on the alluvial, well-drained clayey loams of Punjab-Haryana and western Uttar Pradesh, but adapts well to varying types of soils excepting the sandy or the waterlogged. Nearly one-half of the country's wheat lands are found in these areas, which account for 60 per cent of the total production.

The area under wheat more than doubled between 1950 and 1995 (10 to 26 million ha), but production increased almost ten-fold (6.5 to over 65 million tons a year) during this period. Significant increases between 10 and 15 per cent a year were recorded in Punjab, Haryana and western Uttar Pradesh (Figure 9.7). Extension of wheat cultivation to Bihar, West Bengal and Orissa as a dry, winter crop has also been attempted. Yields are still low by world comparisons, averaging between 1,000 and 2,000 kg/ha, despite the introduction of HYV.

*Millets* as cereal crops are intermediate between rice and wheat lands in terms of location and environmental needs (climate and soil conditions). Several types, principally *jowar*, *bajra* and *ragi*, are extensively grown, collectively accounting for more area than is given to rice cultivation. Both production and area have increased modestly during the last twenty years. A long-standing prejudice against these crops as inferior cereals fit for consumption only by the poor has traditionally relegated their cultivation to the poorer soils. This prejudice stems from the fact that millets are a coarser grain, although in nutritional content they rank higher than either wheat or rice.

*Bajra*, a staple food of the poor, tolerates drier conditions and sandier soils than *jowar*, and is grown best in Rajasthan, south Haryana, western Uttar Pradesh and Gujarat. Its cultivation also extends to most of the northwestern sections of the peninsula. It is primarily a *kharif* (summer) crop.

*Jowar* (sorghum) is grown in an extensive area stretching from Punjab to Tamil Nadu, with the greatest concentration of cropland in the northwestern and central sections of the Deccan plateau where it is a leading crop and accounts for over 30 per cent of the cropland. In Tamil Nadu, it is grown as a *rabi* (winter) crop, its growing period coinciding with the winter monsoons. Elsewhere it is a summer crop.

*Ragi* is a *kharif* millet grown in a wide-ranging area from the Himalayas in Uttar Pradesh to the uplands of Tamil Nadu. It is a secondary foodgrain crop to rice in Tamil Nadu, and occupies generally heavier, poorer upland soils. Its main value lies in its relative dependability during the lean famine years, particularly in the poorer sections of the Deccan. The principal of production lies in the uplands of southern Karnataka where it occupies one-third of the cropland,

and is consumed locally as a staple food. It is nutritionally superior to rice and has a high calcium content, but the grain is coarse like other millets.

*Maize* (corn) and *barley* are two cereals grown typically under temperate conditions in north and central India. Maize has a wider distribution and extends well into south Rajasthan, Madhya Pradesh and Andhra Pradesh. It has the status of a coarse, secondary cereal crop, but its cultivation covers a large area (nearly 6 million ha) because it is also used as a fodder crop. A deep-rooted, leafy crop, it grows best on well-drained, deep, heavy loams in areas of assured rainfall or irrigation. The nucleus areas are in the hills, principally in the Kashmir valley and the western Himalayas where it occupies 15 to 30 per cent of the cropland, and in the uplands of southeast Rajasthan. Yields are very low by world standards, and hybridization has been recommended for the improvement of strains. Barley is grown as a minor cash crop in the Uttar Pradesh Himalayas, southern Haryana, and in eastern Rajasthan.

Among other staple crop cereals, *pulses* of various kinds are grown extensively, covering over 23.2 million hectare of cropped land. Grown ubiquitously over the country, their cultivation exhibits a few areas of concentration in central Rajasthan, Bihar, Orissa, Kutch (Gujarat) and Madhya Pradesh. For the diet of millions of Indians these provide a major source of protein. Although ecologically most responsive to areas of low moisture and light soils, pulses can be grown in widely varying environmental conditions. Both *rabi* and *kharif* varieties exist; best known among these are: *gram* (chick-pea), *moth* (brown grain), *tur* or *arhar*, *moong* (green beans) and *masur* (lentils). Regional distributions vary with different types, which are adapted to local soil and moisture conditions. In most parts of the country the cultivation of pulses forms a necessary activity in the agricultural calendar for they also provide a useful leguminous rotation crop for retention of soil fertility. *Gram* and *tur* account for over 10 per cent of the total cropland. Most pulses are used as *dal*, a puree complementing bread, rice or curry dishes or are roasted, fried for snacking.

### *Specialized Cash Crops*

Both *refined and raw sugar* have long been used in India. Raw cane, probably native to the country, grows ideally in tropical climates. The largest producing areas lie outside the tropics in the North Indian Plains from Punjab to the middle Ganga basin. Its total cultivated area amounts to a mere 3.8 million hectares which yields nearly 25.8 million tons of raw sugar (*gur*) annually. Extension in its cropland occurred during the First and Second Plans (1951-56/1956-61) in Maharashtra and Bihar. After reaching self-sufficiency and providing a modest exportable surplus in the 1960s, sugarcane cultivation expanded only slightly.

Indian yields are low because its areas of greatest production do not coincide with the climatically most-suited region in south India. The crop is highly demanding, requiring rich soils and abundant moisture during the growing season. In the western Uttar Pradesh plain and the *tarai* areas, cane occupies over 10 per cent of the cropland. Nearly one-half of India's cane cultivation area lies in Uttar Pradesh, the remainder is shared mainly among the states of Punjab, Maharashtra and Bihar, leaving only minor acreage in the deltas of the Godavari, Krishna and Kaveri, despite the existence of ecologically optimal conditions of rich alluvial soils, cheap irrigation and year-round high temperatures. In the North Indian Plains where rich alluvial soils have been traditionally utilized, a long growing period, widely-practiced irrigation, and cheap labor partially offset the disadvantages of a less than ideal environment.

Ranking after Cuba and Brazil in output, India is the third most important sugarcane producer in the world. Its cultivation area is the largest among growers but the native variety is of thin cane containing little juice and yieid is low particularly in the sub-tropical north Indian growing region. In the peninsula (Maharashtra, Andhra Pradesh and the Kaveri delta) where most favorable ecological conditions prevail, yields are two times higher than in the south.

*Oilseeds*, used both as cooking media and as animal feed, are widely grown in the country. Five major varieties, namely, linseed, rape-seed/mustard, sesamum seed, castor seed and groundnuts collectively occupy 26.3 million hectares of the cropped area. Rapeseed/mustard and sesamum are the principle cool season (*rabi*) crops, utilizing about one-fifth of the area of oilseed cultivation. Sesamum is mostly a *rabi* crop in southern India, but *kharif* in the north. Major cultivation areas are central India, the Malwa plateau, east Rajasthan, Madhya Pradesh and Kerala, which account for about 15 per cent of the area given to oilseeds. Groundnut ("peanut" in the U.S.A.) is a leading crop in Gujarat and parts of Maharashtra, but is also grown extensively in Tamil Nadu, Andhra Pradesh, Karnataka, Punjab and western Uttar Pradesh. It is an important cash crop, grown widely in warm, tropical lands and is well-adapted to poor, light soils. India is the leading producer in the world, accounting for nearly one-third of the world's total output. Newer high-yielding strains have recently been tried but with little success.

Spices, coconuts, arecanut and chilies are grown chiefly along the Western Ghats coast, especially in Malabar, south of Goa. *Coconut* is a major source of vegetable oil and its fiber is the basis of the coir industry. Total output is over 600 million nuts. Kerala, Tamil Nadu and Karnataka produce over 85 per cent of the total crop. India's coconut fiber-based coir industry is the largest in the world.

### *Fiber Crops*

*Cotton* holds a premier position among fibers and occupies a large cultivation area, i.e., 7.9 million hectares. With an annual output of over 12 million tons, India ranks high among the world's producers. The quality of lint is poor, and yields are low. Cotton is a demanding crop: moisture restricted to a limited growing period within the 60 to 86 cm rainfall zone, warm temperature and rich clayey loam soils. Indian production is, therefore, limited to a few areas including the ideal Deccan lava region of Maharashtra, Madhya Pradesh and Gujarat where rich moisture-retentive lava soils, warm temperatures combine to make it the leading producing area and the alluvial, irrigated Punjab area where cotton occupies 30 per cent of all cropland. Elsewhere, areas of significant production are Haryana, Karnataka and Tamil Nadu with 26 to 30 per cent of the cultivated area under it. Yields, low by world standards, are highest in Punjab, Haryana and Tamil Nadu, made possible by excellent irrigation facilities, rich alluvial soils and the introduction of better crop strains (Figure 9.8).

*Jute*, a minor crop in terms of cultivated area, is an important cash crop and a foreign exchange earner. Despite expansion of cultivation outside the traditional area in the Ganga delta, the crop shows marked

**Figure 9.8**
*Cotton and Jute*

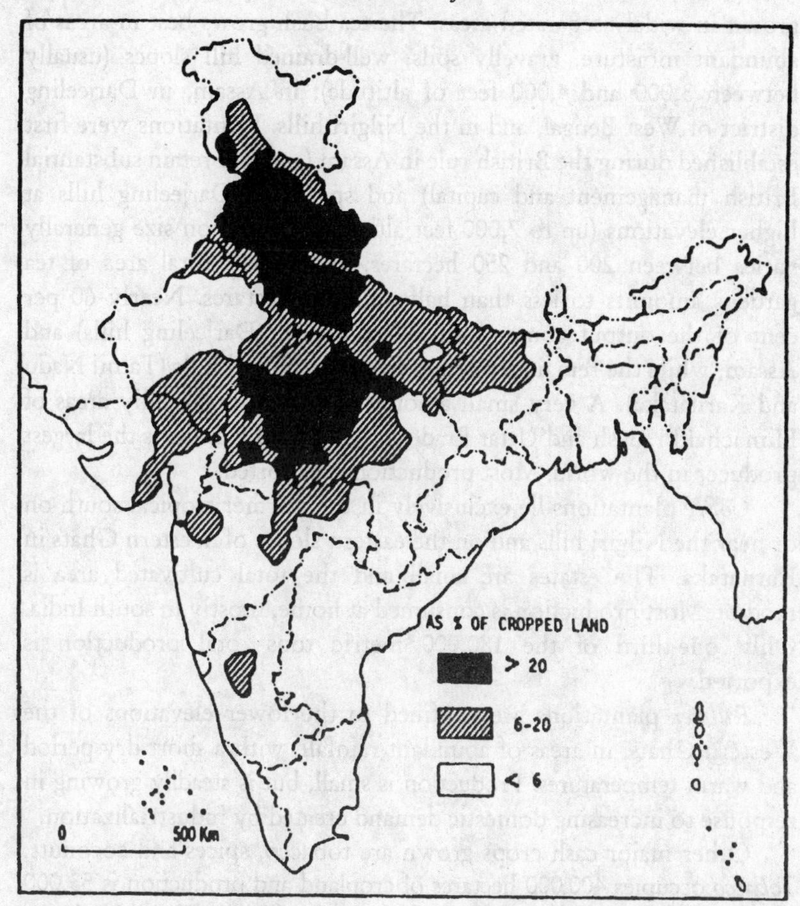


regional concentration. Partition of the subcontinent in 1947 deprived India of its major source area in Bangladesh for the mills around Hooghlyside near Calcutta. Rapid expansion of cultivation followed in the Mahanadi delta in Orissa, in the Assam Valley and in *tarai* areas of Uttar Pradesh and Bihar. Diffusion of a newly-developed jute variety in the *tarai* helped meet requirements of the Hooghlyside jute industry. India is now nearly self-sufficient in its needs, importing only a modest amount of very refined varieties from Bangladesh (Figure 9.8).

### *Plantation Crops*

Tea, coffee, rubber and cashewnuts are the principle plantation crops, grown in widely separated areas. The tea bush grows best in areas of abundant moisture, gravelly soils, well-drained hill slopes (usually between 3,000 and 4,000 feet of altitude), in Assam, in Darjeeling district of West Bengal, and in the Nilgiri hills. Plantations were first established during the British rule in Assam (and still retain substantial British management and capital) and spread to Darjeeling hills at higher elevations (up to 7,000 feet altitude). Plantation size generally varies between 200 and 250 hectares, and India's total area of tea gardens amounts to less than half-a-million hectares. Nearly 60 per cent of the output comes from West Bengal (Darjeeling hills) and Assam, while the remainder comes from the Nilgiri hills (Tamil Nadu and Karnataka). A very small amount is grown in the hilly areas of Himachal Pradesh and Uttar Pradesh. After China, India is the largest producer in the world. Most production is exported.

*Coffee* plantations lie exclusively in the warmer tropical south on or near the Nilgiri hills and on the eastern slopes of Western Ghats in Karnataka. The estates are small and the total cultivated area is modest. Most production is consumed at home, mostly in south India. Only one-third of the 180,000 metric tons otal production is exported.

*Rubber* plantations are confined to the lower elevations of the Western Ghats, in areas of abundant rainfall, with a short dry period and warm temperatures. Production is small, but is steadily growing in response to increasing domestic demand created by industrialization.

Other major cash crops grown are tobacco, spices and coconuts. *Tobacco* occupies 400,000 hectares of cropland and production is 53,000 tons annually. Nearly one-third of its cropland lies in Tamil Nadu. In the north it is grown widely in the silty, alluvial Ganga plains. Smoking is widely practiced in the country; thus most production is consumed at home, although some is exported to Britain.

Many varieties of *spices* are also grown in various parts of the country. Chilies are widely used and are grown primarily in south India, with the states of Tamil Nadu, Andhra Pradesh and Maharashtra accounting for 60 per cent of the production. Pepper production is concentrated along the moist slopes of Western Ghats. India is the largest exporter of pepper. Gingerroot and cardamom are grown mostly in Kerala and large quantities are exported. Coriander

Figure 9.9
*First-Ranking Crops*

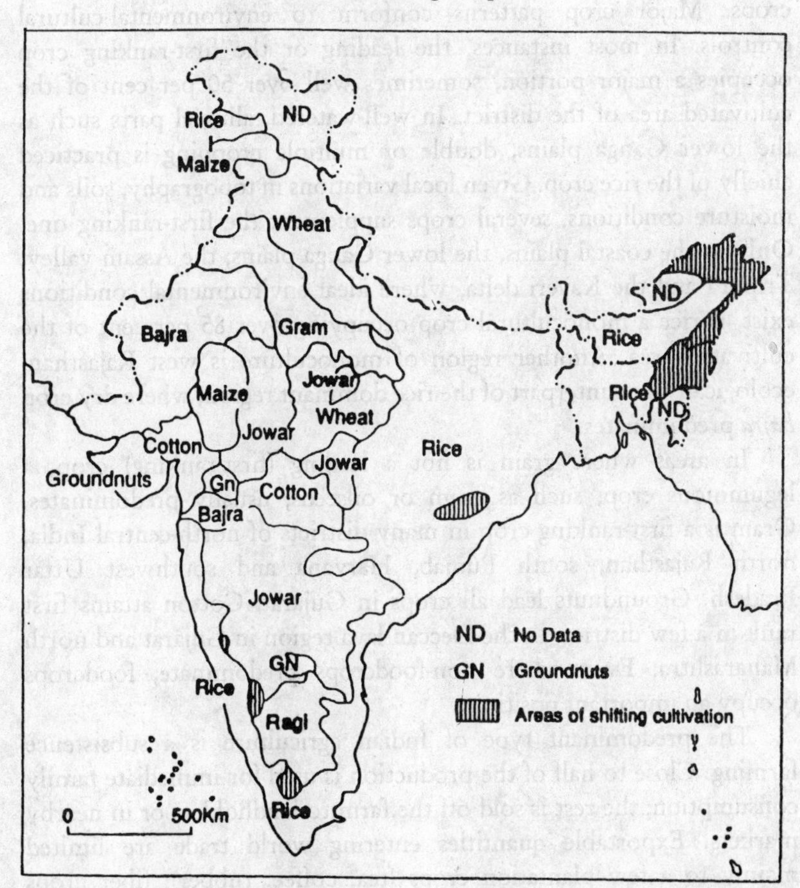


and turmeric are more widely grown, mainly in south India. The most favored area for coconut production is the Kerala coast. Spices and coconuts are grown on small holdings. Unlike rubber, tea and coffee these two are largely non-plantation crops. In addition, a variety of fruits and vegetables are grown which enter the domestic and international markets (like potatoes and mangoes).

## *Crop Association*

Figure 9.9 identifies areas of leading (first-ranking) crops. The predominance of foodcrops is apparent occupying over 70 per cent of the cropland. Cash crops (jute, tea, coffee) occupy a secondary position. Another characteristic is the regional distribution of several crops. Major crop patterns conform to environmental-cultural controls. In most instances, the leading or the first-ranking crop occupies a major portion, sometimes well over 50 per cent of the cultivated area of the district. In well-watered, alluvial parts such as the lower Ganga plains, double or multiple cropping is practiced chiefly of the rice crop. Given local variations in topography, soils and moisture conditions, several crops supplement the first-ranking one. Only in the coastal plains, the lower Ganga plains, the Assam valley, Tripura and the Kaveri delta, where ideal environmental conditions exist, is rice a monocultural crop occupying over 85 per cent of the cultivated area. Another region of monoculture is west Rajasthan, ecologically a counterpart of the rice dominant region, where dry crop *bajra* predominates.

In areas where grain is not a leading (first-ranking) crop, a leguminous crop, such as gram or oilseeds, usually predominates. Gram is a first-ranking crop in many districts of north-central India, north Rajasthan, south Punjab, Haryana and southwest Uttar Pradesh. Groundnuts lead all crops in Gujarat. Cotton attains first rank in a few districts in the Deccan lava region in Gujarat and north Maharashtra. Even where non-foodcrops predominate, foodcrops occupy an important position.

The predominant type of Indian agriculture is a subsistence farming. Close to half of the production is used for immediate family consumption; the rest is sold off the farm to landholders or in nearby markets. Exportable quantities entering world trade are limited mainly to a few plantation crops (tea, coffee, rubber), fiber crops (cotton, jute). Cash crops meant for nearby or distant markets occupy only a secondary place in the crop regions. Crop associations within the crop regions often have complex distributional patterns. As many as four and eight crops may compete for cropland and be termed secondary crops in a crop region, but such crops, nevertheless, tend to have widespread distribution. Pulses, oilseeds and fodder crops are examples of secondary crops with extensive distribution. Cotton, groundnuts, maize, bajra and barley are more widely distributed.

Secondary crops in the rice region are jute, oilseeds, pulses and tea, all of which, excepting tea, are associated with low-lying alluvial plains. Tea is important in the hills adjoining the Assam valley and in Darjeeling and Jalpaiguri districts of West Bengal. A rice-coffee combination is restricted to the plantations in the hills of Coorg and Karnataka. The pulse-*jowar*-rice combination is particularly noteworthy in the Krishna delta, where rainfall is less than 125 cm annually. Tobacco is also locally important there. Oilseeds are widely grown as a second- or third-ranking crop in the rice region of southern Bihar and the Mahanadi delta. Maize and wheat in association with rice occupy a major proportion of cropland in the Kashmir valley and the *tarai* region in Uttar Pradesh. Cotton is a second crop in Meghalaya and along the northern coast of Maharashtra. Gram is important in the lower Ganga plains and in western Uttar Pradesh.

Within the wheat region, maize and barley are second crops in the hills of Himachal Pradesh, while cotton is significant in areas of irrigation in western and central districts of Punjab as a *kharif* crop. *Bajra* holds a secondary position in the drier districts of south Punjab and south Haryana. Wheat is an important second crop in the *jowar* region which extends from east Rajasthan and the Malwa plateau to the drier, central parts of the Deccan plateau and as far south as north Karnataka. *Jowar* and cotton are important rainfed *kharif* crops in the Deccan lava region, whereas oilseeds and groundnuts occupy drier portions of the *jowar* region. Cotton is a leading crop in north Maharashtra and southern Madhya Pradesh. Groundnuts lead in Gujarat and one district in Andhra Pradesh, whereas maize in the Aravalli hills of southern Rajasthan with barley as a second crop. In the southern part of the Deccan peninsula, major crops are *ragi* on the uplands of south Karnataka and north Tamil Nadu. Oilseeds and groundnuts are secondary crops in the *ragi* region.

## Trends in Agricultural Development

Given the importance of agriculture in the Indian economy and the complexity of its underlying problems, the overall record of its achievements since Independence appears promising. Future directions would depend on how effectively past experience is utilized. A major initial disappointment was in the area of foodgrain production, the growth rates of which were either slow to fluctuating to generally bleak during the first three Five Year Plans. Performance has been satisfactory since then, a result primarily of breakthroughs in productivity made possible by the introduction of high-yielding varieties (HYVs) of foodgrain crops.

Until the Third Plan (1961-65) growth performance of most crops remained weak, but major acceleration occurred during and since the Fourth Plan (1966-71). The watershed of 1966 in agricultural production is coincident with the diffusion of agricultural modernization (introduction of HYV, mechanization and chemical fertilization) in selected parts of the country. Since then, the growth in the areas utilizing HYV has increased dramatically.

Contrary to the prevailing stereotypes, Indian farmers demonstrated willingness to work and accepted challenges and opportunities to improve the conditions. This was particularly true in regions where Intensive Agricultural Area Programme (IAAP) of 1965 and High-Yielding Varieties Programme (HYVP) of 1966 were launched by the government. These two programmes were initially concentrated in those geographic areas where these were most likely to succeed rapidly. Dramatic results were obtained in Punjab, Haryana and western Uttar Pradesh where dwarf Mexican wheat varieties were introduced and in Tamil Nadu, Andhra Pradesh, Kashmir and Kerala by the use of new strains of rice. Since 1970 these programs have been gradually diffused to other sections of the country. A new HYV of wheat as a dry *rabi* crop was also diffused to the predominantly rice producing areas of West Bengal, Assam, Orissa and Andhra Pradesh with promising results. For example, little wheat was grown in West Bengal before 1972, but 1 million tons of wheat was produced almost entirely of the HYVs in 1972.

### *Green Revolution*

For several years prior to Independence, growth of foodgrain production was virtually stagnant, averaging only 0.11 per cent annually, whereas population had been growing at about 1.4 per cent. Colonial policy primarily emphasized the production of export-oriented crops with the result that the country's capacity to feed itself had steadily deteriorated as population pressure mounted. Since Independence, despite substantial gains made in food production, the tempo of population pressure increased, posing a major threat to foodgrain self-sufficiency.

Since 1951, when the First Five Year Plan was launched, the

country has been importing foodgrains. The volume of imports fluctuated from less than 1 million tons in 1955 to over 10.4 million tons in 1966. Since then these have oscillated between 3 and 4 million tons annually, fluctuations responding to variations in monsoon behavior and resulting harvests. In 1976, a near self-sufficiency was reached, and attainment of food self-sufficiency became a major development objective.

India's foodgrain production has fluctuated widely from year to year since 1947, but four distinct phases are discernible. During the pre-planning period (1947-50) foodgrain production ranged between 55 and 60 million tons annually, growing at about 2 per cent a year, barely keeping up with population increase. In the second period, which synchronized with the First Plan (1951-56), India experienced foodgrain production increase at an average rate of 4 per cent a year, thus outpacing population growth. This was made possible by huge investments in irrigation projects and the Community Development Programmes. Considerable gains, both in area and in production of foodcrops, resulted in a decline of food imports, from over 3 million tons in 1951 to 700,000 tons in 1955. Production during the First Plan period peaked at 63 million tons. The third period coincided roughly with the decade of 1955-65 (Second and Third Plans). Production initially grew at 3.4 per cent a year, then slackened to nearly 2 per cent annually in 1965 and 1966, the years of disastrous droughts. Population growth averaged over 2.6 per cent annually during the period. In output, production of foodgrains had declined to a level at 65 million tons in 1966 (the level of 1955), after having risen to 70 to 71 million tons in the early 1960s. Consequently, foodgrain imports rose substantially, from 700,000 tons in 1955 to over 10.4 million tons in 1966. By the late 1980s a near self-sufficiency in foodgrain requirements had been achieved.

It became clear that by the end of the Third Plan the government had not succeeded in its stated objective of reaching foodgrain self-sufficiency. In fact, during the late 1960s there were serious food shortages, while the foreign exchange situation had also reached a crisis level in view of large food imports. Nearly 30 million people in the major cities lived under mandatory rationing; another 200 million, or over a third of the entire population, were under partial rationing. Failure of the monsoons in 1965-66 further aggravated the shortfalls in production and drastic, revolutionary efforts were clearly needed to improve the foodgrain situation.

The traditional approach to agricultural development had relied heavily on such methods as expansion of cultivable area and efficient utilization of the underemployed farm labor. These had failed to produce sufficient results during the first three Five Year Plans. During the 1960s some policy changes were elaborated which rested on two basic principles: first, that the farmers have a larger role in transforming the environment; and second, there are greater regional variations in agricultural productivity. The second principle was demonstrated by a joint Indo-American study, known as the Intensive Agricultural District Programs (IADP) which emphasized the need for a fresh approach to agricultural development. The IADP used a package program of raising foodgrain production by an intensive effort concentrated in 17 selected districts which were most likely to yield good results. The results of this experiment were mixed, but it led to the development of a new strategy based on the recognition of the need for provision of technological inputs into agriculture particularly in areas which can yield optimal results. In a sense, then, the Green Revolution was not an overnight phenomenon but was based, in part, on the experience of the IADP study.

In 1965, the administration shifted its focus to the new technological applications for agricultural development. The goal of the new strategy was to bring India to self-sufficiency in foodgrains by 1971 by raising production from 89 million tons in 1964 to 110 million tons by 1975. The approach was three-pronged: introduction of new seeds and modern agricultural techniques, concentration of these new inputs in the irrigated areas, and the provision of price and cheap credit incentives to the farmers so as to draw them into the participation process of the development program.

Indian government announced a "New Strategy" in 1965 with its basic objective of maximizing agricultural production by directing state effort in the first instance to those areas which were best endowed for food production. The key element of this maximizing process was the injection of large capital and technological inputs, first in the selected areas and subsequently to enlarge its coverage to other parts of the country. The introduction of HYV of foodgrain crops and mechanization were to be the primary inputs; inputs of chemical fertilizers and regulated water supply formed the remainder of the package.

This bold, new strategy was based on three major assumptions: first, the new varieties of crops to be introduced would nearly double per hectare productivity for the major foodcrops; second, maximum potentials would be attained by concentrating HYVs of seeds and modern inputs in irrigated lands; and finally, cultivators would respond to scientific practices if price and credit incentives were provided. The IADP was extended later on to include 114 districts (approximately a third of total), and its name was changed to IAAP (Intensive Agricultural Area Program).

A major element in increased foodgrain production was the introduction in 1966 of the HYV of wheat and rice developed in Mexico by Norman Borlaug for the Rockefeller Foundation. Dwarf strains of these crops were introduced in areas of favorable environmental conditions, particularly in Punjab, Haryana and western Uttar Pradesh, and became an instant success. Between 1966 and 1971 wheat production rose by over 90 per cent, recording a growth rate of 19 per cent a year. Close to 60 per cent of all foodgrain production between 1955 and 1971 was attributed to the new HYV of wheat. The foodgrain situation became so promising that by 1966 analysts proclaimed that the "Foodgrain Revolution" or the "Green Revolution" had arrived in India.

Yields increased phenomenally in the initial stages in favored areas. For example, in 1968, farmers who had adopted the new technology in Ludhiana district of Punjab increased their average yields in wheat from 2,100 lbs to over 4,200 lbs per hectare. Nationally one-fifth of wheat land was switched to the new Mexican strains, which produced in 1971 about 35 per cent of the total wheat crop. Yields did not rise as dramatically elsewhere as in the Punjab and Haryana districts since the new technological innovations were not accompanied by adequate supporting facilities for irrigation, fertilization and mechanization.

Regionally, the diffusion of HYV was spotty. From the "core" area of Punjab-Haryana, where about 80 per cent of the area was devoted to it, the HYV of "miracle wheat" quickly spread to other parts of the North Indian Plains from western Uttar Pradesh to districts as far east as Bihar, and southward to Rajasthan, Gujarat and

Maharashtra. Between 1965 and 1972, the area given to HYV of miracle wheat increased by 120 per cent nationally. Initial growth was rapid in the "core" area, its diffusion to east Uttar Pradesh, Bihar, West Bengal, the Brahmaputra valley and Orissa as a second *rabi* crop followed.

HYVs of rice were simultaneously introduced in several parts of the country: in Bihar, and West Bengal in the north, and Tamil Nadu, Kerala and Karnataka in the south. Yields of rice rose appreciably after the introduction of these new varieties, but not as dramatically as in the case of HYVs of wheat. Areas of notable success, with rice yields showing gains of over 30 per cent between 1965 and 1970, were limited to Malabar, Tamil Nadu and West Bengal. New rice strains were particularly susceptible to damage by pests and diseases. Rice farmers were generally poor and could ill afford the expenses of chemical fertilizers and pesticides used for new seeds and, unlike the Punjabi farmers, they did not possess reliable irrigation and drainage facilities. Furthermore, new rice strains were short-grained, tasted unfamiliar and lumped together in cooking. Public acceptance was weak. Rice, therefore, remained the "orphan" of the Green Revolution.

As regards other foodgrain crops, substantial success has been achieved in developing high-yielding strains of maize, which are responsive to good irrigation facilities as well as in millets for the dry regions. But actual production gains in both cases have remained insubstantial primarily on account of the limited area originally devoted to these crops, and also because of ineffectual irrigation facilities, lack of fertilization and weak enthusiasm among maize and millet growers.

The accomplishments of the Green Revolution have been highly selective among farmers and unevenly distributed in space. Its backbone has been the phenomenal success in dwarf Mexican wheat strains. Performance of new rice strains has been less satisfactory. Its continuing success will depend on how effectively the remaining serious problems of diffusion and management are overcome. A major challenge is, of course, the expansion of its benefits to cover remaining sections of the country and the farming community.

Most gains have been achieved where farmers possessed larger landholdings, had access to improved farm equipment and good irrigation facilities and could muster a reserve of capital through savings or credit loans. Studies in Ludhiana, a particularly advanced district in Punjab where new dwarf wheat varieties were introduced in relatively larger landholdings and where farmers had access to irrigation facilities, gains produced through diffusion of HYV have been truly remarkable. But Ludhiana is not a typical wheat-growing district. In Bihar and Uttar Pradesh, both wheat-growing areas, over 80 per cent of all cultivating holders have farms of less than 3 hectares. In the rice-growing regions, the average farm size is even smaller, between 1 and 2 hectares. On such small holdings, the impact of new technology has proved to be inadequate for future investments in agricultural development.

In general, big farmers maintaining holdings of 15 hectares were the largest beneficiaries. They were able to muster capital reserves for the purchase of farm machinery and equipment, install electric wells and utilize chemical fertilization. The result was that economic disparities between large farmers and the majority of small farmers generally widened. These widening disparities were clearly counter-productive to the objectives of the Five Year Plans of enabling small marginal farmer and agricultural laborers to "participate in the process of development". A Small Farmers Development Agency (SFDA) was set up during the Fifth Plan to help 3.5 million small farmers to grow HYV of foodgrains. This agency was responsible for the provision of irrigation facilities and technical advice to small farmers through existing channels of credit and cooperative societies. Yet, there was always an element of risk for the small farmer in obtaining loans, and he feared loss of his land to repay heavy debt in the event of a bad monsoon. As a result, agricultural modernization failed to have any serious impact on smaller farmers and might even have intensified polarization among farming classes.

Crucial to the dramatic rise in foodgrain production have been the two pillars of the Green Revolution: an adequate and controlled supply of irrigation and the application of chemical fertilizers to the crops during the growing season. Studies in the 1970s (Frankel, 1971; Mellor, 1976) have demonstrated that the big farmer was the major and quick beneficiary of the Green Revolution because he could make capital investments in irrigation and chemical fertilizers. A quarter of India's cultivated land currently enjoys an assured water supply from irrigation; most of this is held by big farmers. In consequence, insofar as the majority of the small farmers adopted the new practices, they can be expected to do so slowly, cautiously and partially. However, the meager infrastructural and marketing facilities, created during the Third and Fourth Plans, also tended to help the big farmer. Big farmers could afford to maintain their own transport or buy new tractors through credit loans; small farmers were considered poor credit risks and left without adequate facilities. Moreover, their crop would be generally too small to warrant provision of credit or other facilities. As a rule they would be left to use more expensive modes of transporting crops to market.

The legislative record in regard to land redistribution has been adequate, but implementation has flagged. Big landowners have ways to evade the land ceiling orders. One method is to assign portions of their land among relatives on legal documents thus evading the "land to the tiller" and land ceiling orders which enjoin each cultivating household to adhere to a fixed ceiling on his farming landholdings. This has bred discontent among small farmers, and has led to political turmoil, particularly among the landless scheduled castes (*Harijans*) who generally work as hired laborers. No wonder *Harijans* have been involved in recent political unrest and violence in Bihar and Tamil Nadu.

Indian planners are currently faced with the challenge of meeting fertilizer and new-strain seed needs incumbent on the introduction of HYV foodcrops. To serve 40 million ha of irrigated area placed under new strains of foodgrain crops, the country would require 2.4 million tons of chemical fertilizers and 144,000 tons of high-quality foodgrain seeds (combined for wheat, paddy, maize, *bajra*, *jowar*). Over 100,000 hectares of additional cultivated land will be needed to produce the required amount of seeds. Potentials for augmenting fertilizer production also appear restricted particularly because the escalation of world prices since 1973 of imported oil has resulted in oil shortages. Pesticides, another key input in the Green Revolution, are in critically short supply.

Despite setbacks and shortcomings, the outlook for foodgrain production appreciably improved in 1974 when a record of 104 million tons were produced. Production reached the level of 120 million tons in 1978 and 200 million tons by 1994-95. Since 1973, the country has virtually attained foodgrain self-sufficiency, and imports of foodgrain crops were reduced to a trickle. On the average, foodgrain production grew at a rate of a little over 4 per cent annually during the Fourth Plan. To run a winning battle with population, a sustained growth rate of 4.5 per cent a year in foodgrain production is required. Prospects for such increases for some years appear promising. Whether India will win this foodgrain-population race, is difficult to predict.

In sum, the introduction of HYV of foodgrains has brought a foodgrain revolution by stimulating agricultural modernization and commercialization. Accomplishments have been uneven in farming classes from one region to another. What is needed is a carefully designed and vigorously implemented strategy of agricultural modernization encompassing institutional, economic and political aspects of the agricultural problem applied universally and over all agricultural classes.

## *References*

Blyn, G., *Agricultural Trends in India*, 1981-47. Philadelphia, 1966.

Brown, D.D., *Agricultural Development in India's Districts.* Cambridge, Mass, 1971.

Chakravarti, A.K., "Green Revolution in India," *Annals of the A.A.G.*, Vol. 63, 1973, pp. 319-30.

Chatterjee, S.R., *National Atlas of India* (Hindi Edition). Calcutta, 1957.

Dayal, E., "A Measure of Cropping Intensity," *Professional Geographer*, Vol. 300, 1979, pp. 289-96.

Eitenne, G., *Studies in Indian Agriculture.* Bombay, 1968.

Farmer, B.H., *Agricultural Colonization in India Since Independence.* Oxford, 1974.

Frankel, F.R., *India's Political Economy, 1947-77.* Princeton, 1979.

Frankel, F.R., *India's Green Revolution: Economic Gains and Political Costs.* Princeton, 1971.

Government of India. *India, 1997: A Reference Annual.* New Delhi, 1990.

Jannuzi, F.T., *Agrarian Crisis in India.* Austin, 1974.

Khusro, A.M., *Economics of Land Reform and Farm Size in India.* Madras, 1973.

Ladejinsky, W., "Ironies of India's Green Revolution," *Foreign Affairs*, Vol. 48,1970, pp. 758-68.
Lewis, J.P., *Quiet Crisis in India*. Washington, D.C., 1962.
Mellor, J.W., *Developing Rural India*. Ithaca, N.Y., 1968.
Mellor, J.W., *The New Economics of Growth*. Ithaca, N.Y., 1976.
Muthiah, S., et al (eds.), *A Social and Economic Atlas of India*. Delhi, 1987.
Rao, K.L., *India's Water Wealth*. New Delhi, 1975.
Sengupta, R. (ed.), *Census Atlas*. Vol. 1 (ix). Delhi, 1970.
Singh, J., *An Agricultural Atlas of India: A Geographical Analysis*, Kurukshetra, 1974.
Singh, S.K., *The Indian Famine of 1967*. New Delhi, 1975.
Streeten, P.L. (ed.), *The Crisis of Indian Planning*. London, 1968.
Thorner, D. and Alice, *Land and Labor in India*. New York, 1962.

# 10

# Mineral and Power Resources

## Mineral Resources

India is fairly well-endowed with major minerals basic to the development of industrialization. It ranks high among world producers of iron ore, coal, mica and manganese. Its high-grade iron ore (*Fe* content over 55 per cent) reserves are among the largest in the world; manganese reserves are extensive and production of mica provides nearly 75 per cent of the world's output. Reserves of chromite, bauxite, kyanite and limestone are also considerable. A major deficiency is that of petroleum which must be imported in substantial amounts, particularly to meet its rising demand, created in a large measure by fertilizer manufacture. Discovery of offshore oil along the Gujarat-Cambay coast, and offshore sites near Bombay (now Mumbai) known as Bombay High, with seemingly large potential, is likely to ease the situation.

Major mineral-bearing areas (Figure 10.1) are dispersed primarily south of the Northern Plains in the Deccan plateau; a few also lie in the Himalayas. The northeastern section of the Deccan plateau is country's major mineral belt. Principal mineral concentrations lie in close association with rocks of the Palaeozoic era. Mineral areas outside the Chota Nagpur plateau are in Karnataka and in east Rajasthan-Madhya Pradesh.

India ranks seventh in the world (after the United States, the Soviet Union, China, the United Kingdom, Poland and West

**Figure 10.1**

*Non-Fuel Minerals*

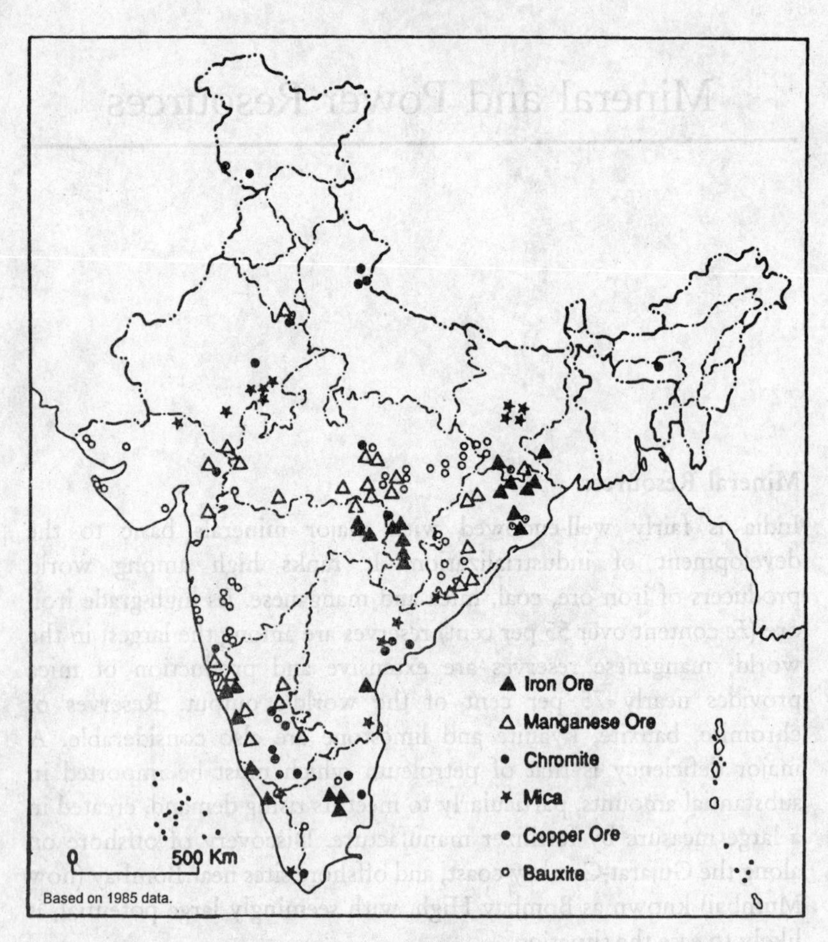


Germany) in the production of coal. The production is between 280 and 300 million tons annually, or 3 per cent of the world production. Consumption of coal was 299 million tons in 1996-97 and the projected demand for 2001-2 is 386 million tons. Production has gone up rapidly since Independence (nearly four and a half times), and to achieve self-sufficiency, an annual growth rate of 13 per cent needs to be targeted. Proven reserves are moderate, placed at 206 billion tons by the Geological Survey of India, but about a quarter of these lie outside easy reach below the surface. Quality is also poor.

High-quality near-surface coking coal, suitable for the metallurgical industry is obtained in small quantities from the Jharia coal fields of Bihar. At the present rate of consumption India's near-surface coke could last only for 25 years. The reserves of non-metallurgical coal are larger, currently estimated at 114,000 million tons, but their production costs are high.

Electricity generating stations, steel mills and railroads consume 65 per cent of the total coal production. The two states of Bihar and West Bengal supply nearly three-fourths of the total production. Major fields are Jharia, Raniganj, Bokaro and Giridih in the Damodar valley. Mines have thick, near-the-surface, easily extractable seams of moderate quality. Fields of Madhya Pradesh, the Godavari valley and at Singreni in Andhra Pradesh together contribute 124 per cent of India's production. Minor fields are Umaria and Talchar in Madhya Pradesh and in the lower Mahanadi valley.

Some coal of tertiary age is mined in Assam and utilized in tea factories. Lignite is also extracted from Neyveli mines near Madras (now Chennai) and is used for the production of electricity and in the fertilizer industry. The Eighth Plan aimed at raising lignite production to 20 million tons by 1995. By 1996 India was producing 24 million tons of lignite.

The location of major fields in the Chota Nagpur plateau has helped the growth of Jamshedpur as a major iron and steel center, but that city is distant from such other industrial areas as Mumbai, Hyderabad, Chennai, and the growing light engineering industries in Punjab and Haryana. Nearly 40 per cent of the coal needs of the states of Punjab and Tamil Nadu remain unmet largely because of the shortage of hopper cars and the inability of the railroads to make the necessary transfers. Indian reserves of high-grade iron ore are estimated to be over 880 million tons or nearly 4 per cent of the world's total. Four other countries—the Soviet Union, Canada, Brazil and Australia—have larger reserves. Current production is 61 million tons annually, which is three times that of 1960, of which 30 million tons are mined by private companies and the remainder is state-controlled. Fields are scattered in the Deccan plateau; many are located in the relatively inaccessible interior. The major producing region extends from Singhbhum in Bihar to Mayurbhanj in Madhya Pradesh and Keonjhar in Orissa. Large reserves occur in fairly close proximity to the coal fields and the iron and steel manufacturing centers. Other reserves lie in more inaccessible locations in Karnataka. Large scale exploitation has so far been hindered by inadequate transport system.

Domestic demand for iron ore is weak; less than a half of the production (69 million tons) is exported. About 80 per cent of the exports go to Japan and the remainder to European countries, West Germany and the United Kingdom. Ore is exported through Vishakhapatnam and the newly developed port of Paradeep near Cuttak in Orissa.

Petroleum reserves are limited. Its annual production of 35 million tons is capable of satisfying only one-third of the nation's demand. Consumption has been rising at a faster rate than production during the last three decades. In 1951 production was less than half-a-million tons but could meet a quarter of India's requirements. The Eighth Plan estimated that by the end of 1995 domestic demand would have escalated to 40 million tons annually and that increased imports will place an "excessive burden" on the economy. The projected demand for 2010 will be 197 million tons and if this is to be met through imports, the import bill could be something like US $ 26 billion per year.

Potential oil-bearing areas are located in Assam, West Bengal, the Ganga valley, Punjab, Himachal Pradesh, Kutch (Gujarat) and the west coast. Structurally, the most promising region is the Indus-Ganga-Brahmaputra valley and the synclinal tracts marginal to the peninsula in the Kaveri and Godavari basins. Recent exploitation has been limited to off-shore sites near Gujarat-Kutch-Cambay coast and of Bombay High. The Cambay field, developed since 1965 with Russian-Romanian technical collaboration, is currently yielding 5 million tons annually. The inland fields of Kalol and Ankleshwar in Gujarat are being now developed. Reserves of this region are estimated at 60 million tons. Refineries are located at Trombay near Mumbai, and Koyali in Gujarat. Since 1980, the country has been making a determined effort at oil exploration and exploitation and has begun laying a network of pipelines connecting production sites with refineries. The network is expected to extend to the Kaveri and Godavari basins when oil is struck there. Production in Assam started in 1892 in the upper Brahmaputra valley. Major fields are at Digboi and Sibsagar. Reserves are probably as large as those of the Gujarat fields; current production is a little larger. Refineries are located at Tinsukia and Nunmati, both close to the producing fields.

Refineries are also set up at Haldia near Calcutta, Vishakhapatnam, Chennai and Cochin, each with capacities ranging between 4 and 5 million tons annually, and close to the major importation points or to the producing fields. The Barauni refinery has a capacity of refining nearly 32 million tons annually, and is located close to the industrial region of south Bihar-West Bengal. The Barauni refinery serves Kanpur and other industrial centers in the North Indian Plains, and is connected with Haldia port. In all, the current capacity of the Indian refineries is around 35 million tons. Currently, the Bombay High contains the greatest potential in both crude oil and gas reserves.

Estimates of the reserves of natural gas yet remain to be determined; its usage is very limited, ranging between 7 and 7.5 billion cubic meters annually. Production is closely associated with the three major petroleum areas of Assam, Cambay coast, and Bombay High from where it is piped to the consuming centers in the country.

### *Other Minerals*

Manganese, extensively used in the iron and steel industry, has been produced in exportable quantities since 1891. Among minerals, it is second to iron-ore as a foreign exchange earner. Production is nearly 1.7 million tons annually or roughly 8 per cent of the world's output. India ranks after the Soviet Union, South Africa and Brazil in production. Price fluctuations in the world market and increasing competition from South Africa have adversely affected India's position. Mining centers are widely distributed in the states of Madhya Pradesh, Andhra Pradesh, Orissa, Karnataka, Maharashtra, Rajasthan and Goa. The principal producing area lies in the Keonjhar hills of Orissa, which is close to Vishakhapatnam—its export outlet. Total reserves are estimated to range between 165 and 175 million tons, of which 24 per cent are of high grade.

Among the metals used in the engineering industries (nickel, lead, zinc, copper, chromite and kyanite) only chromite and kyanite are abundantly found in the country; the remainder are imported or are inaccessible for extraction. Lead and zinc are found in Rajasthan in small quantities. The former is refined in Bihar and the latter processed in Udaipur (Rajasthan). Production of the ore is limited to 50,800 tons of lead concentrate annually. Total reserves *in situ* of chromite (used largely in chemical industries) are estimated at 135 million tons, although refractory grade are meager. Annual production has risen sharply from 107,000 tons in 1960 to over half-a-million tons, most of which is exported. Major fields are near Cuttak and Keonjhar in Orissa; other fields lie in Singhbhum in Bihar, Bandara in Maharashtra and Hassan in Karnataka. Magnesite is also produced in exportable amounts from the fields near Salem in Tamil Nadu and Karnataka. Asbestos is, however, imported. India's kyanite deposits are probably the largest in the world. Annual production has increased from 20,000 tons in 1960 to four times that amount since then. The major producing area lies near Jamshedpur.

Bauxite, mica, uranium, thorium and graphite, the chief minerals used in the production of electricity, are produced in fair to abundant quantities. Producing three-fourths of the world's mica, India faces competition from the Soviet Union and Brazil which have recently stepped up their production. Production has declined from 30,000 tons in 1950 to 25,000 tons. The country's major mica belt lies in the Hazaribag area of the Chota Nagpur plateau which accounts for about one-half the production, the remainder comes from Nellore in Andhra Pradesh, and Bhilwara in Rajasthan. India is the world's leading producer of sheet mica, and accounts for about 50 per cent of the world's mica trade. Bauxite production has nearly leaped five-fold since 1960, from 387,000 tons to over 4 million tons in 1995, most of which is exported to Japan, Australia and Europe. Domestic demand is meager but growing fast. West Bihar, Orissa, Madhya Pradesh, Maharashtra and Tamil Nadu are its major fields. Ore is processed and aluminum is produced at places near the hydro and thermal electricity installations at Hirakud in Orissa, Asansol in West Bengal and Alwaye in Kerala. The alumina content of Indian bauxite ranges between 50 and 80 per cent.

Minerals used for the production of atomic energy exist in relatively large quantities. Thorium reserves, estimated at 500,000 tons, are among the largest in the world, and are located in the sands of the Kerala coast. The uranium-bearing formations have been found in Singhbhum district of Bihar, and in the Himalayan regions of

Himachal Pradesh and Uttar Pradesh.

Minerals used in the glass, ceramic and fertilizer industries are limited. Rock salt production is about 3,500 tons annually. About 90 per cent of the annual salt output is, however, derived from evaporated salt-water along the Kutch, Maharashtra and Tamil Nadu coasts. A small amount is exported to Japan. Gypsum is obtained in adequate amounts, i.e., one million tons annually, chiefly from Rajasthan. The Sindri fertilizer factory utilizes most of the output.

Building materials such as sandstone, slate, limestone and marble are in adequate amounts. The Deccan peninsula is a rich storehouse of several of these. The Himalayas also yield slates and limestone; the slate quarries of Himachal Pradesh are particularly noteworthy.

Gold occupies an important place in Indian culture and mythology. Holdings, considered essential in Indian homes, are substantial, although estimates are hard to make. Nearly 70 per cent of the annual production of about 4,000 kilograms comes for the Kolar gold fields in Mysore in Karnataka.

## Power Resources

Basic to India's economic modernization is the development of an adequate supply of energy. Traditional energy sources such as wood, dried cattle dung and vegetable wastes are not only primitive and uneconomic, but also scarce and unsuited to modernization. The use of coal, hydroelectric power, petroleum and nuclear energy, started essentially in the twentieth century, has remained underdeveloped. Presently, 90 per cent of the domestic fuel used in the rural sector and 75 per cent in the cities is derived from the traditional sources with fuelwood accounting for 70 per cent of all the sources and the remainder from dried cattle dung and vegetable wastes. Fuelwood has become very scarce, as the forest cover had been removed continuously since medieval times while demand for wood has grown unabated.

The installed power generation capacity of India as on March 1998 was 89,166 MW comprising 64,150 MW from thermal (coal and gas), 21,891 MW from hydel, 2,225 MW from, nuclear and the remaining 900 MW from wind. Thermally-derived energy is focused primarily on the coal-rich section, particularly in the Damodar valley region. The hydro-based energy is potentially large and fairly well distributed. The Central Water and Power Commission estimated it to be about 51 million KW at 60 per cent load factor, of which only 3.6 per cent had been developed by 1961. By the end of the Seventh Plan 19 per cent of the potential had been developed, generating nearly 46 million KW in 1979.

Potentially, several areas are well endowed for the development of hydroelectric power: (a) a belt along the Himalayas from the Sutlej to the Brahmaputra valley, which is estimated to account for about 60 per cent of India's total potential; (b) hills in the southern part of the Deccan, the area of Nilgiri hills; and (c) the region of Western Ghats' eastern slopes in Maharashtra and Karnataka. Demand is minimal in the Assam section of the Himalayas, where potential is greatest (30 per cent of Indian total), but little development has taken place. In the Deccan plateau development has generally been hampered by seasonal and annual fluctuations in rainfall, lack of perennial rivers and problems of reservoir construction in hard rocks. Inter-state rivalries and water disputes have occasionally prevented the exploitation of hydro-resources for power generation. In the Himalayan region, where rainfall, terrain and water storage conditions are ideal, problems of inaccessibility, construction and distance from demand areas have caused hydroelectric development to lag.

Tamil Nadu and Maharashtra lead in actual production with over 3 million KW each in development of installed generating capacity, followed by Uttar Pradesh, West Bengal, Karnataka, Gujarat, Punjab and Orissa, and the DVC region as a unit, each with over 2 million KW installed capacity of hydro-energy. In per capita consumption, however, Punjab leads, followed by Maharashtra, Tamil Nadu and Gujarat. Currently there are over 100 operating power stations, of which 45 are hydroelectric and the remaining mostly thermal.

Generation of power through nuclear sources is still in its early stages, although India is one of the few countries in the world that can design, construct and operate a nuclear station all on its own. Its uranium resources are small, estimated at 70,000 tons, whereas the long-range potential of nuclear power depends largely on thorium which has an estimated reserve of 360,000 tons. The generating capacity of the three operating nuclear-powered stations, located at Tarapur near Mumbai, Kalpakkam in Tamil Nadu and Kota in Rajasthan represent only 2.6 per cent of the country's total installed power capacity. The Seventh Plan (1985-90) envisaged commissioning of two more units, one at Narora in Uttar Pradesh and the other at Kakrapur in Gujarat which will add a cumulative capacity of 4,070 MW. A total capacity addition from all sources was projected at 30,538 MW in the Eighth Plan but only 16,422 MW (46 per cent) could be added during this period.

In sum, power utilized in Bihar, West Bengal and Gujarat is primarily thermal, urban-industrial, market-oriented, and with the exception of Gujarat, close to the coal fields. In Punjab, Jammu & Kashmir, Karnataka, Kerala and Tamil Nadu, it is both thermal and hydro. In Gujarat, Assam and Maharashtra gas turbines and diesel are also used for power generation in addition to hydro and thermal sources.

Both industrial progress and the success of the Green Revolution have been accomplished through the utilization of electric power. Demand for it is likely to accelerate despite current shortfalls. The principal users are: industry (68 per cent), electric irrigation pumps (10 per cent, up from 4 per cent in 1950), domestic household consumption (11 per cent), and railroads (3 per cent). Electrical power utilization for irrigation pumps has dramatically increased in Tamil Nadu, Punjab, Haryana and western Uttar Pradesh, where high-yielding varieties of foodcrops have become important. Between 1941 and 1995 nearly 11 million electric pumps were installed.

Electricity has been an instrument of industrial decentralization. It has helped pull the agro-based small industries into the rural sector in the Haryana-Punjab-Delhi industrial corridor. The Fifth, Sixth, Seventh and Eighth Plans envisaged extension of electricity to the backward and rural sections under the Minimum Needs Programme (MNP) by the provision of electric power to health centers, irrigation pump sets, drinking water and construction of roads. By the end of 1989, 450,000 of India's 567,000 villages were provided with electricity, although less than one quarter of the households in the electrified villages were using electric power. At present, all the villages of Haryana, Punjab, Kerala, Andhra Pradesh, Karnataka, Gujarat, Himachal Pradesh and Tamil Nadu have electricity. Areas poor in rural electrification was the hilly states of Manipur, Meghalaya, Tripura, Arunachal Pradesh, and Orissa. The Seventh Plan (1985-90) outlays for rural electrification is Rs 20,920 million (approximately $1,747 million in 1992). It also envisaged a 100 per cent electrification of Maharashtra, Jammu & Kashmir, Assam and Nagaland.

The Fourth Plan (1969-74) recognized the need for the development of an integrated regional network of electric grid lines (Regional Grid) in order to provide an economic and efficient flow of power to the places of need. So far integrated grids limited to a few industrial areas like the DVC region, the Hooghly-side, the Greater Mumbai industrial complex and localized areas in Kerala and Tamil Nadu. The DVC grid is the only large integrated system of transmission of power utilizing both thermal and hydroelectric energy. By 1972, the total length of the integrated grid of transmission lines was 11,800 kms. In 1990, the network totaled 145,000 km of high voltage, spanning all over the states. The Seventh Plan (1985-90) aimed at extension of high voltage lines and standardization of power transmission, and the establishment of a National Power Grid. A major step towards power sector reforms had been the setting up Central Electricity Regulatory Commission in July 1998 to set tariffs and regulate inter-state power exchange, licensing, planning and other functions for all central generation and transmission utilities.

### *References*

Chaudhuri, M.R., *Power Resources of India*. Calcutta, 1970.
Government of India, *India 1997: A Reference Annual*. New Delhi, 1990.
Government of India, *Seventh Five Year Plan*, 1985-90, Vol. II, 1985.
Government of India, *Economic Survey*. 1989-90. New Delhi, 1990.
Muthiah, S. et al (eds.), *A Social and Economic Atlas of India*. Delhi, 1987.
Rao, K.L., *India's Water Wealth*. New Delhi. 1975.
*Statistical Outline of India*. 1989-90, Tata Services Limited, Bombay, 1989.

# 11

# Industries

India has been a traditional home of refined consumer-oriented manufactured items since pre-Christian times, although its modern large-scale industry is of recent growth. A base for industry had been laid during the pre-Independence period, but the colonial authority did not push vigorously for industrial modernization. Despite its potentially large market and resource endowment, industrial development of the country remained limited. The railroads established by the British were geared to administrative convenience and exploitation of raw materials with export potential. Industrial growth remained spotty and limited.

At the present time (of a labor force of nearly 300 million) only 8.7 million are employed in about 120,000 registered factories. Additionally, about 25 million work in cottage industries of handloom spinning or village crafts. The latter have a long and distinguished heritage.

In general, Indian industry is labor-intensive. Nearly 80 per cent of the establishments are small units employing less than 50 workers. Capital-intensive and power-based enterprises, with an efficient scale of operation, have been largely in the public sector which accounts for 6,200 factories, a quarter of the industrial workers. A relatively under-developed technological-managerial base, comparative capital deficiency, dependence on imported machines (much reduced since the 1960s) have resulted in low per worker production.

These shortcomings apart, industrial production has nearly quadrupled since Independence, the largest expansion being in the manufacturing of heavy machinery, engineering goods and chemicals. Textiles, heavy metallurgical and engineering products dominate industrial production, others occupying a minor position. Table 11.1 indicates the relative significance of the various industries. Industry in the private sector accounts for over 70 per cent of the industrial labor force. The government enunciated Industrial Policy in 1956, demarcating areas of jurisdiction. Heavy metallurgical and engineering, petro-chemical, pharmaceutical, aircraft, defense, nuclear and shipbuilding industries were placed in the public sector. Manufacturing of sugar, textiles, paper, cement, and light engineering goods were to be developed in the private sector.

**Table 11.1**

*Industrial Employment and Output, 1992-93 (Percentage Distribution)*

| *Industries* | *Employment* | *Value Added by Manufacturing* |
|---|---|---|
| Food Products and Beverages | 20.0 | 9.4 |
| Textiles (Cotton, Woolen and Jute) | 17.4 | 9.4 |
| Wood | 0.8 | 0.3 |
| Paper | 3.5 | 3.0 |
| Leather | 1.3 | 0.9 |
| Rubber | 3.2 | 7.5 |
| Chemicals and Chemical Products | 7.3 | 15.1 |
| Non-Metallic Minerals | 5.2 | 3.7 |
| Basic Metals and Metal Products | 7.6 | 9.0 |
| Electrical Machinery | 10.2 | 13.8 |
| Transport Equipment | 5.8 | 17.1 |
| Electricity | 10.2 | 0.4 |
| Repair Services | 0.6 | 0.4 |
| Others | 6.9 | 10.0 |

*Source:* Statistical Outline of India, 1997-98, Mumbai, 1997, p. 84.

## Growth of Manufacturing

Indian manufactured goods ranked with those produced in China and the Byzantine Empire during ancient and medieval times. The fifth century A.D. wrought-iron pillar near Delhi and the sheer Dacca muslins are testimonials to the quality of products manufactured in India. Industrial decline, however, set in by the early nineteenth century in the face of competition from cheap British goods. The colonial policy of the British favored the development of "home" industries in England, and kept India merely as a provider of raw materials or as a consumer of British products.

As early as the fourth century B.C. north India was dotted with flourishing manufacturing centers. Kashi (Varanasi), Mathura, and several cities in Vanga (now Bengal) and Kalinga (present-day Orissa) were important cotton textile centers. Magadh (Bihar) and Pundra (Bangladesh) were famous for the woolen fabrics. Silverware, gold ornaments, furs, skins and perfumes were other important products. Smelting technology was well advanced. Domestic and intercontinental trade in gold, silverware and iron products existed. An efficient and elaborate economic organization, dealing with the production and trade of crafts and mineral products, was in operation.

Contemporary Greek historians chronicle the flourishing status of trade and manufacturing during the first four centuries of the Christian era. Both Ptolemy's *Geography* and the *Periplus* (of anonymous authorship) mention that the coasts of India were studded with trading ports, chief among those were Barygaza (Broach) and Muziris (Cochin), carrying on trade with the Greek and Roman ports. Major items of export were cotton, silk, pepper, ivory, aromatics and precious stones. Imports consisted of wines, clothing, glass and coins.

Manufacturing and trade continued to flourish during the medieval times in India. During the period of Mughal rule (sixteenth-seventeenth centuries) Indian art, culture and crafts reached their zenith. New manufacturing techniques were developed by artisans and craftsmen. By the middle of the sixteenth century, the Europeans had developed commercial contacts with India and had established trading posts or "factories" along the coasts. By the early nineteenth century they had consolidated themselves as traders-rulers not only along the coasts, but in the North Indian Plains as well. The European posts were heavily fortified. Their commercial activities consisted mainly of the production of ornaments and textiles. These fortified posts, in due course, became foci for inland commerce and military penetration. Principal among the manufactured items during the eighteenth and nineteenth centuries were textiles whose production had become a highly specialized activity. Fabrics such as taffeta, muslin, percale, gingham and satin were produced for the domestic market. Other items produced were brassware, carpets, silk, silverware and pottery.

As late as 1850 manufacturing, however, was still practiced as a set of handicrafts largely in the rural areas. Initially, the urban centers along the railroad lines (established by the British) became collection and transit points for domestic trade. As communication links were improved, for example, building of roads, and supply lines were facilitated. Government establishments mainly produced gunpowder, ammunitions, arms and military uniforms. Major manufacturing centers during the early twentieth century using factory techniques were confined to a few locations such as Calcutta, Varanasi, Lucknow, Agra, Delhi, Jaipur, Bombay (now Mumbai), Madras (now Chennai), Pune, Hyderabad and Jamshedpur.

Even as these commercial industrial centers were developing, the focus of economic, social and political organization was the village-centered craft manufacturing based on the traditional *jajmani* system. In a typical self-sufficient rural organization, craftsmen, carpenters, smiths, tailors, oilcrushers, potters, weavers and others were supported by a fixed annual payment in kind by the landlords or clients. The producers belonged to certain castes and depended on the client-castes for livelihood. Little was produced beyond local demand. Luxury items, such as silk fabrics, jewelry and ivory carving, were manufactured in urban centers and the capitals of the princely states. The village-produced wares were immune to the price fluctuations of the cash-structured urban society.

With increased urbanization and commercialization, cities began to produce many of the items produced in villages and the demand for such items was also growing in the cities. The villages slowly became unable to cope with the increased demands from the growing cities, nor could these compete in price with the urban-made goods. Urban factories began to pour forth large quantities of cheap products. Rural crafts soon became uneconomic. The *jajmani* system was weakened, resulting in the flight of craftsmen to agricultural activities and newly developing industries.

Most of the railroad system was laid between 1870 and 1920. This facilitated the movement of such raw materials as tea, cotton and jute to the ports for processing. British colonial policy particularly favored the import of manufactured cloth from Britain to boost the Lancashire textile industry. Between 1850 and 1880 Indian exports of raw cotton increased from £25 to £55 million, which accounted for 50 per cent of Indian exports. Finished cloth imports from Britain surged from £7 to £25 million during the same period.

As domestic demand for textiles increased, cotton mills were opened in Mumbai, Ahmedabad, Nagpur and Sholapur in or near the country's major cotton-producing region. With the exception of iron and steel manufacturing, cotton textiles remained the only large scale industry in the country. In 1900 it employed nearly one-third of all factory workers (of a total of about 1 million).

In the intervening years between World War I and World War II, factory employment increased to 1.5 million, reflecting textile industry's expansion and also the expansion of war-based industries (gunpowder, arms, ammunition). Other industries were growing slowly. In 1939 India's steel production was 1 million tons, and coal production 28 million tons. Industrial employment had risen to 1.8 million.

### *Recent Industrial Growth*

In addition to the traditional manufacturing of sugar, leather products and rural crafts, at least two modern industries—textiles and iron and steel—were established during the British rule. The foundations of industrial institutions had also been laid, and there existed a small class of experienced entrepreneurs. As of 1950 the industrial sector, however, was a very small one. Registered and regulated industries utilized a mere 2 per cent of the labor force, and contributed only 6 per cent to the net domestic product. Textiles continued to dominate the industrial sector.

Since Independence, Indian planners have considered industrial growth to be a catalyst for the economic transformation of the country, and accorded the development of an industrial base a central place in the Five Year Plans. Central to the creation of a sound industrial base were two premises: first, the development of heavy industry would lay a foundation for the development of other industries; and second (related to the first), only a sound development of the industrial sector would make the country economically independent of foreign sources for capital goods and machinery production. For such an industrial development, government planning rested on three basic assumptions. First, the country, with its vast potential market base, would provide an adequate level of effective demand. Secondly, large industrial undertakings demanding technical knowledge would be domestically available. Domestic demand was weak in the 1950s and 1960s, and grew at a slower pace than anticipated. In sophisticated items requiring the use of modern technology, the country was obliged to depend on imports. Thirdly, Indian planners were confident of building adequate domestic capital formation required for developmental programmes. Such was not the accomplished. The creation of domestic capital continued to lag and capital supplies had to be replenished by large foreign aid infusions. Large imports of heavy machinery and foodcrops during the 1960s undoubtedly aggravated the foreign aid situation. Moreover, domestic production of such exportable goods as tea, jute, iron ore, mica, coconut fiber, cotton goods and sugar did not experience the large targeted expansion which would have created a surplus in the balance of trade. Despite restrictive import policy, import substitution and export stimulation, negative trade balances continued to create problems of foreign aid and lack of hard currency.

Industrial production was slow during the early 1950s; growth rates oscillating between 3 and 6 per cent per year. It rose to 8 per cent between 1953 and 1956, reverted to a little over 4 per cent in the late 1950s. Development picked up after that, averaging a healthy 10 per cent a year during the early mid-1960s. Index numbers of most manufacturing items jumped substantially, particularly in the production of heavy machinery and electric machinery. Industrial employment rose at an annual rate of 6 per cent. Production of consumer goods and light engineering items, however, registered poor growth performance. The tea and sugar industries prospered, but domestic demand for these items had been rising faster. Meanwhile, foodgrain imports had been rising rapidly, particularly after the serious droughts of 1966 and 1967. Recurrence of droughts in 1972, and the wars with China and Pakistan during the 1960s led to an industrial stagnation. By 1973, it had become clear that emphasis on heavy industry had to be relaxed, and that a reorientation toward the development of a diversified and integrated system of graded industry, broadly based to cover rural and urban needs and capable of mobilizing a vast, idle rural and urban labor force was required. This reorientation was reflected in the policy statements of the recent Five Year Plans.

The need to create an industrial base for the rural and medium-sized urban centers recognized by the Fourth and Fifth Plans was largely obscured by the emphatic importance given to heavy industry (heavy metallurgical products, locomotives, heavy machine manufacture, etc.) in the earlier plans. All these items are capital-intensive, rather than labor-intensive. Although the Fourth Plan did not entirely forsake the development of petrochemicals (new projects were set up at Mumbai and Koyali near Vadodara in the state of Gujarat), steel manufacture, and the production of cement, paper and fertilizer all recorded moderate growth. Emphasis on these was relaxed to further the growth of chemicals, fertilizer and light engineering goods, with greater private sector participation. Domestic demand for these items was rising fast in response to the modernization process which accompanied the "Green Revolution". While these industries were not directly labor-intensive, they did create a base for the development of ancillary, small-scale agro-industries which created additional employment opportunities. As a result of the growth of small-scale engineering and metal industry an industrial-urban corridor (Figure 11.1) between Delhi and Amritsar in Haryana and Punjab states to include the industrial centers of Karnal, Ambala, Ludhiana, Jullundur and Amritsar, oriented to the production of such items as bicycles, sewing machines, tractors and automobile parts, was developed.

In an effort to diversify and decentralize the industrial base, several industrial estates were set up during the 1950s and 1960s. These were located close to cities in order to make use of the existing infrastructure. Industrial estates contain small-scale industries and produce diverse goods ranging from consumer products such as textiles, light engineering goods and small machines, watches, bicycles and handicrafts. These mainly private, industrial estates, low-capital enterprises were occasionally housed in shed-like structures and made use of the local semi-skilled labor force. In 1962, there were 71 estates

**Figure 11.1**
*Industrial Areas*

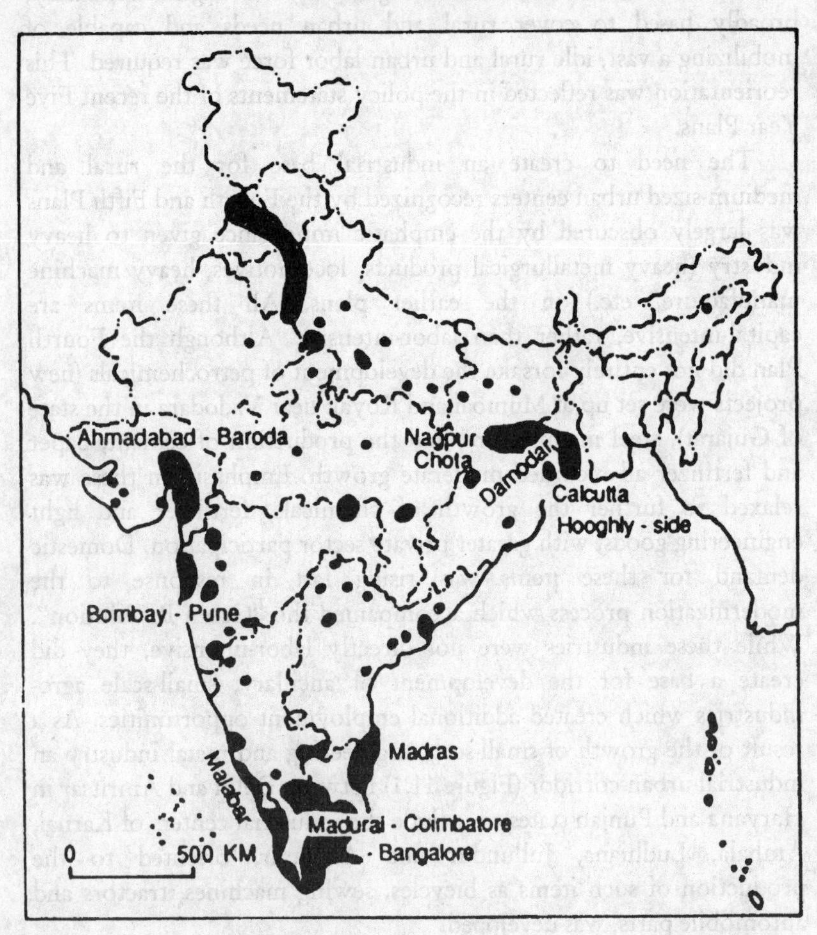


containing 138 factories located in different parts of the country. These employed 19,000 persons and producing $20 million worth of goods. By 1972 the number had gone up to 140 estates employing 23,350 persons. The Fifth Plan's outlay for expansion of estates amounted to Rs 300 million, and increased during the subsequent plans.

The policy of industrial decentralization was mainly limited to setting up industrial estates. It hardly affected the spatial pattern of industry, which remained concentrated in a few foci. Some spatial dispersal has recently started, especially in sugar, textiles, cement and paper industries. A special feature of the Fifth, Sixth and Seventh Plans was the development of industrially backward areas. As many as 229 districts in the country were identified as backward and eligible for grant of loans. Not much attention was, however, given to draw a coherent, national plan of industrial decentralization based on a policy of spatially-integrated development of industrial expansion lessening the rural-urban dichotomy.

## Distribution of Industry

At the close of colonial rule, industrialization was localized in a few major ports and scattered inland urban-commercial nuclei. The post-Independence period, particularly since the mid-1950s, witnessed a process of intensification of industrial growth in the existing areas and the emergence of a few new industrial-urban nuclei. Large industrialized stretches of landscape profoundly affecting the economy, transport and lifestyles of the inhabitants, were only beginning to be formed. In scale, degree of manufacturing activity and the impact on landscape, such areas could be termed as "nascent industrial regions". These could hardly be compared to the manufacturing regions of the U.S.A. or U.K. where industrialization overwhelmingly affects large tracts of land. In India, within the nascent regions, large stretches of rural countryside remain virtually unaffected by the industries of urban centers. Major and other nascent industrial areas which come closest to a typically western analogue of an industrial region are shown in Figure 11.1. Calcutta-Hooghly side employs about 750,000 workers, Bombay-Pune some 670,000 and Ahmedabad-Baroda 330,000 while the Chota Nagpur region in south Bihar, Madurai-Coimbatore-Bangalore and the southern Malabar coast manufacturing region are also significant. In all, these areas account for 3.8 million or one-half of the country's industrial labor force.

Calcutta-Hooghly side area is a sprawling 50-mile long industrial conurbation along the Hooghly river (a distributary of the river Ganga). Besides Calcutta, other major industrial centers located in this area are Howrah, Kanchanpura, Bhatpara, Chandernagore, Naihati, Barrackpur, Titagarh, Serampur, Dumdum, Bally and Alipur. Jute textile manufacturing is the main industry, absorbing nearly 40 per cent of the industrial workers, followed by engineering, chemical, food processing and paper industries. The English and other European powers selected lower Hooghly in the seventeenth century because of its convenient access from the Bay of Bengal to the North Indian Plains. Soon it became a focal point of Indo-British trade, although it lost a part of its hinterland to Bombay after the opening of Suez Canal in 1869. In 1912 it further lost some of its importance when the capital of India was moved inland to New Delhi. Its primary functions as the premier jute manufacturing center persisted, however, based on its location near the world's leading jute producing region. The development of railroads across the Ganga plains as the British administration fanned out across north India during the nineteenth century further enhanced Calcutta's connections and commerce.

Partition struck hard at the jute industry by removing its major source of raw jute in the Ganga-Brahmaputra delta. Problems of refugees from East Bengal further aggravated the economic situation. Problems multiplied as the original site became increasingly prone to inundation and silting. Population congestion, poor communication facilities, power and water supply breakdowns and political turmoil all contributed to the region's industrial stagnation in the early 1950s. Recent improvements in port facilities include channel deepening and the construction of a deep water access port at Haldia near the mouth of the Hooghly river, both designed to allow access by oil tankers. An oil refinery for processing imported oil has been constructed near Haldia. In 1976 an upstream barrage was built at Farakka to relieve the problems of silting and water shortages. Recovery since the 1950s has also been a result of its inherent locational advantages, such as proximity to India's chief mineral belt of Chota Nagpur, and to the power and navigation resources of the Damodar Valley Corporation. The country's three foreign exchange earners—tea, jute and jute textiles—pass through the port.

Mumbai-Pune area includes the industrial centers of Mumbai, Pune, Ambarnath, Pimpri, Trombay, Nasik and Tarapore. Textile industries, long dominant, now account for 41 per cent of the manufacturing labor force. Diversification since 1945 has led to the growth of other industries, particularly engineering, chemicals, pharmaceuticals and petroleum-based products. Ship-building, electrical, automobile, film and food-processing industries are also important. Coal-hauling from the distant Damodar valley fields in south Bihar and West Bengal has been a major problem, but is now largely mitigated by the development of hydroelectricity in nearby power plants located on the steep western edge of the Western Ghats. India's first major nuclear plant was established at Trombay. Oil refining is located in the nearby Cambay region where oil has been recently discovered.

Mumbai's advantage as a port closest to Europe has been enhanced because of the city's proximity to Middle Eastern oil and to market for finished textile and electrical goods in the Middle East and Africa. The entrepreneurial experience and skill of Parsi merchants and industrial families, the Tatas being one of these, and other entrepreneurial groups have transformed the area into a leading manufacturing region of the country.

The industrial area of Ahmedabad-Vadodara lies north of the Mumbai-Pune region and specializes primarily in cotton textile manufacturing, which consumes over 75 per cent of the industrial labor force. Chief centers are Ahmedabad, Jamnagar, Surat, Vadodara, Rajkot and Kaira. Lying in a cotton-growing region, the textile industry enjoys a decided locational advantage. Entrepreneurial experience is amply provided by several Gujarati mercantile groups (Hindu, Muslim and Parsi). Surat is an important silk manufacturing center as well. Non-textile industries include the manufacture of chemicals and light engineering goods. Vadodara is an important center of chemicals and pottery manufacture. The development of Kandla, although somewhat removed from the region, as a new port, has contributed to the recent expansion of industry for the region as it has for Ahmedabad which lies closer to it.

Chota Nagpur – Damodar valley industries owe their location chiefly to their proximity to India's major coal-fields of Jharia, Raniganj, Bokaro and Burnpur. The major iron and steel center of Durgapur and Asansol are located in the Damodar valley, as are also the plants of Kulti, Burnpur, and Bokaro. Sindri has a major fertilizer plant. The country's oldest and largest iron/steel plants (Tata Iron and Steel Works) at Jamshedpur and that at Rourkela developed during the Five Year Plans are in the Chota Nagpur plateau.

The Damodar Valley Corporation supplements coal-based power by provision of hydroelectric energy. A variety of industrial minerals, such as bauxite, iron, copper and manganese, are found in this leading mineral belt of India.

In south India, the area of Madurai-Coimbatore-Bangalore forms an important manufacturing region. Nearly one-half of manufacturing labor force is engaged in cotton textile industries. Bangalore has diversified industries like aircraft, automobile, telephone and locomotive manufactures—all primarily in the public sector. High grade coal is hauled from the distant Chota Nagpur plateau, but hydroelectric power is accessible within easy distance in the Nilgiri hills. Iron ore is available nearby. Industrial location is favored by relatively mild climate at altitudes up to 3,000 feet. Most production is consumed at home but exports to the Middle East and African countries are steadily growing. Chennai acts as the major port for the goods.

Malabar coast contains several industrial centers like Cochin, Ernakulam, Alwaye, Kozhikode, Quilon, Trichur and Kalamasary. Food processing industries, based on the local products of the plantations and of fisheries, such as rice milling, coconut oil extraction, soap making and shrimp canning, predominate. Oil refining and fertilizer manufacturing are other industries. Cochin is a major port.

## Major Industries

### *Iron and Steel Industry*

Prior to Independence, the colonial administration had practically ignored the creation of strong industrial sector in India despite the adequate resource base for the development of modern metallurgical and chemical industry which would compete with the growing industries in the United Kingdom. The British authorities held the view that India was unlikely to develop a large-scale industrial base, and that its minerals (iron ore, coal, manganese and bauxite) could be more profitably exported, whereas finished iron and steel goods and machinery could be more competitively imported than produced at home. At the turn of the twentieth century, the total production of pig iron in the country amounted to a mere 75,000 tons, produced by a company started in 1874. The Tatas established the first large-scale plant to manufacture iron and steel site which became the city of Jamshedpur in Bihar in 1907. Backed by the Tata entrepreneurial skill, financing and determination, and favored by a set of ideal geographic circumstances, the plant rapidly prospered.

At the time of Independence in 1947, there existed only three large, modern plants in the country: two in the private sector (Tata Iron and Steel Company at Jamshedpur and Indian Iron and Steel Company at Kulti-Burnpur in West Bengal) and the third was at Bhadravati in the state of Karnataka. Total production was a little over one million tons of crude steel and one million tons of pig iron. During the First and Second Plans, emphasis on heavy industry led to rapid growth of the iron and steel industry. Three new plants were set up in the public sector, each with an initial capacity of one million tons production of iron and steel ingots at Rourkela in Orissa, Bhilai in Madhya Pradesh and Durgapur in West Bengal. Expansion resulting from these plants pushed the production of crude steel from 6.5 million tons in 1966 to a little over 22 million tons in 1995-96. India is now the tenth largest producer of steel.

The concentration of all major plants, with the exception of the one at Bhadravati was in the Chota Nagpur region in the northeastern section of the Deccan plateau, reflecting the constellation of ideal circumstances in this region. The industry tends to be raw material oriented. Studies in 1962 revealed that transport costs in raw material assembly for the steel plants was the lowest for Jamshedpur and highest for Bhadravati. The iron ore fields of Singhbhum in Orissa are within 75 miles at Jamshedpur and the Damodar valley coal within 125 miles, the molding sand from the Subarnarekha river and several alloys such as manganese within 50-60 miles of Jamshedpur and within easy reach of Durgapur, Burnpur-Kulti area. The engineering works at Hooghly-side near Calcutta, offering the largest market for steel ingots, are only 150 miles away by a connecting railway link with Jamshedpur and closer for the Durgapur and Kulti-Burnpur industries. Limestone and manganese come from the nearby fields in Orissa. The Bhilai industry is a little closer to the manganese and limestone supply base. These centers are conveniently connected by a railroad with the Vishakhapatnam shipyard on the east coast and lie on the main railroad link between Calcutta and Mumbai. The three public sector plants set up during the Second Plan were established with foreign aid and collaboration: Durgapur with the United Kingdom, Rourkela with West Germany and Bhilai with the Soviet Union Public sector undertakings were subsequently expanded at

Bokaro during the Fourth Plan with Soviet assistance. The Bokaro factory is blessed with advantageous location, including proximity to mineral resources and transport connections with the existing foci of the iron steel industry in Chota Nagpur and was targeted to develop a production capacity of 10 million tons by 1979. Both Bokaro and Bhilai plants, however, fell considerably short of the targets despite their expansion.

The Bhadravati plant in southern India is located within 25-35 miles of iron ore, manganese, limestone and timber areas, but is remote from coal fields. Timber is used as a base fuel in the industry but is being rapidly depleted. The transport cost of hauling coal by rail from the distant Damodar valley area are the highest among major plants. The nearby industries of Bangalore utilize its production in metallurgical, engineering and ancillary industries of the Hindustan Machine Tools factory.

Steel production, which recorded substantial progress during the 1960s flagged later on for a number of reasons. Emphasis on the development of heavy industry, which was capital-intensive but which generated little employment, was shifted in favor of employment-oriented industries. It was widely feared that India could not afford capital-intensive enterprises as these tended to create foreign exchange pressures and did little to aid the unemployment situation. Furthermore, it was estimated that investment of capital in larger firms from 1957 to 1965 may have led to a net loss of 2.9 million potential jobs. Analysis of Indian industrialization generally favored the creation of an industrial base of smaller, consumer-oriented, employment-rich manufacturing units capable of utilizing the idle segment of the rural workforce in agro-based machinery and agricultural goods.

Several developmental problems have also plagued the growth of industry, notably inadequate transport linkages, underutilization of mill capacities, shortages of coal supply, and inefficient power supply. The Eighth Plan (1990-95) has set up a production target of 23 million tons of steel ingots, and nearly 15 million tons of finished steel (more than twice the figure of 1975). Those who favored the development of an employment-oriented, capital-conserving, small consumer-based steel industry of the view that these targets could be reached easier by the creation of a rural agro-based sector of the industry (e.g., by manufacture of agricultural implements, engines, small machinery, etc.). However, economy became sluggish during the 1980s, and the talks of implementation lagged. Government conceded that the Seventh Plan would be a period of consolidation and not expansion.

As per the new industrial policy of July 1991, the iron and steel segment is no longer reserved for the public sector. Private entrepreneurs are now free to set up steel plants of any capacity. Iron and steel industry is now on the high priority list and automatic approvals are available for foreign equity up to 74 per cent.

### *Engineering Industries*

Insignificant and confined to a few large industrial centers such as Calcutta, Jamshedpur, Bombay (now Mumbai) and Madras (now Chennai) before Independence, engineering industries have recently acquired special significance. The country is now self-sufficient in such engineering goods as bicycles, diesel locomotives, and several types of light electrical and electronic goods. Despite substantial progress made in the last 40 years, engineering manufactures form over 25 per cent of the import bill, a major portion of which is paid by foreign-aid receipts.

During the last two decades substantial progress has been gained in the manufacture of machinery for sugar, tea, cement, jute, cotton and paper. Printing, rice milling and machine tools manufactured also became significantly mechanized. Calcutta, Mumbai, and Chennai are the leading centers. Calcutta specializes in the production of machinery for textiles, tea, chemicals, sugar and paper, while in the machinery for cotton textiles, chemicals and paper. Chennai is more diversified in manufacturing. Light electrical machines and electronic instrumentation and machine tool manufacturing areas are also located in all these centers. During the post-Independence period, Coimbatore, Ahmedabad and Vadodara have emerged as centers producing electrical, textile and chemical machinery. Indian electrical and electronic component industry experienced a phenomenal increase during the last twenty years, registering a ten-fold increase between 1970-90. Maharashtra, Gujarat, Tamil Nadu, and Karnataka are the major producing states. The major impetus for the growth was the establishment of 15 State Electronic Development Corporations in the 1980s which also enabled the regional dispersal of the industry.

The Indian government has recently been encouraging

decentralization of industry with a view to achieving a more balanced regional development. In 1970, Hindustan Machine Tools Ltd., a public sector undertaking, established its major plants at Bangalore, Pinjore (Haryana), Hyderabad and Kalamasary (Kerala). Bangalore is the largest machine-building center in the south and has attracted a concentration of industries like watch-making, production of telephone equipment and aircraft manufacture in the public sector. Its private sector industries include ceramics, electric goods, and the manufacture of soap and textile machinery. Pleasant climate, planned urban development and government encouragement have been important factors in the growth of these industries there. Elsewhere, public sector machine tool plants are located at Ajmer (Rajasthan), a machine plant for the production of weapons for the Defense Services at Secunderabad (Andhra Pradesh) and a precision-instrument plant at Kota, also in Rajasthan.

The manufacture of heavy metallurgical engineering machines during the last three decades also progressed substantially, particularly in the production of mine-pumps, conveyors, shuttle cars, and haulers, paid for largely through foreign aid. Production of these items is over one million tons annually making the country practically self-sufficient in coal mining machinery requirements. Major plants are located in the Chota Nagpur plateau area in Ranchi where the Heavy Engineering Corporation is located, at Dhanbad in Bihar and at Durgapur in West Bengal. Private sector heavy machinery plants are located at Naini (near Allahabad, set up in 1965) and in the south at Tungabhadra, started in 1967 to manufacture transmission power machinery and ancillary items.

The public sector heavy electrical equipment production has also made good progress during the 1960s and 1970s. Major plants are located at Bhopal (Madhya Pradesh), New Delhi, Hardwar (Uttar Pradesh), Tiruchirapalli (Tamil Nadu), Hyderabad and Rupnarainagar (West Bengal). These plants manufacture electric transformers, steam turbines, boilers and telecommunication equipment. Ship building is chiefly in the public sector and is located at Vishakhapatnam, Cochin and Malgaon.

Light industries are regionally dispersed, but tend to be concentrated close to the centers of heavy industry and have a larger private sector participation. Manufactured items include bicycles, watches, typewriters, sewing machines, diesel engines and electric pumps. Production is growing rapidly, in pace with the increasing domestic demand and has even shown recently a healthy export performance. Indian manufacturers have established contacts with the Middle East, Africa and Southeast Asia to assist them in setting up these industries in their countries. Although small industrial units specializing in light industry have sprung up near the major industrial cities all over the country, particularly noteworthy is the emergence of a light industries urban corridor (Figure 11.1) along a 200 km belt in Punjab-Haryana-Delhi-West Uttar Pradesh including the urban centers of Ludhiana-Ambala-Rohtak-Panipat-Delhi-Ghaziabad-Meerut which contains numerous small private sector industrial units producing agricultural machinery, scooters, bicycles and sewing machines.

Jamshedpur industries also manufacture a wide variety of light engineering goods. Lucknow (Uttar Pradesh) and Jabalpur (Madhya Pradesh) are the important centers of automobile equipment and parts. Delhi, Calcutta, Mumbai and Chennai, including their suburbs, are the other centers. Production of commercial automotive vehicles has steadily grown since 1960, but domestic demand has been rising faster and is now half-a-million units per year.

### *Metal-Based Industries*

With the growth of heavy industry, domestic demand for such metals as copper, nickel, zinc and lead has risen. Although current production is still small (51,000 tons of lead, 253,000 tons of zinc, and 47,800 tons of copper), the Ninth Plan has ambitious targets for stepping up their production. While the production of copper, lead and nickel was to be developed exclusively in the public sector, aluminum and zinc were also to be encouraged in the private sector. The non-ferrous metals were allocated over $60 million for their development in the Seventh Plan—a figure representing a sizable amount of all industrial development programs.

Major aluminum plants are located at Jaykaynagar (West Bengal) close to the bauxite fields of the Chota Nagpur plateau, in Salem (Tamil Nadu) and near Goa, also proximal to bauxite areas. These plants make use of the thermal power obtained from nearby generating stations. Another plant is located in Alwaye (Kerala). New smelters are planned at Korba (Madhya Pradesh) and Ratnagiri (Maharashtra). Copper smelting is located close to the mines at Khetri in Rajasthan and Gharala in Bihar. Zinc smelters are located at Debri (Rajasthan) near the mines and at Vishakhapatnam and Alwaye, based on imported concentrates. Lead and nickel smelting plants are also located near mines in Rajasthan, Orissa and Andhra Pradesh.

### *Textile Industries*

Producing between 12 and 13 per cent of the world's raw cotton, India ranks behind the United States, the Soviet Union and China in its production. In jute it rivals Bangladesh as the leading producer, and in the production of raw silk it is behind Japan, China and the Soviet Union. Textile industries based on raw cotton and raw silk have ancient roots in the country. India was renowned in the western world for the production of fine silks and sheer Dacca muslins in ancient and medieval times. Textiles formed its leading exports even before the Christian era. This fame remained uneclipsed until the nineteenth century when the imported, cheap and plentiful cloth manufactured by the Lancashire industry virtually wiped away domestic production.

Both in terms of employment and output, however, the cotton textile industry still occupies a pre-eminent position among industries. India is not only self-sufficient in its textile requirements, but exports textile fabrics, and readymade garments. It employs over 1.9 million persons, or nearly 24 per cent of the total industrial labor force, in its nearly 8,000 factories and contributes over 20 per cent of the value added by manufacture. The first organized cotton textile factory was established in Calcutta in 1818, whereas the first jute mill was set up in 1854, also near Calcutta. The modern industry picked up momentum much later, during the 1930s and 1940s after the grant of government protection to appease the *swadeshi* (national) movement. In 1939, there were 389 registered cotton textile mills. Since then, the growth has been steady. Between 1950-51 and 1974-75 the production of cotton yarn nearly doubled, commensurate with increased domestic requirements. The peak was reached in 1961 when exports to Third World countries became an essential part of foreign trade.

The start of the cotton textile manufacturing in 1850 in Mumbai was undoubtedly favored by its geographic location near the cotton growing region, its accessibility to a wide variety of cotton grades and by its superior transport facilities. Industry was initially spurred by its nearness to British industrial centers for the import of machinery and export of yarn. Besides Britain, early importers included China and Japan, but the Chinese market dried up as Japan and China experienced a manufacturing upsurge by the early twentieth century. Meanwhile, the home market was expanding particularly of the urban elite. The *swadeshi* movement, launched against imported British manufactured cloth, helped the growth of the fabric manufacturing industry substantially. Mumbai and, later, nearby locations such as Ahmedabad became important centers, followed by the development of numerous smaller urban constellations surrounding the two foci. Freight costs of cotton lint hauling were competitive at these places as was the case also at some inland locations such as Nagpur and Sholapur. Nagpur benefited from fuel supplied by Damodar coal fields, whereas Sholapur was close to hydropower produced in the Western Ghats.

In North Indian Plains, good transport facilities, a growing home market and the local cotton-growing tracts, all contributed to the emergence of such centers as Kanpur, midway between Calcutta and Delhi on an important railroad route, and others farther west at Delhi and beyond it in Punjab. Cotton being a non-perishable commodity, the location of manufacturing centers was determined on the basis of demand, transport facilities and availability of power. Calcutta's cotton industry started in 1905 by which time the jute industry had already been well-established. Despite competition from the jute industry such advantages as proximity to Damodar valley coal, port facilities and a large local market fostered the rapid development of cotton textiles. Industrial centers in Punjab (Ludhiana, Amritsar) grew up following the development of Mandi hydroelectricity.

In south India, industry came around 1870, first at Chennai (then Madras) and then at Coimbatore, Bangalore and Madurai. Major development occurred since 1930s, following the arrival of hydroelectric power from Pykara (Tamil Nadu). Although the industry is regionally well-distributed in areas of consumption, the heart of the cotton mill industry is a pentagon formed by the cities of Mumbai, Ahmedabad, Ujjain, Nagpur and Sholapur. These five centers account for over one-half of the mill consumption of raw cotton, and three-fourths of the number of looms in the industry. In addition to the initial advantages which led to industry's location here, a number of factors helped sustain and expand the industry. The Gujarati and Parsi entrepreneurial experience, well-regulated markets for collecting and distributing raw cotton and excellent internal transport facilities all helped in maintaining the pre-eminence of this region.

In addition to the mill sector (registered factories) discussed thus far, a much larger (in terms of employment) cottage industry sector of decentralized handloom cloth-making has traditionally existed in the country. This has recently expanded more rapidly than the "organized" sector because of the administration's special encouragement provided through funding under Five Year Plans. Expansion of this sector has been rapid largely because the hand-woven cloth is inexpensive and oriented toward local, rural markets. There are thousands of small handlooms spread throughout the country. The total production of cotton cloth in 1973-74 was 7,800 billion square meters, mainly by expanding the decentralized handloom sector.

India is now next only to Japan as an exporter of cotton fabrics. In value, cotton yarn and goods rank after tea and jute manufactures among Indian exports. Southeast Asia, the Middle East and Africa are the major markets. Trade statistics of the last two decades, however, indicates that the industry has suffered in its relative contribution to the country's earnings. Government policy since the First Plan has favored capital-intensive heavy industry, particularly that of iron and steel. In 1951, cotton textile produced 41 per cent of the nation's value-added through manufacturing, and nearly one-third of the textile output was sent abroad, representing 38 per cent of India's exports. Since that year foreign shipments have been continually declining. In 1974, textiles' share of manufacturing value-added was 21 per cent (about one-half of the 1951 level), and in export value, cotton fabrics represented 25 per cent of the total. The administration is now considering a more rapid expansion of the decentralized sector.

Jute manufacturing is one of the important industries, accounting for 10 per cent of textile exports. Before 1947 India held a pre-eminent position as a producer of jute manufactures as all the industrial units (70,000 looms) were concentrated in the Hooghly-side region near

Calcutta. Partition in 1947 deprived India's jute industry of a major supply area of raw jute. In order to retain its supremacy as the largest jute manufacturing country, India had to extend its jute growing area outside the traditional jute heartland of the Ganga-Brahmaputra delta, a major portion of which became East Pakistan. Jute cultivation was, therefore, extended to Orissa, west Bihar and the Tarai region of the Ganga plain, thus restoring the supply base to the industry. Domestic raw jute supply is now sufficient to meet domestic demand; only small quantities are imported from Bangladesh.

The highly localized distribution of the industry along a belt 60 miles by 3 miles adjacent to the Hooghly, began in 1859, under ideal geographic conditions. The Ganga-Brahmaputra delta supply region of jute cultivation was connected by excellent river-borne and railway links. Hooghly-side lay roughly midway between the Damodar valley coal fields and the jute cultivation region. Transportation economy virtually dictated the selection of a Hooghly-side location, for if the mills were to be located in East Bengal trans-shipment costs would have been prohibitive, first in carrying coal and mill stores, and to the mills, and again in trans-shipping the finished goods to Calcutta for export. Following the emergence of Bangladesh in 1971, India and Bangladesh lacked enough mills to process all of its raw jute.

Current production of jute manufacture is 1.4 million tons of which 16 per cent is exported. Production has remained nearly steady or has declined slightly in view of competition from Bangladeshi industry and the growth of a synthetic fabric industry.

India's woolen industry is relatively small. It received a setback in 1947 when the major supply base fell to West Pakistan after partition. Production of yarn is about 200,000 tons, and that of fabric 125,000 meters annually. Domestic demand is large. Major manufacturing centers are Kanpur (in Uttar Pradesh), Dhariwal (near Amritsar in Punjab), Beawar (in Rajasthan) and Srinagar (in Kashmir). Elsewhere, the handloom cottage woolen industry is well established in Himachal Pradesh based on the local, fine sheep wool.

The silk industry is even smaller in terms of total employment, although it has traditionally held a pride of place among Indian textiles. Karnataka, Tamil Nadu, West Bengal and Kashmir are the major producing areas. The cottage industry sector has expanded rapidly during the last 20 years under government encouragement.

Rayon, and the synthetic fabric industry has also grown since 1950, with annual production of yarn jumping from 2,000 to 13,000 tons. The industry is located at the large cotton-textile centers, such as Mumbai, Ahmedabad and Amritsar.

### *Sugar Industry*

India produces nearly 25.8 million metric tons a year or 10 per cent of the world's sugarcane. It is world's leading producer of unrefined country-sugar or jaggery. Next to textile manufacturing, sugar is the country's major agro-industry.

At the time of Independence it was one of the well-established industries. There were over 100 factories producing less than one million tons of sugar annually. It was world's largest sugar producer. This was an eight-fold increase since 1932. It had also emerged as a modest exporter by the early 1940s. Cuba and Brazil, however, soon took over as the largest producers and world's principal sugar exporters. During the 1960s, despite a remarkable increase in domestic production, export was meager, roughly 2 to 5 per cent of domestic production. Since then, export has averaged between 10 and 15 per cent of production, made possible by the four-fold increase in the domestic output since Independence. Sugar exports increased dramatically since 1987.

The traditional "heartland" of the industry has been the Ganga plain, which accounted for over 80 per cent of the production until 1930, when a dramatic shift to the peninsula began to take place. In 1959-60, the Ganga plain (mainly the states of Uttar Pradesh and Bihar) contributed 65 per cent of the country's production, whereas the current production of the region is about 50 per cent. The dispersal was oriented toward the tropical region in the peninsula where the four states of Maharashtra, Andhra Pradesh, Tamil Nadu and Karnataka now account for about 45 per cent of the total production. Between 1950 and 1965 twenty new factories were set up in Maharashtra, eleven in Andhra Pradesh, eight in Tamil Nadu and four in Karnataka. Mills are raw material oriented and lie close to the sugarcane areas. They are generally small, clinging to the railroad or roads for easy hauling of the bulky raw material. The Ganga plain, the peninsula, west coastal plain and the delta districts of the east coast contain most of the factories, utilizing the good transport facilities of the areas.

In the northern states cane quality suffers from low sucrose content, a short crushing season and low productivity. The tropical southern states are ideally suited for cane production from the climatic standpoint, capable of producing cane with higher sucrose content. Extension of irrigation facilities, growth of cooperatives and administrative regulatory incentives and improvements in transport, both for cane production and for the sugar manufacturing industry, helped the growth of industry during the 1960s in the southern states. This process of decentralization of the industry is likely to continue. Factories in the south also enjoy a comparative cost advantage in respect of the sugarcane supply. The Sugarcane Inquiry Commission of 1965, after examining the comparative cost-benefits of sugarcane cultivation in the various parts of the country, concluded that Maharashtra ranked first in cost-benefit advantages (one measure of which was gross value of sugarcane per unit area).

A problem which the industry has chronically faced is the recurring cycles of international pricing and demand fluctuation affecting the domestic production and its exportable amounts. Domestic demand has, however, risen appreciably during the last two decades and has been met partly by the production of unrefined sugar (*gur* and *khandsari* and palm sugar) especially in the rural areas. Statutory controls over refined sugar for domestic consumption have occasionally been applied in order to meet the export demands.

### Chemical and Related Industries

Drugs, pharmaceuticals, paints, varnishes, synthetic fibers, DDT and other pesticides, industrial gas and chemical fertilizers were all virtually entirely imported until early 1960s. Only caustic soda, phosphates, sulphuric acid, soaps and synthetic detergents were produced at home in limited quantities. So phenomenal has been the growth of chemical and related industries during 1960s and since then these industries rank only after textile manufacturing and iron and steel in value added. Production is still below domestic needs, but the targets of the Sixth, Seventh and Eighth Plans were kept high. Previous achievements during the Third, Fourth and Fifth Plans fell 20 to 30 per cent short of the targets.

Before 1960 sulphuric acid manufacturing plants existed near the textile and steel centers at Asansol, Jamshedpur and Alwaye, whereas caustic soda and soda ash were produced near the areas of availability of limestone and sea salt in Gujarat. Sindri (Bihar) and Alwaye (Kerala) were the two ammonium sulphate manufacturing centers. Demand for chemical fertilizers has increased dramatically following the introduction of HYV in agriculture and shortfalls have been critical since 1965. The consumption is still among the lowest, i.e., nearly one-third, in the world. Production of the working plants in the public sector, located at Sindri, Nangal (Punjab), Trombay (Maharashtra), Rourkela (Orissa), Alwaye, Neyvelli (Tamil Nadu), Namrup (Assam), Gorakhpur (Uttar Pradesh) and Chennai (Tamil Nadu) is 13 million tons of nitrogenous fertilizer and phosphatic fertilizers annually. Private sector factories are located at several places in Rajasthan, Uttar Pradesh and Orissa. The Fifth Plan projects included new plants at Durgapur, Cochin, Barauni, Goa and Kota, in an effort to decentralize and expand the fertilizer industry since the transportation cost of fertilizers for farmers appears to be a critical factor. Scarcity of raw materials (rock phosphates, sulphur, petroleum) is a stumbling block in the decentralization programme. The escalating price of oil in the international market is a limiting factor in raising production. From a long term view it will be necessary to explore the possibility of producing fertilizers from other locally available sources, such as coal. An expected increase in oil production from the Bombay High raises some hope of utilizing domestic oil for fertilizer as well.

Soda ash and caustic soda industries are located in Gujarat, Maharashtra, Tamil Nadu, Delhi and Kerala. Cotton and synthetic textile and paper industries utilize caustic soda and the industry tends to be close to those centers. The present production is 1.5 tons of soda ash and equal amount of caustic soda.

India's drug and pharmaceutical manufacturing industry is concentrated at three major centers: antibiotics at Rishikesh (Uttar Pradesh), vitamin manufacturing at Hyderabad and sulfa drugs at Chennai. Insecticides are manufactured at New Delhi, Mumbai, and nearby Kolaba, Pimpri (near Pune) and Alwaye.

The petrochemical industry is of comparatively recent growth, the first plant was established in 1966. Production of plastics is slightly over one million tons, and of synthetic fibers about 130,000 tons. Potentials for growth appear limited, primarily a result of the paucity of raw materials. The synthetic manufacturing industry tends to be located near the existing textile centers of Mumbai, Ahmedabad and Coimbatore as are the industries of soap, plastics and cosmetics, all deriving benefits of the established markets, banking and capital resources from those centers. In the south these industries are dispersed in such small and medium-size cities as Ernakulam (Kerala), Bangalore, Ramnathapuram (Tamil Nadu) and Chennai, close to the sources of raw materials (coconut, vegetable oils, rubber, sea salt, soft wood), labor supply and hydroelectricity.

### *Other Industries*

Cement and paper manufacturing are two other major industries. Both were established before Independence. Cement production capacity at the time of Independence was nearly 2.5 million metric tons, as against 15 million metric tons in 1975. Paper production also registered similar gains since 1947. As in most developing countries, both products are critically needed in independent India for the construction of dams and buildings, and for the dissemination of news. Both are in short supply, despite the remarkable progress already made. Occasional shortages and black marketing during the 1960s led the government to monitor distribution in the private sector. Production is now close to 75 million tons.

Both industries are raw material oriented, because of heavy transportation costs. Ideal locations for the industries are near fuel supply and markets since the products are low priced and bulky. Distance from markets increases freight costs substantially. For cement, raw materials include limestone, gypsum, clay, coal and water; for paper, woodpulp, grass, coal and water. The cement industry is located in Bihar (close to the coal fields and areas of limestone), Orissa, Gujarat, Madhya Pradesh, Tamil Nadu and Rajasthan. Since 1950 dispersal has been to coastal locations in Kerala and Gujarat where plants could utilize sea sand, marl and sea shells. The Mumbai market is now served by Gujarat factories, with a grinding and packing plant located in Mumbai. Important centers of the industry are Dalmianagar (Bihar), Katni (Madhya Pradesh) and Lakheri (Rajasthan).

Major paper mills were located in Calcutta, Raniganj, Dalmianagar (in Bihar) and Brajrajnagar (Orissa) until 1950. Since then mills in Bellarpur and Nepanagar (for newsprint) in Maharashtra, Sirpur and Kaghaznagar in Andhra Pradesh, Dandeli and Bhadravati in Karnataka, and Kaveri in Tamil Nadu have been set up, all dependent on the supply of bamboo from nearby forests. Smaller plants were set up in the Mumbai, Pune and Bangalore areas; these centers utilize nearby hydroelectric and thermal power. Dalmianagar in Bihar has been recently utilizing sugarcane pulp (bagasse) instead of woodpulp for paper making. Production of 3.8 million tons of paper and 198,000 tons of newsprint in 1995 fell much short of the country's requirements.

## Village and Small-Scale Industries

Small-scale industry forms an integral sector of Indian economy. It accounts for about 50 per cent of employment in the industrial sector. A small-scale industry is defined as an enterprise which is operated mainly with hired labor, usually employing 10 to 50 persons. Village industries are those which cater primarily to local markets. The village industries are labeled as "traditional". Although Indian planning has been oriented to large-scale industrialization, it has recognized the continuing and special role of small-scale industry in the development process. Between 1950 and 1970 nearly $78 million were spent for the development of this sector of industry. In addition, a vast unorganized, unregistered segment remains virtually outside the orbit of government planning. Absence of reliable statistics make it difficult to assess the dimensions of small-scale and rural industry.

Such industry, as the label suggests, is composed of small firms, often deriving their labor force from the owner's household, and is primarily based in villages or small towns. The Five Year Plans sought to encourage household (cottage) industries by the provision of credit facilities, access to machinery and technical assistance.

The Eighth Plan allocated an outlay of Rs 1,600 crores for the development of rural and small-scale industries, the largest slice of which was assigned to the development of the handloom industry. With a million persons utilized therein, it currently ranks next only to agriculture in employment, fulfilling about 46 per cent of country's domestic needs in cloth and art silk. Processing of cereals, vegetable oils and *gur* (jaggery) and manufacture of soap, pottery, carpets, wood-carvings and iron-smelted objects are some other small-scale (basically rural) industries.

The coir (coconut fiber) industry is primarily export-oriented. The annual production of coir yarn is 150,000 tons, 90 per cent of which is produced in Kerala. Coir goods' export earnings average about $3-4 million annually and about half-a-million persons are engaged in their production. India produces nearly 10 million kg. of silk yarn annually, most of which is obtained from small-scale, rural industrial units. Two-thirds are provided by Karnataka alone, the remainder comes from West Bengal, Assam, Jammu & Kashmir, Bihar and Madhya Pradesh. Nearly 1.6 million persons are employed in silk textile industry, and exports of silk have been rapidly increasing.

In the Indian context, where a large reservoir of rural labor exists, village and small-scale industries (by producing goods and providing services) can play a vital role in development. Therefore, the development of a rural employment-oriented strategy of growth based on labor-intensive rather than capital-intensive industries has been advocated by several economists.

## Problems of Industrial Development

In the last four decades the main objects of industrial development have been to increase national income, speed modernization, reduce dependence on imports, and raise labor productivity. The pace of industrial development since Independence has been more rapid than under British colonial rule, but has, in, general, fallen short of planning targets. A complex set of forces responsible for slow progress of industrialization in the country is reviewed briefly in the following paragraphs.

Industrialization started in India roughly a century later than in the developed countries. Its development coincided with other nations' development when they had reached a mature status of technology. India has not only to integrate modern means of development but also try to narrow the gap between it and the developed world in establishing the basic structures of industrial development. In other words, it had to tackle greater tasks than other industrial nations at a similar stage of development. To attain world standards of production and labor efficiently, India had to develop not only the metallurgical and engineering industries, but also the nuclear, electronic and computer capabilities used by other nations. The pace of Indian industrialization was, therefore, bound to be initially slow. A base of "intermediate goods" (such as steel plates, cement, machines, electrical goods) manufactured domestically to help create heavy industry was first to be created.

Given its meager sources of financing and capital accumulations, Indian ruling classes could not muster domestically the large investments needed to create a base for heavy industry. Over a period of planned development large amounts of foreign aid were obtained which resulted in problems relating to balances of foreign exchange. Heavy reliance on foreign aid and investment constrained domestic initiative and compelled the country to borrow foreign technology and goods at higher than competitive prices in the international market. Foreign exchange balances eased off only in the later 1970s when the heavy industry and intermediate goods sector had measurably matured but deteriorated sharply in the late 1980s when the foreign exchange reserves dried up and foreign debt escalated. It also reduced dependence on foreign electric, electronic, engineering and metallurgical goods. During the 1960s foodgrain imports had further aggravated the foreign trade situation and enormously constrained imports of capital-intensive goods.

Indian economy has had to work under conditions labeled as the "economics of scarcity". This scarcity was not only in the matter of capital reserves, but in terms of infrastructural deficiencies as well. Shortages of power (electricity, coal, gas), transport facilities, technical knowledge and machinery all impeded industrial growth. An inadequate pace of investment and widespread shortages of construction material resulted in the scarcity of buildings needed for modernizing and expanding the industrial base. Scarcity of administrative, management and technical capacities to sustain development processes was particularly critical during the early phases of industrialization. Import substitution and strict industrial licensing as a means of expanding domestic manufacturing resources initially helped to depress the availability of consumer and intermediate goods. The Fifth Plan had to concede that the initial difficulties experienced in the process of indigenization through reliance on domestic technology became an important factor in impeding the pace of investment.

A major development thrust during the Five Year Plans was toward the establishment of a vigorous public sector developed hastily without the creation of a base of administrative machinery adequate to undertake this enormous task. Preparatory work for such tremendous institutional reorganization was poor. High performance was rarely insisted on even after the construction of an administrative base. The result was non-achievement of targets. During the Fourth, Fifth and Sixth Plans, achievement levels fell short of targets by 15 to 18 per cent.

Political and social agitation, fueled in part by the class, and caste inequalities, often marred industrial relations. Labor-management relations, never properly nurtured before Independence, remained strained. Factory legislation was slow to take shape. The quality of labor and of the Indian market has remained notoriously antiquated. The development of skilled, technical manpower for industry has lagged primarily to placate traditional caste prejudices against manual and technical work. Deep-routed prejudices prevail, hampering the creation of an efficient labor force. The domestic market remained chronically underdeveloped through lack of enthusiasm generated by the middle and upper class segment who did not terribly need to raise their standards. The result: Indian living standards remained among the lowest in the world.

Among the major geographical problems that faced planners in promoting industrialization was the inadequate infrastructural and locational base of the industries. Industrial locations, in several instances, were established without reference to cost-effective points. Each state clamored for the establishment of major industries in the public sector within its boundaries, and the locational decisions have thus been often politically motivated. The plans set out to reduce regional inequalities by a process of regionally distributing major public industrial units. This resulted in high transport costs, inefficient operations, power shortages and poor infrastructural amenities for the industries, which in turn led to marketing problems. The Indian railroad system has been working under heavy stress to cope with increasing industrial demands. Pressure has become so heavy that the movement of raw materials to industrial sites and of the finished products to the ports and hinterland is uneconomically slow.

One approach toward the development of a national, comprehensive industrial policy was suggested by the Planning Commission in 1955. The underlying idea was that industrial development could start at the village and small town level, progressively building a pyramid of industry with the rural economy as a base. Influenced by this approach rural industrial estates were set up in the various states, using districts as an operational area for developing rural industry. Another approach later advanced by the Planning Commission, envisaged the creation of urban-nucleated industrialization in towns of 20,000 to 50,000 population. It was argued that towns of this size would be technologically ill-suited for heavy industrialization. More promising results could be reached if the selected population range varied between 20,000 and 300,000 which could "offer the most congenial physical setting" since these were places likely to maximize the proximity of agriculture to centers of industry and commerce, and to supply the necessary framework for "the whole network of development sequences, linkages and feedbacks upon which the successful transformation of the Indian countryside so largely depends" (Lewis 1962). This approach has been strongly debated in favor of a rural-led employment-oriented strategy which suggests the development of a graded system of market centers from the village level to the larger urban nuclei of the size of 300,000. A graded system of market towns would draw upon the local resources of raw materials and labor force. A prerequisite to development is the creation of spatial integration by the provision of adequate infrastructure, communication linkages and banking-marketing facilities. Implicit in the above approaches is the idea that locational policies of economic growth should be aimed at reducing the rapid migration of people from the farm to urban areas. The creation of employment-oriented agro-based industries in villages and medium-sized towns would provide alternatives for the rural labor force, thus helping reduce its urban-ward migration. Such a policy of industrial decentralization, it is argued, will help restore the regional balance in industrialization.

## *References*

Alexander, P.C., *Industrial Estates in India*. Bombay, 1963.

Bhagwati, J.N. *et al*, *Planning for Industrialization and Trade Policies Since 1951*. London, 1970.

Chaudhuri, M.R., *Industries, Development and Location*. Calcutta, 1970.

Chaudhuri, M.R., *Power Resources of India*. Calcutta, 1970.

Economic and Scientific Research Foundation, *Changes in the Locational Pattern of Select Indian Industries*, 1950-65. New Delhi, 1969.

Gandhi, M.P., *Major Industries of India*, 1970-71. Bombay, 1972.

Government of India, *India, 1991: A Reference Annual*. New Delhi, 1990.

Government of India, *Seventh Five Year Plan, 1985-90*, Vol. II. New Delhi, 1985.

*India: A Country Study*. Government Printing Press, Washington, DC, 1985.

Karan, P.P., "Indian Industrial Change," *Annals of the A.A.G.*, Vol. 54,1964.

Karan, P.P. *et. al.*, "Geography of Manufacturing in India", *Economic Geography*, Vol. 35,1959, pp. 269-278.

Lewis, J.P., *Quiet Crisis in India*. Washington, DC, 1962.

Mellor, J.W., *The New Economics of Growth*. Ithaca, N.Y, 1976.

Muthiah, S. (ed.), *A Social and Economic Atlas of India*. Delhi, 1987.

Pattanshetti, C.C., *Dimensions of India's Industrial Economy*. Bombay, 1968.

Rao, K.L., *India's Water Wealth*. New Delhi, 1975.

Sinha, B.N., *Industrial Geography of India*. Calcutta, 1972.

Schwartzberg, J.E. (ed.), *A Historical Atlas of South Asia*. Chicago, 1978.

Wilber, C.O. (ed.), *The Political Economy of Development and Underdevelopment*. New York, 1973.

# 12

# Communications, Transport and Foreign Trade

The key to India's economic development lies in the improvement and extension of its infrastructural base, particularly the development of its communication and transport systems. For a long time rural and urban segments of the economy have remained separate and unlinked. It has been argued that as long as rural India remains a bystander rather than a participant in country's development process, it cannot be the beneficiary of, nor contribute to, country's development. Basic to country's economic planning, therefore, is the creation of a process which is capable of linking and regenerating its long-stagnant economy. Only by drawing rural India within the orbit of India's economic expansion, can the development process be accelerated.

In absolute numbers, the country's infrastructural base appears impressive. The railroad system, with a total track of 63,000 kilometers, is the fourth largest in the world, ranking after those of the United States, the Soviet Union and Canada. It employs about 2 million persons, has a fleet of over 5,000 locomotives, and operates 14,000 trains carrying daily over 4 million passengers and 4 million tons of freight. The trans-India road system dates back to the Mauryan times and its 63,000 km track is one of the largest in the world. But these and other infrastructures (airways, inland navigation, shipping, post and telegraph operations, banks, etc.) are, generally speaking, out-moded, and inefficient, despite considerable infrastructural improvements through the Five Year Plans. The extent of transport deficiency can be indicated by comparisons with the transport resources in the rest of the world. India accounts for 16 per cent of the world's people but only 4 per cent of its improved roads, 1 per cent of its rail freight, and 1 per cent of its trucks and buses.

The railroad system, inherited from the British, reflects the pattern of British control and administrative convenience. The track was extended inland starting from the three foci along the coast, viz., Calcutta, Bombay (now Mumbai) and Madras (now Chennai), where trading and military posts had been established. In north India, the Ganga plain, a commercially productive area, was suitable for laying railroad tracks. The major British thrust of annexation in the Ganga plain followed the establishment of a Calcutta-Delhi link between 1757 and the early nineteenth century. Meanwhile, coastal and inland penetration resulted in British expansion along the established routes in central and southern India. Thus, a skeletal railroad system was developed from the coastal ports, radiating inland to link the ports with the hinterlands. Delhi, a long-time economic node in the country, became a major inland focus of internal communications. Gradually, feeder links were attached to the major railroad routes.

The road system, despite its ancient history, has remained underdeveloped. Major roads parallel the railroad system. Feeder links joining the interiors are in poor condition, in places passable only during fair weather. Water transport has been only regionally significant, especially in the northeast parts of the country and along the coasts. Air traffic was initiated only a few years before Independence.

Transport development during the British rule was based on two key considerations. First was to facilitate the movement of cash crops such as cotton, tea and jute to the ports for export to Britain or for re-import into the country after processing into finished products (e.g., tea, gunny-bags, textiles). Secondly, transport links were developed with an eye on strategic points for effective administrative control and movement of military forces from the cantonments (garrison towns) established along the railroad lines. Transport development did help, however, in a number of ways. In times of famine, foodgrains could be dispatched from surplus areas to the needy regions, while commercial centers sprang up along the railroads.

The Five Year Plans accorded the development of transport facilities a priority status and investments ranged between 20 and 24 per cent of total outlays in the public sector. A significant expansion of railroad and road transport took place during the Third and Fourth Plan periods. The volume of passenger and freight traffic increased from 88 to 319 million tons, and 78 to 269 million passenger km between 1961 and 1988. Road traffic increased even faster, registering gains particularly in volume of freight carried. In 1988 road transport accounted for more than one-half of all passenger traffic (up from 41 per cent in 1961), and approximately one-third (31 per cent, up from 16 per cent in 1961) of the freight traffic. Growth of commercialization and increased spatial integration between the rural and urban markets during the period largely accounted for the relatively rapid growth of road transport. Inland navigation and coastal shipping remained underdeveloped. The Fifth Plan aimed at improvement in railroad freight transport of major bulk commodities (coal, iron ore, cement, fertilizers), as well as extension of rural feeder road links. It envisaged linking all villages with a population of 1,500 or over by all-weather roads by the end of 1979, but by 1995 nearly 75 per cent of these had been connected by a road.

The railroads are nationalized, their operations are controlled by the Ministry of Railways. Trucking and buses are handled largely by private firms regulated by state legislation. Surfaced-roads are estimated to be approximately 1.5 million km, nearly one-half of the total road length. The national highway system covering over 40,000 km, is the responsibility of the union administration. Major highways include: (1) the Grand-Trunk Road from Calcutta to Amritsar via Varanasi and Delhi; (2) Kanpur to Agra; (3) Agra to Mumbai; (4) Mumbai to Chennai; (5) Chennai to Calcutta; and (6) Calcutta to Mumbai via Nagpur. A major future project is to link the national highways to an international highway network, principally with Pakistan and Burma. In 1960, a Border Roads Development Board was set up for accelerating the development of border transport along the Tibet-Burma-Pakistan borders in strategic frontier areas.

Despite steady improvement in transport facilities during the forty years of planning, still many parts of the country remain inaccessible. Spatial segmentation, particularly of the rural areas, is still a major problem. Great regional disparities in the availability of services foster spatial segmentation and are a direct outcome of it. Four major concentrations of railroads are: (1) Calcutta-Dhanbad industrial belt in West Bengal and East Bihar; (2) Kanpur-Lucknow commercial belt in eastern Uttar Pradesh; (3) Agra-Mathura commercial belt in western Uttar Pradesh; and (4) Ahmedabad-Vadodara industrial region in Gujarat. The Ganga plain from West Bengal to Punjab contains high railroad concentrations. Elsewhere, in the peninsula, the distribution is spotty, concentrated around chief commercial industrial centers (Nagpur, Hyderabad, Chennai, Bangalore, Mumbai, Madurai, Vijayawada and Coimbatore). Road densities generally exhibit similar patterns, but contain several deviations, particularly in south India where road densities are higher than in the North Indian Plain.

Accessibility patterns, i.e., distance from surfaced roads, are essentially similar to the above patterns. The North Indian Plain, Gujarat and Tamil Nadu belong to the highest category containing areas less than 10 kms from roads. Among the inaccessible parts (those which lie more than 10 kms from roads) include the hill regions (Kashmir, Himachal Pradesh, Arunachal Pradesh, Meghalaya, Nagaland, Manipur), the western sections of Rajasthan Desert and the forests of Chota Nagpur, Baghelkhand and Dandakaranya.

## Problems and Prospects of Infrastructural Development

Foremost among infrastructural needs is the extension of surfaced roads to villages. In 1988, only 40.7 per cent of the villages were connected with any type of roads—surfaced or unsurfaced. Only in three states, viz., Punjab, Haryana and Kerala, nearly all the villages had access to a dirt or surfaced road. A study in Uttar Pradesh concluded that such transport inadequacy not only hampered the economy of the area in general, but contributed to foodgrain deficits in a potentially rich farming area. Large sections of Uttar Pradesh (and indeed all the Middle Ganga Plain) suffer from low productivity because better seeds and technology cannot be easily supplied to the isolated villages. Crop yields are 75 per cent of those prevailing in other states with better communications. In the late 1960s nearly one-third of the Uttar Pradesh villages were without any road connections, one-half were located 5 to 10 miles from an important market and over 20 per cent were located more than 10 miles from any road. Nearly one-third of the villages located on a surfaced road were remote from any market. Problems of transport are compounded where the means of communication are mainly in the form of bullock carts, camels or human carriers. Poor transport facilities create difficulty of crop marketing and of diffusion of technology to farmers. The first step toward spatial integration is the provision of a surfaced rural road network.

The railway network, like the surfaced roads, is not distributed in proportion to the service area or the population size of various sections of the country. Excepting Tamil Nadu, Gujarat and parts of the North Indian Plain, one-half of the country lies more than 10 miles distant from a railroad track. Furthermore, the effective usage of railway track is also ill-proportioned, for 15 per cent of the track carries 60 per cent of the traffic.

Related to road and railway inaccessibility is the question of producing equipment parts for such transport carriers as buses, freight cars and locomotives. After the World War II shortages of vehicles and freight cars have grown. Shortages often led to industrial shut-downs and stoppages of coal movements; road transport and the use of oil reflected such shortages. Near self-sufficiency has now been claimed in the domestic production of locomotives, freight cars and other vehicles. But the rise in oil prices since 1973 has aggravated the transportation situation.

Growth of railroads, also of crucial importance to economic development, is beset with a number of problems. It depends, at least partially, on the domestic production of locomotives. Production of rail locomotives, currently amount to 200 diesel and electric engines annually at Chittaranjan workshops, the diesel locomotive plant in Varanasi, the coach factories at Chennai and Bangalore, are barely sufficient to meet the domestic needs. Demand for these is increasing with attendant strain on their present capacity. In this regard, readjustments in freight rates of road haulage to competitive levels with those of the railroads can release pressure on railroad freight traffic.

The low operating efficiency of Indian railroads stems from a combination of factors: one, the multiple track-gauges, ranging from 610 to 1,676 mm. (the narrow, meter and broad gauges) in width in different parts of the country impeding the continuous flow of goods;

two, a poorly-conceived freight rate structure; and three, the uncompetitive edge given to freight rates in comparison with those of passenger traffic. Passenger movement rates have been found at times to be below the investment levels, whereas freight rates are unreasonably higher. This has led to frequent overutilization of existing facilities and unnecessary congestion on parts most traveled by passengers, and underutilization of services less frequented by passengers. A more judicious rate structure, based on actual costs, distance, freight volume and passenger amenities has yet to be devised. In trans-shipment of high value commodities, road transport needs greater encouragement.

Road transport experiences different problems. Truckers complain of complex administrative constraints by state and central authorities. Road maintenance is appallingly poor. Most trucking and passenger busing are in private hands and are prone to complex administrative regulatory hurdles. There has been a phenomenal increase in private trucking and passenger busing in several areas, notably Punjab, Haryana and Gujarat. During the British rule, road transport was meager. Railways received major attention at the expense of road transport, because colonial interests lay in the large-scale movement of military forces and export-oriented goods for which railroads were most suitable. Such a policy greatly inhibited road construction. This trend has been reversed since Independence.

Development of surfaced roads is more cost-effective for the movement of passengers and agricultural commodities. Already, in Tamil Nadu, Maharashtra, Haryana, Punjab and Kerala, the rural landscape has begun to undergo some transformation due to the construction of all-weather roads. New market centers have developed and agricultural technology is being diffused. Farmers have begun to shift their production emphasis from subsistence crops to cash crops, yielding them better investment returns. By promoting the inflow of new technology (fertilizer, high yielding varieties of crops, power-driven farm implements) roads are acting as agents of modernization and progress (schools, banks, medical clinics, credit bureaus) to rural areas. Studies point out that areas of newly constructed surfaced roads received many economic and social benefits.

## Foreign Trade

During the British rule India's foreign trade was characterized by the traditional colonial pattern of agricultural exports (jute, tea, cotton, hides and skins) and imports of manufactured items. India's trading world was confined to the United Kingdom and a few other countries. Following Independence it sought to expand its trade territory and diversify items of trade as it embarked upon economic reconstruction and development. Industrial expansion was, however, below expectation during the first four Five Year Plans (1951-56 to 1966-71), and only improved a little since then. Dependence on the industrial countries for machinery, manufactured items and foodgrains steadily increased during the 1960s and 1970s. After 1960 India was also able to make inroads into the markets of the Soviet Union, the Eastern European countries and Japan. Trade with other countries, particularly with the United States, the Middle East and Southeast Asia also increased considerably, particularly during the 1970s and 1980s during the period of industrialization and need for oil imports.

In statistical terms (value and number of trading items and countries) progress since 1947 appears deceptively promising. Despite a significant enlargement of the trading area and substantial increase in dollar amount of trade was achieved, yet India's share in the world's total trade remained low, less than 1 per cent in 1974 and 0.5 per cent during the 1980s and 1990s, a drop from the figure of 2.4 per cent in 1948. More than 60 per cent of trade was still conducted with only six countries in 1974. Moreover, the nation's balance of trade steadily deteriorated, because of increased imports of foodgrains, machinery and petroleum which were needed for developmental programmes.

Nearly one-half of India's international trade is contributed by six countries, viz., the United States, Britain, Japan, the Soviet Union, West Germany and Canada. In 1973-74 these nations absorbed about 55 per cent of Indian exports and supplied almost 52 per cent of its imports. Despite India's widening trade territory, these countries, as the tables indicate, strengthened their share of trade. The United States retained its leading position, although its primacy was threatened by the Soviet Union and Japan after 1970-71. Since the increase in oil prices in 1973, Iran and Saudi Arabia superseded Canada and West Germany in import value. Since a large part of India's imports were comprised of foodgrains, annual fluctuations in harvest affected the relative position of the main suppliers. By 1970-71 internal production had also increased, markedly reducing foodgrain imports and weakening Canada's position as one of India's major suppliers. Meanwhile, Japan, the Soviet Union and West Germany had emerged as major exporters of a diversified group of technological and manufactured items. Competition from these new exporters also led to a progressive and substantial decline in the value of imports from the United States (from 37.8 per cent of total imports in 1965-66 to 16.7 per cent in 1973-74, and 10.5 per cent in 1995-96).

In 1947 India ranked among the ten leading trading nations, but in 1995 it ranked below the 40 largest traders, contributing less than 0.5 per cent to world trade. This decline in its international trade position corresponded ironically with the period of economic planning and export stimulus. In Asia, its position as a trading nation suffered during the 1950s and 1960s as the newly developed countries such as Singapore, Hong Kong, Malaysia, Taiwan and Indonesia emerged as new traders, although the total volume of trade remained low.

**Table 12.1**

*Imports, Exports and Trade Balance, 1950-51 to 1995-96*
*(Selected Years)*

(in millions of Rs)

| | *Imports* | *Exports* | *Total Value* | *Balance* |
|---|---|---|---|---|
| 1950-51 | 6,080 | 6,061 | 13,141 | -19 |
| 1960-61 | 11,216 | 6,423 | 17,640 | 4,792 |
| 1970-71 | 16,342 | 15,351 | 31,693 | -990 |
| 1980-81 | 125,491 | 67,107 | 192,598 | -58,384 |
| 1985-86 | 196,576 | 108,945 | 305,522 | -87,631 |
| 1986-87 | 200,957 | 124,519 | 325,477 | -76,438 |
| 1987-88 | 223,989 | 157,412 | 381,402 | -66,577 |
| 1988-89 | 281936 | 202,951 | 484,888 | -78,985 |
| 1995-96 | 1,216,466 | 1,064,648 | 2,281,115 | -151,818 |

*Source:* *India: A Reference Annual*, New Delhi, 1997, p. 441.

*Note:* The value of a rupee compared to that of a dollar has been sliding since 1950. In 1950 a dollar was worth Rs 4.75, in 1988 Rs 18.75, and in 1998 Rs 42.30.

Statistically, India recorded modest gains in foreign trade following initiation of the Five Year Plans in 1951. The total value of foreign trade increased in the 1960s and fell during the 1970s, but has been increasing since then. Exports expanded more steadily, as a trend of continued and expanding unfavorable balance of trade. Table 12.1 summarizes the trade statistics in terms of the value of India's exports, imports, total value of trade and trade balance for the selected years between 1950-51 and 1995-96.

The export trade has long been dominated by three "traditional" commodities, namely, tea, cotton-based goods and jute products. These items represented over half the value of all exports until the 1970s, after which the domestic production and export of such items as cashew kernels, coffee, leather goods and iron ore progressively reduced their share until it reached 27 per cent in 1974. In 1988-89 it was 20.1 per cent. The structure of exports and imports for selected years between 1950-51 and 1995-96 is given in Tables 12.2 and 12.3.

The three leading exports in the 1950s were jute manufactures, tea and cotton goods which suffered from competition with synthetic substitutes as well as from Bangladeshi products. The decline in tea exports stemmed from two basic sources, viz., price-fluctuations in the world market and expansion of tea production in East Africa and Latin America. Labor costs on Indian plantations rose significantly during the 1960s, while African tea prices became competitive. Tariffs imposed on Indian tea imports by British rose when Britain entered the Common Market in 1973.

A dramatic decline in the share of cotton-based goods during the 1960s and 1970s in Indian exports to world markets primarily reflected the emergence in the 1960s of such new producers as Hong Kong, Singapore, Japan and Taiwan. Their product was superior and was in greater demand in international markets. Britain's import policies regarding cotton yarn and woven material, resulting from British commitments to the Economic Community, also affected Indian exports. Expansion of the domestic market for cotton textiles did not help the export situation either. In the 1980s exports however registered significant increases as the quality of the cotton fabrics at home improved. All exports excepting jute goods registered gains, nearly doubling in value since Independence, although their contribution to export trade was drastically reduced. Meanwhile,

**Table 12.2**

*Major Exports for Selected Years, 1950-51 to 1995-96*
*(As Percentage of Total)*

| | 1950-51 | 1970-71 | 1984-85 | 1995-96 |
|---|---|---|---|---|
| Jute Manufactures, Raw Jute | 18.8 | 12.3 | 2.9 | 1.1** |
| Tea | 9.9 | 9.7 | 6.5 | 1.2 |
| Spices | * | * | 1.7 | 0.7 |
| Cotton Manufactures (including readymade garments) | 19.9 | 6.3 | 8.1 | 12.0 |
| Textile fabrics (except Cotton and Jute) | 0.8 | 0.8 | 14.6 | 8.2 |
| Textile Yarn | 4.3 | 2.2 | 5.2 | 8.2 |
| Ore (non-ferrous) | * | 1.1 | 3.9 | 1.6 |
| Leather | 4.3 | 4.7 | 6.1 | 5.7 |
| Raw-Cotton | 0.8 | 0.9 | 0.5 | 0.1 |
| Fruits, Vegetables, Nuts (Cashew Kernels) | 4.1 | 5.7 | 2.8 | 0.7 |
| Raw Wool | 1.3 | 0.3 | * | *** |
| Rice | * | * | 1.4 | 3.2 |
| Sugar | * | 1.9 | 0.2 | *** |
| Fish | * | * | 3.2 | 1.8 |
| Iron Ore | * | 7.7 | 3.4 | 3.0** |
| Tobacco | 2.3 | 2.1 | 1.5 | 0.5 |
| Vegetable Oils (Oil Cake) | 2.1 | 0.5 | 1.1 | 1.5 |
| Crude Minerals (except Coal) | * | 1.4 | 3.0 | 1.0 |
| Woolen Carpets | 0.9 | 0.6 | 4.4 | 3.5 |
| Iron and Steel | 0.5 | 5.9 | 8.1 | 10.4 |
| Coffee | 0.2 | 1.5 | 1.1 | 1.4 |
| Hides and Skins | 1.6 | 1.2 | 0.7 | *** |
| Petroleum Products | * | * | 15.5 | * |
| Coal, Coke | * | * | * | * |
| Gems and Jewelry | * | * | 10.5 | 5.7 |
| Chemicals and Allied Products | * | * | * | 10.5 |

* Data not available.
** Estimated.
*** Negligible.

*Source: India 1997: A Reference Annual*, New Delhi, 1997.

**Table 12.3**

*Major Imports for Selected Years, 1950-51 to 1995-96*
*(As Percentage of Total)*

| | *1950-51* | *1970-71* | *1984-85* | *1995-96* |
|---|---|---|---|---|
| Iron and Steel | 2.2 | 9.0 | 5.4 | 5.6 |
| Machinery (except electrical and electronic) | 10.3 | 15.8 | 11.2 | 25.8 |
| Petroleum and Products | 8.3 | 1.8 | 0.7 | 19.2 |
| Transport Equipment | 5.4 | 4.1 | 2.1 | 2.9 |
| Electric Machines | 3.4 | 4.3 | 4.2 | 5.5 |
| Raw Cotton | 15.4 | 6.1 | 0.2 | * |
| Cereal (mostly wheat and rice) | 23.5 | 12.4 | 1.4 | 0.1 |
| Chemicals | 1.4 | 4.2 | 5.0 | 10.1 |
| Fertilizers | * | * | 6.1 | 4.9 |
| Metals Manufactures | 2.1 | 0.6 | 2.4 | ** |
| Textile Yarn | 2.3 | 0.2 | 0.2 | ** |
| Non-Metallic Minerals (includes aluminum and zinc) | 2.7 | 5.4 | 6.0 | ** |
| Medicines | 1.5 | 1.5 | 0.7 | 1.1 |
| Fresh Fruits and Nuts | 1.5 | 2.2 | * | * |
| Raw Wool | 0.9 | 1.0 | 0.4 | * |
| Paper and Paper Board | 1.6 | 1.8 | 1.1 | 1.4 |
| Coal, Dyes and Tar | 1.8 | 0.2 | 0.2 | * |
| Pearls and Precious stones | * | * | 6.1 | 6.1 |
| Milk and Cream | 0.5 | 0.4 | * | * |
| Crude Rubber | * | * | 0.5 | * |
| Vegetables | 0.5 | 1.4 | * | * |
| Edible Oils | * | 0.2 | 5.3 | 0.5 |

* Negligible amounts.
** Data not available.
*Source:* As for Tables 12.1 and 12.2.

items such as sugar, leather, raw cotton, nuts, vegetables, coffee, engineering goods and petroleum products recorded gains.

In the 1970s engineering goods such as electrical machinery, transmission lines, railroad wagons and chemical plants, which Indian firms produced more competitively, found their way to markets in Africa, the Middle East and Southeast Asia. The government's policy of import substitution had one clear effect. It helped reduce the widening trade deficit during the 1960s and 1970s but sharply widened again in the 1980s. Meanwhile, exports of neither primary goods (jute, tea, cashew kernels, mica) nor of manufactured goods (textiles cloth, sugar and consumer goods) registered dynamic expansion during the 1980s.

In 1970-71 over three-fourths of India's imports were in machinery, transport equipment, foodgrains, petroleum products, chemicals and manufactured goods. Between 1950-52 and 1970-71, these items accounted for over 80 per cent of imports. Machinery and transport equipment led the group among imports. After 1973-74 oil and petroleum-based products became the second largest category of imports (24.6 per cent of all imports, up from 7.8 per cent in 1984-85). Paper and paper board accounted for 1 to 2 per cent of imports. Chemicals and fertilizers represented 4 to 6 per cent of import value. Other major imports were raw cotton, wool and medicines.

The most noticeable shift in import structure since the mid-1960s was the large reduction in foodgrain imports, from 23.5 per cent in value in 1950-51 to 12.4 per cent in 1970-71. In 1988-89 foodgrains accounted for 2.2 per cent of the total imports (Table 12.3).

Overall, two notable shifts in the structure of imports were the dramatic increase in petroleum and petroleum-based products and sharp decline in foodgrains during the 1970s and 1980s. Changes indicated growing industrialization on the one hand and the dramatic improvement in foodgrain production largely resulting from the "Green Revolution". Demand for unfinished stones, gems and pearls also markedly grew.

## *Direction of Trade*

India's trading partners are shown in Tables 12.4 and 12.5. Despite fluctuations and decline, the United States has remained the major supplier of foodgrains, a situation which over the years pushed India into a position of heavily unfavorable trade balance and caused a deterioration in foreign exchange reserves. Most of the foreign exchange needed to cover the import costs was made possible by foreign aid extended by the United States to India. Between 1950 and

**Table 12.4**
*Imports from Principal Suppliers, 1950-51 to 1995-96*
*(As Percentage of Total)*

| | 1950-51 | 1970-71 | 1984-85 | 1988-89 | 1995-96 |
|---|---|---|---|---|---|
| U.S.A. | 18.3 | 27.7 | 9.9 | 11.4 | 10.5 |
| United Kingdom | 20.8 | 7.8 | 5.4 | 8.5 | 5.2 |
| Federal Republic of Germany | 3.3 | 6.6 | 7.5 | 8.7 | 2.2 |
| U.S.S.R. | 0.2 | 6.5 | 10.5 | 4.4 | 2.3 |
| Japan | 1.6 | 5.1 | 7.2 | 9.3 | 6.7 |
| Canada | 3.4 | 7.2 | 2.9 | 1.5 | 1.0 |
| German Democratic Republic | 3.3 | 6.6 | 7.5 | 8.7 | ** |
| France | 1.7 | 1.3 | 2.0 | 2.9 | 2.3 |
| Belgium | 1.4 | 0.7 | 4.6 | 7.2 | 4.6 |
| Netherlands | * | * | 2.0 | 1.9 | * |
| Italy | 2.5 | 1.8 | ** | ** | ** |
| Switzerland | 1.1 | 0.7 | ** | ** | ** |
| Australia | 5.1 | 2.2 | 1.2 | 2.4 | 2.8 |
| Iran | 5.7 | 5.6 | 2.8 | 0.4 | 1.6 |
| Iraq | * | * | 3.9 | 0.6 | n.a. |
| Saudi Arabia | 0.1 | 1.5 | 7.3 | 6.7 | 5.5 |
| Kuwait | * | * | 2.1 | 1.8 | 5.4 |
| Czechoslovakia | 0.4 | 1.2 | ** | ** | ** |
| Egypt | 5.1 | 2.4 | * | * | * |
| Kenya | 2.8 | 0.6 | * | * | * |
| Sudan | 1.1 | 1.3 | * | * | * |
| Malaysia | 0.1 | 0.3 | * | * | * |
| Burma | 0.9 | 0.6 | * | * | * |
| Asia (all countries) | ** | ** | 13.9 | 13.2 | ** |

* Negligible.
** Data not available.

*Sources:* 1. *India 1990 Economic Survey*, 1988-89.
2. *Statistical Outline of India*, 1997-98, Mumbai, 1997.

**Table 12.5**

*Export Destinations, 1950-51 to 1995-96*

*(As Percentage of Total)*

| | *1950-51* | *1970-71* | *1984-85* | *1988-89* | *1995-96* |
|---|---|---|---|---|---|
| U.S.A. | 19.3 | 13.5 | 15.0 | 18.4 | 17.4 |
| United Kingdom | 23.3 | 11.1 | 5.2 | 5.7 | 6.3 |
| U.S.S.R. | 0.2 | 13.7 | 16.0 | 12.8 | 3.3 |
| Japan | 1.7 | 13.2 | 8.7 | 10.6 | 7.0 |
| Australia | 4.9 | 1.6 | 1.1 | 1.3 | 1.2 |
| Federal Republic of Germany | * | 2.1 | 4.1 | 6.1 | 6.2 |
| Belgium | * | * | 1.6 | 4.3 | 3.5 |
| Netherlands | 1.7 | 0.7 | 1.5 | 1.9 | ** |
| France | 1.5 | 1.2 | 1.5 | 2.1 | 2.3 |
| Canada | 2.3 | 1.8 | 1.1 | 1.1 | 1.0 |
| Italy | 2.5 | 0.9 | 1.7 | ** | * |
| German Democratic Republic | * | * | 0.7 | 0.9 | * |
| Sri Lanka | 3.3 | 2.1 | ** | ** | * |
| Burma | 4.1 | 0.9 | 1.1 | * | * |
| Bangladesh | * | 0.3 | * | * | * |
| Singapore | 5.1 | 1.1 | 1.1 | ** | ** |
| Egypt | 1.0 | 3.7 | * | * | * |
| Iran | * | 0.9 | 1.1 | 0.4 | 0.5 |
| Iraq | * | * | 0.4 | 0.7 | ** |
| Kuwait | * | * | 0.9 | 0.7 | 0.4 |
| Saudi Arabia | * | * | 2.3 | 1.6 | 1.5 |
| New Zealand | 0.6 | 0.3 | ** | ** | ** |
| Nigeria | 1.0 | 0.6 | * | * | * |
| Romanio | * | * | 0.7 | 0.1 | * |
| Argentina | 1.8 | 0.2 | * | * | * |
| Sudan | 0.7 | 2.4 | ** | ** | ** |
| U.A.E. | * | * | 2.3 | ** | ** |
| Kenya | 1.1 | 0.5 | * | * | * |
| Malaysia | 1.0 | 0.7 | ** | ** | ** |

* Negligible.

** Data not available.

*Sources:* As for Table 12.4.

1970, India received over $9,517 million from the United States, 20 per cent of which was in the form of outright grants, and 33 per cent was refundable over long periods partly in rupees, and partly in dollars at low interest rates. The United States Export-Import Bank oversaw these arrangements. PL 480 Title I agreements enabled India to purchase foodgrains. In addition to foodgrains, India imported metals, machinery, transport equipment, fertilizers and dairy goods from the United States. Increased production of foodgrains at home and expansion in industrial production (machinery, farm equipment) during the last two decades narrowed the trade balance deficit in some years.

The United Kingdom, West Germany, France, Belgium, Italy and Switzerland were the principal Western European sources of Indian imports. Britain remained the leading supplier of machinery and industrial plants, although after 1965 West Germany became a serious competitor. Imports from West Germany included machinery, machine tools, electric goods, chemicals, nitrogenous fertilizers and ships. High-cost, sophisticated, manufactured items, machinery and dairy products have been the chief imports from the West European countries with whom the trade balance has been normally unfavorable to India. Most imports were financed through long-term aid and loans arranged through agencies like the Aid India Consortium, of which most of these countries were members. The comparative decline in India's purchases from Western Europe in recent years stemmed from these countries' strict foreign exchange policies silks, knitted garments, cigarettes and iron ore.

Although raw materials continued to form a major part of India's exports to these countries, several compositional changes during the last few years have occurred. Domestic production of small electric and non-electric machinery increased, leading to the exports of electric fans, refrigerators, canned fruits, electric transformers, electric water heaters and air conditioners to the Soviet Union. Between 1968 and 1970 the Soviet Union also purchased large quantities of Indian steel and steel products, notably some 1 million ton of steel and 4,000 railroad wagons.

Indian imports from these countries consisted mainly of industrial plants, construction equipment and machinery for irrigation and power projects, and oil prospecting and drilling. Political convulsions in 1990 and 1991 may radically alter the trading postures of these countries. Looking for hard currency themselves, these countries would be seeking market expansion in the United States and Western Europe. India's reliance on the Soviet Union and East Europe as trading partners has dramatically reduced since 1990.

India's trade with the rest of the world was very small until 1960, but has increased steadily since. Specifically, trade with Japan has increased dramatically in the recent years. In 1972-73 Japan became the leading supplier of Indian imports, surpassing the Untied States, the United Kingdom and the Soviet Union.

In the Middle East, Iran and Saudi Arabia have become important sources of Indian imports accounting for between 2 and 5 per cent each. Major imports from this region are petroleum, fruits and fine-quality cotton. Imports increased substantially in value after 1966 as a result of greater demand for petroleum by a growing industrial sector, particularly in the manufacture of fertilizers. India's trade balance with this region has always remained negative and received severe setbacks in 1972 when the OPEC countries increased oil prices, and since the Gulf War of 1991.

In Africa, Egypt, Sudan, Uganda, Tanzania, Zaire and Zambia have been the main sources of India's imports. Raw cotton, cloves, wattle bark, sisal and asbestos were the principle imports from these countries.

In Southeast Asia, Thailand, Burma, Malaysia and Singapore have been the major suppliers. In East Asia, Japan is a principle supplier of such items as machinery, yarn and synthetic fiber. As the Indian industrial sector expanded during the 1960s, dependence on Japan decreased.

Imports from neighboring countries (Nepal, Burma, Pakistan, Ceylon, Afghanistan) have always been small because these underdeveloped countries largely duplicated India's own production and could not offer machinery or foodgrains, the items India needed. Trade with Pakistan and China, never substantial, came to a virtual standstill during periods of armed conflict in the 1960s. Major imports included tires and hides from Ceylon, dry fruits from Afghanistan, and *ghee* (clarified butter) from Nepal. India's imports from Latin

America have remained insignificant. The major trading partners are Brazil and Argentina, the latter as a principle supplier of foodgrains.

As regards exports, the Middle East consumed 7 to 8 per cent of India's export value during the 1970s. Iran, Saudi Arabia, Iraq, Kuwait, Qatar and the Yemen People's Democratic Republic were the major customers. Electric and engineering goods were the principle items of export accounting for over 25 per cent of total export value. Despite competition from Japan and the United States for manufactured goods and from Ceylon and Pakistan for primary, the region remained an important customer for Indian textiles, jute, iron and steel, electric machinery (fans, sewing machines), bicycles, tea, fruits, spices, air conditioners, agricultural implements, sugar, tobacco and footwear.

The countries of Africa normally consume less than 5 per cent of India's exports. The main exports are cotton textiles, jute goods, spices, sugar, vegetable oils and light engineering goods. Egypt, Sudan and Libya are the chief customers. Indian exports to this region have been steadily increasing and the Indian government has been striving to push exports to the large potential markets in Africa for the last two decades.

Exports to Latin America are small, and have been declining in value and relative position since the 1960s. The principle customer is Argentina. Major exports include tea, spices, and jute manufactures.

In Asia, Japan has been an important customer since 1950 when it absorbed 1.7 per cent of India's exports. In 1988-89 Japan accounted for 10.6 per cent of India's exports. Principal exports to Japan include raw cotton, iron ore, iron ore concentrates, and iron and steel scrap, manganese, sugar, groundnuts and cashew kernels. Japan's purchases of India's exports increased rapidly in the 1960s, and India's adverse balance of trade with Japan became favorable. In 1988-89 the trade balance surplus with Japan amounted to over $200 million. Japan's rapidly growing iron and steel industry and economy afforded incentives for Indian iron ore production.

In Southeast Asia, Burma, Singapore, Thailand and Hong Kong have been the principal customers; their main items of purchase are jute goods, iron and steel products and engineering goods. There appears to be scope for expansion of markets for such materials as electric and engineering goods and inexpensive semi-manufactured items in this region, especially in Thailand and Malaysia.

India's exports to Australia remained relatively small, amounting to 1.3 per cent of the total in 1988-89 (decline from 5 per cent in 1950-51). Scope for expansion undoubtedly exists, especially for exports of jute products, cashew kernels, hides and skins, woolen carpets and vegetables.

During the 1980s India's policy focused primarily on export diversification and market expansion in the developing countries. Exportable amounts of fairly high quality manufactured items such as electric fans, bicycles, radios, refined textiles, sewing machines, electric transformers and copper sheets for markets in several countries of West Africa, East Africa, the Middle East, Southeast Asia and Latin America were the primary targets of Indian exports in these items. Expansion of such exports as tea, jute, spices and cotton were also tapped for markets in Australia and New Zealand. Trade with the Soviet Union and East Europe has mostly stabilized, while future expansion to quality conscious markets of the Western European countries and the United States would appear to depend on tariff considerations. Indian government's policy of liberalization of tariffs and granting investment inducements to foreign establishments adopted in 1991 is a step in this direction.

### *References*

Bhagwati, J.N. *et al*, *Planning for Industrialization and Trade Policies*. New York, 1966.

Bhardwaj, R., *Structural Basis of India's Foreign Trade*. Delhi, 1978.

Dayal, E., "The Changing Patterns of India's International Trade", *Economic Geography*, Vol. 44, 1968, pp. 240-269.

Dutt, A.K. *et al*, "Dimensions of India's Foreign Trade", *International Geography*, 1972, Vol. 1, Toronto, 1972.

Epstein, T.S., *Economic Development and Social Change in South India*. Manchester, 1962.

Government of India, *A Reference Annual 1997*. New Delhi, 1997.

Government of India, *India 1977 and 1978: A Reference Annual*. New Delhi, 1997.

Malenbaurn, Wilfred, *Modern India's Economy*. Columbus, 1971.
Muthiah, S. *et al* (eds.), *A Social and Economic Atlas of India*. Delhi, 1986.
Owen, W., *Distance and Development, Transport and Communication in India*. Washington DC, 1968.
Schwartzberg, J.E. (ed.), *A Historical Atlas of South Asia*. Chicago, 1978.
Stein, A., *India and the Soviet Union*. Chicago, 1960.
*Times of India Directory and Yearbook*. Bombay, 1987.

# 13

# Economic Development Patterns

We have already discussed in brief India's economic development in Chapter 9 and Chapter 11. Similarly, in Chapter 5, we have examined the dimensions and complexity of the population problem, and emphasized that a key factor in the development process has been the spiraling population growth in the country during the last six decades. This chapter examines the regional distribution of poverty and economic development in the country.

Studies of the World Bank and the statistics released by the Indian National Sample Surveys indicate that nearly 480 million persons were illiterate in India, 210 million or 46.6 per cent of the working age population was unemployed, and over 310 million fell below the "poverty line" in the late 1980s. The census of 1991 as well as some economic surveys reveal that only modest gains in the fight against illiteracy and poverty had been achieved by early 1990s. In 1991, 405 million persons were counted as illiterates and 280 million fell in the unemployed category.

Measuring poverty and its world comparisons are beset with a number of problems. First, the rich Indians, much like their counterparts elsewhere, do not report their incomes accurately, often because of the illegality of the incomes. Secondly, the incomes of the poor are difficult to determine. As an example, the urban pavement dwellers, the beggars and the migrant rural workers elude such a determination. India's National Sample Survey, one of the world's major surveys recording household incomes and expenditures, failed to design a meaningful sample frame to assess the poverty of these groups in India. Thirdly, the famous "poverty line" adopted by the Indian administration as a part of the Third Five Year Plan (1961-66) of Rs 20 per capita per month (in 1960-61 prices) became quickly invalidated because of ever-increasing price index. The poverty level was based on minimum monthly per capita income linked to per capita calorie requirement, and had to be constantly revised. The poverty line kept on shifting the base-line. In 1977-78 it was computed at Rs and Rs 75 per month for rural and urban areas with a daily calorie intake of 2,400 and 2,100 respectively. In 1984-85, the poverty line was revised to Rs 107 and Rs 122 for the rural and urban areas respectively with calorie intake remaining constant. For the 1993-94 price level, it was revised to Rs 229 for rural area and Rs 264 for urban area. Fourthly, comparison of India's poverty with other countries, particularly with the United States suffers due to an ever-declining value of a rupee as compared to that of a dollar since the sixties.

Significantly, however, India became the first democratically elected government in the Third World to make "reduction of poverty" as an announced planning goal. Comparisons with China, where revolutionary strides were made between 1949 and 1960 in wealth redistribution and elimination of poverty, are inappropriate, because the remarkable achievements in China were carried out under an undemocratic and coercive environment.

The status of poverty in India for 1970, 1980, 1988 and 1993-94 is given in Table 13.1. The trends are worth noting. Between 1970 and 1988, the poor increased in numbers both in rural and urban areas, although that represented a substantial, if not significant, percentage decline of nearly 10 per cent points. Urban poverty, however, did not register as much decline as the rural component of the population, a situation reflecting the migration of the rural poor to the rural areas. The trend continued in the 1990s.

In per capita income of Gross Domestic Product (GDP), Punjab and Haryana were in the forefront with their annual per capita incomes of Rs 3,835 and Rs 3,296 respectively for the year 1984-85, resulting mainly from their agricultural productivity. Several of the more industrialized states, like Maharashtra (Rs 3,232), Gujarat (Rs 2,997) and West Bengal (Rs 2,231), on the other hand, had

**Table 13.1**

*Rural and Urban Poverty in India*

(in millions)

| | *Rural* | *Urban* | *Total* |
|---|---|---|---|
| 1970 | 236.8 | 50.5 | |
| | (53.0) | (25.6) | (51.6) |
| 1980 | 252.1 | 64.6 | |
| | (44.9) | (17.7) | (42.9) |
| 1988 | 252.3 | 70.1 | |
| | (41.7) | (15.8) | (32.3) |
| 1993-94 | 244.0 | 76.3 | |
| | (37.2) | (32.3) | (35.9) |

Figures in parentheses indicate percentage.

*Sources:* World Bank, *India: Poverty, Employment, and Social Services*, Washington D.C., 1989, pp. 175-176.
*Statistical Outline of India*, 1997-98, Bombay, 1997, p. 205.

lower income levels. The explanation is simple. Agriculture is the dominant activity in all the states and contributes most to the income generated everywhere. A higher industrial base of states like Maharashtra and West Bengal does not contribute significantly to the total income of these states as does the agricultural sector. In comparison, the agricultural sector in Punjab and Haryana generates much larger total income. The growth rates in income levels of the various states between 1960-61 and 1984-85 also varied greatly. Predictably Punjab and Haryana recorded higher growth rates, but several poorer states like Tripura, Jammu & Kashmir and Manipur recorded even healthier rate of growth during this period, resulting mainly from a large infusion of central funds and assistance advanced to these states. In general, all states showed an increase in per capita income during this period.

The regional distribution of poverty in the country may be described as a broad belt of areas running from east to west from eastern Uttar Pradesh and Bihar to West Bengal, and moving southward to include parts of Orissa, and Madhya Pradesh, Gujarat and Maharashtra. Elsewhere, there are small pockets of poverty in Andhra Pradesh and Tamil Nadu. The proportion of the poor reaches to a peak of 85 per cent in south Orissa, and 67 per cent in south

Bihar, 72 per cent in east Gujarat and 82 per cent in parts of south Rajasthan. As a contrast, some parts of Punjab and Haryana registered ratios of under 5 per cent (World Bank 1989, Heston 1990: 109). Table 13.2 provides data on the poverty of major regions in rural and urban areas for 1970, 1983 and 1988.

**Table 13.2**

*Percentage of Total Population in Poverty*

| | *Rural Areas* | | | *Urban Areas* | | |
|---|---|---|---|---|---|---|
| | *1970* | *1983* | *1988* | *1970* | *1983* | *1988* |
| South | 62.6 | 47.7 | 44.7 | 56.1 | 42.3 | 29.7 |
| East | 65.3 | 60.4 | 57.4 | 41.8 | 43.5 | 40.9 |
| Central | 47.7 | 41.3 | 38.3 | 42.7 | 36.2 | 33.2 |
| West | 49.0 | 41.8 | 38.5 | 39.3 | 31.6 | 19.8 |
| Northwest | 11.0 | 8.5 | 9.1 | 15.9 | 13.9 | 12.2 |

*Source:* World Bank, *India: Poverty, Employment, and Social Services*, Washington D.C., 1989, pp. 175-176.

## Regional Patterns in Economic Development

Both the World Bank and Heston observe the prevalence of sharp regional variations in poverty and income levels within the country. The Centre for Monitoring Indian Economy, based in Bombay (now Mumbai), has developed composite index values measuring the development levels of the 438 districts using nine data series, two from agriculture, three from mining and manufacturing, and four from the service sector by assigning relative weightages for each series (50 for agriculture, 30 for mining and 20 for services) except for the nine urbanized districts, Greater Mumbai, Delhi, Madras (now Chennai), Hyderabad, Ahmedabad, Bhopal, Chandigarh, Yenam, and part of union territory of Pondicherry. The relative weightages assigned for these nine districts were 35 for mining and manufacturing and 65 for services. The composite index values of the districts were classified into eight categories for determining the level of economic development in a district: four categories above the national average value of 100 and four below. Among the indicators (data series) used were: per capita output of 18 major crops (average of 1975-76 to 1979-80), per capita bank credit for the agriculture group carrying 50 points, and for mining and manufacturing workers per 100,000 population, number of households of manufacturing workers per 100,000 population, and per capita bank credit. The indicators for the service group included per capita bank deposits, bank credit to services, percentage of literate population, and urban population as a per cent of total population. Eight categories were then developed taking the national base index value of 100, the values over 100 denoting development level above the nation and below it less developed than the national level. Figure 13.1 summarizes and maps the findings of the Centre for Monitoring Indian Economy. Four broad classes or regions, viz., developed, moderately developed, less developed and least developed (underdeveloped) are identified.

The developed parts of the country with index values of 200 and above consist of two dozen districts scattered all over the country, and therefore, do not form a continuous region. Economic growth in these districts is generated around an urban focus with strong service, agricultural or manufacturing sectors. The urbanized districts of Greater Mumbai, Chennai, Delhi, Calcutta, Hyderabad, Chandigarh and Ahmedabad, with a strong service or manufacturing sector, or other districts with a well-developed agricultural sector like Faridkot, Ludhiana and Patiala in the state of Punjab, and Gurgaon in Haryana have values above 250. Districts with values above 200 but below 250 belong to this category are Coimbatore and Nilgiri in Tamil Nadu, Nashik and Pune in Maharashtra, Vadodara in Gujarat and the district of Dhanbad (Bihar) with a strong mineral-manufacturing base, and Kurukshetra (Haryana).

The districts with values below 200 and above the national average of 100 form a more continuous regional block along the east and west coasts (with a few exceptions in Orissa and North Ratnagiri and Raigarh districts in Maharashtra). These districts are placed in the moderately developed region. Inland, several districts in west Uttar Pradesh, most of the state of Haryana, Nagpur district in Maharashtra, a few scattered districts in the states of Madhya Pradesh, Rajasthan, Bihar (Padra district) and Kanpur district in Uttar Pradesh, also belong to this category. At the state level, the above average

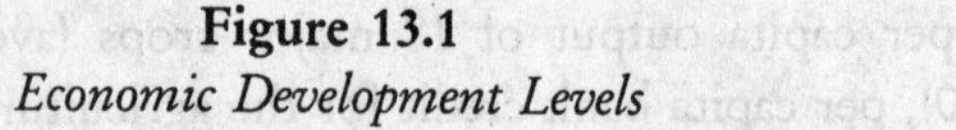

**Figure 13.1**
*Economic Development Levels*

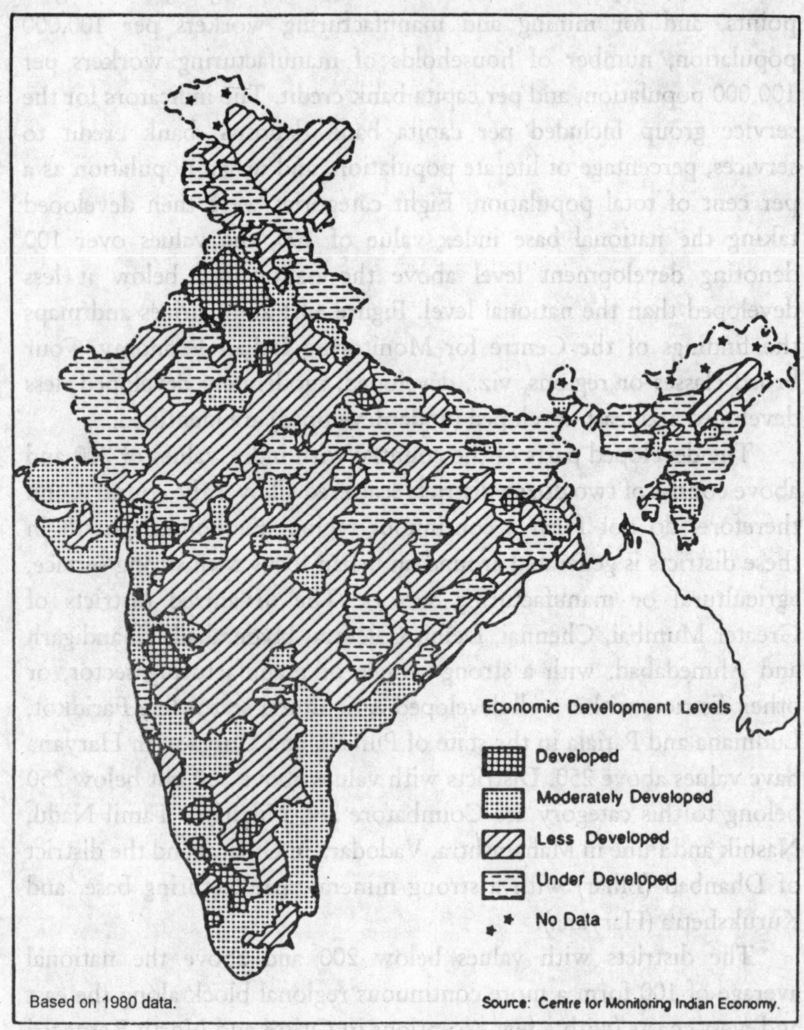


districts are found in the three northern plains states, namely, Punjab, Haryana and western Uttar Pradesh, two west-coast states of Gujarat and Maharashtra, and the four peninsular states of Andhra Pradesh, Karnataka, Kerala and Tamil Nadu.

The rest of the country or nearly two-thirds in area falls below the national average, and is identified as either less developed or underdeveloped. With some exceptions, the two categories identified as lower than average levels of economic development contain districts that occur in the sparsely-populated Himalayan mountain states of Jammu & Kashmir, Himachal Pradesh, Sikkim, Arunachal Pradesh, and the northeast mountain state of Meghalaya, the tribal mountain states of Nagaland, Mizoram, Manipur and Tripura, or in the densely-populated plains of the northern and eastern states (Uttar Pradesh, Bihar, West Bengal, Assam and Orissa), or in drylands (Rajasthan) and tribal hill lands in southern Bihar, and central India (Madhya Pradesh, Orissa and Maharashtra). Within this broad category the least-developed or underdeveloped parts of the country with index values below 50 or one-half of national average are identified as the least developed region which is roughly coterminous with the areas of poverty by the World Bank (World Bank 1989: 175-176). This broad regional belt of underdevelopment lies, in general, from west to east from Himalayan states of Jammu & Kashmir, Himachal Pradesh, and the hill districts Uttar Pradesh, to the plains of east Uttar Pradesh and Bihar, also scattered southwards over parts of Rajasthan and to central India in the states of Madhya Pradesh and Orissa. Elsewhere, this category occurs in a few scattered districts in east Rajasthan, Bastar district in Madhya Pradesh, the coastal districts of Ratnagiri and Sindhudurg in Maharashtra, and Rengareddi district in Andhra Pradesh. The patterns of regional development are remarkably clear. The least developed regions are associated with sparsely-populated mountainous areas, drylands, tribal territories and densely-populated north Indian plains. Comparisons of these regional patterns may also made with those of literacy levels, urban population and population density (Figures 5.3, 7.1 and 5.1).

## *References*

Cassen, R.H., *India: Population, Economy, Society*. London, 1980.

Census of India, 1981. *Census Atlas: National Volume*. New Delhi, 1988.

Census of India, 1991, *Provisional Population Tables, Series 1, Papers 1,2, and 3 of 1991*. New Delhi, 1991.

*Economic Times, Statistical Survey of Indian Economy*. New Delhi, 1991.

Government of India, *India, 1997, A Reference Annual*. New Delhi, 1997.
Heston, A. "Poverty in India, Some Recent Policies" in *India Briefing, 1989*, Boulder, Colo, 1990.
Muthiah, S. (ed.) *et al*, *A Socio-Economic Atlas of India*. New Delhi, 1987.
Schwartzberg, J.E. (ed.), *A Historical Atlas of South Asia*. Chicago, 1978.
*Statistical Outline of India, 1997-98*. Bombay, 1997.
World Bank, *India: Poverty, Employment, and Social Services*. Washington, D.C. 1989.

# IV

## Regions of India

# 14

# The Himalayas

India is a large and spatially varied country, a mosaic of diverse areas. Its parts have been subject to different historical experience in response to their location, ecology, technology and ideology. A knowledge of the various parts will enable us to understand the whole. Chapters 14 to 17 attempt to study the various parts of the country as specific regions which are homogeneous and functionally cohesive.

Several attempts have been made to devise a regional classification of India (Baker 1928, Stamp 1928, Spate 1954, Ginsburg *et al* 1958, and Singh *et al* 1970). Physiography has generally been adopted as the basis of regionalization. Matters relating to location and individual speciality are also taken into account. By comparison, the present study employs three main criteria, namely, natural and cultural landscape, level of socio-economic development, and administrative cohesion, for the purpose. The country is divided into four primary, nine secondary, and 38 tertiary regions. While Table 14.1 lists their names, area and population, Figure 14.1 identifies them.

Though a well-defined geographic unit, the Himalayas contain great regional variety in their physical and cultural patterns. Their physical bifurcation into the western and eastern wings is automatic in the Indian context by the presence of Nepal in between. The Western and the Eastern Himalayas differ from each other on several counts.

**Figure 14.1**
*Regions of India*

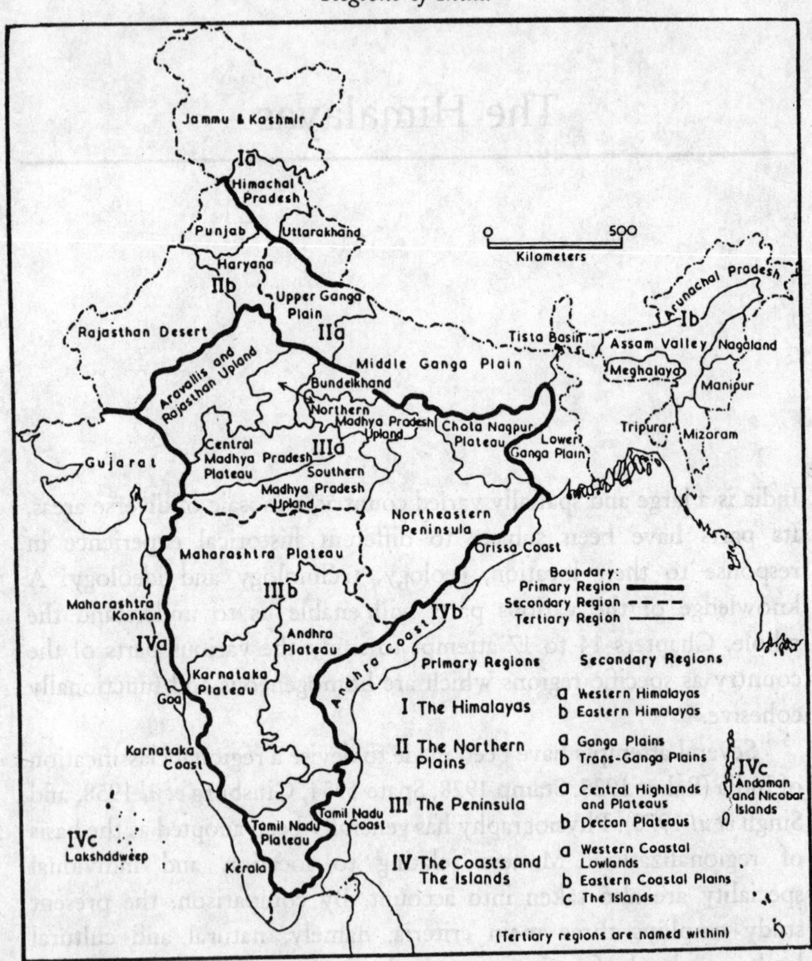


The former is less humid, less tribal, and more integrated with the rest of the country than the latter. The former bears the imprint of the British rule in the form of hill towns, cantonments and tourism; the latter has tea plantations, oil wells and a large Christian population among the tribals. After Independence, in 1947, the eastern wing experienced intense ethnic resurgence and reorganization of new states.

**Table 14.1**

*Regions of India*

| *Primary/Secondary/Tertiary Regions* | *Area ($km^2$)* | *Population (1991)* |
|---|---|---|
| I. The Himalayas | 603,971 | 57,306,848 |
| a. Western Himalayas | 329,034 | 18,730,818 |
| (i) Jammu & Kashmir | 222,236 | 7,718,700 |
| (ii) Himachal Pradesh | 55,673 | 5,111,079 |
| (iii) Uttarakhand | 51,125 | 5,901,039 |
| b. Eastern Himalayas | 274,937 | 38,576,030 |
| (i) Tista Basin | 19,859 | 6,689,119 |
| (ii) Arunachal Pradesh | 83,743 | 858,392 |
| (iii) Nagaland | 16,579 | 1,215,573 |
| (iv) Manipur | 22,327 | 1,826,714 |
| (v) Mizoram | 21,081 | 686,217 |
| (vi) Tripura | 10,486 | 2,744,827 |
| (vii) Meghalaya | 22,429 | 1,760,626 |
| (ix) Assam Valley | 78,438 | 22,294,562 |
| II. The Northern Plains | 677,638 | 313,816,330 |
| a. Ganga Plains | 372,716 | 249,858,965 |
| (i) Upper Ganga Plain | 89,876 | 51,915,485 |
| (ii) Middle Ganga Plain | 206,851 | 136,244,362 |
| (iii) Lower Ganga Plain | 75,989 | 61,699,118 |
| b. Trans-Ganga Plains | 304,922 | 63,957,365 |
| (i) Punjab and Chandigarh | 50,362<br>(114) | 20,190,795<br>(640,725) |
| (ii) Haryana and Delhi | 44,212<br>(1,483) | 16,317,715<br>(9,370,475) |
| (iii) Rajasthan Desert | 208,751 | 17,437,655 |
| III. The Peninsula | 1,504,580 | 300,503,266 |
| a. Central Highlands and Plateaus | 803,009 | 139,887,074 |
| (i) Aravallis and the Rajasthan Upland | 133,488 | 26,442,985 |
| (ii) Bundelkhand | 29,417 | 6,729,885 |
| (iii) Northern Madhya Pradesh Upland | 103,888 | 16,087,912 |

Contd...

Contd...

| | | |
|---|---|---|
| (iv) Central Madhya Pradesh Plateau | 101,691 | 17,880,383 |
| (v) Southern Madhya Pradesh Upland | 102,673 | 14,552,562 |
| (vi) Northeastern Peninsula | 262,045 | 36,884,487 |
| (vii) Chota Nagpur Plateau | 97,714 | 21,348,860 |
| b. Deccan Plateaus | 691,664 | 160,616,192 |
| (i) Maharashtra Plateau | 276,985 | 59,427,189 |
| (ii) Andhra Plateau | 182,139 | 37,653,592 |
| (iii) Karnataka Plateau | 173,059 | 40,895,978 |
| IV. The Coasts and the Islands | 501,074 | 172,697,778 |
| a. Western Coastal Lowlands | 288,652 | 94,847,149 |
| (i) Gujarat, Daman & Diu, and Dadra Nagar & Haveli | 196,627 | 41,414,183 |
| (ii) Maharashtra Konkan | 30,728 | 19,321,026 |
| (iii) Goa | 3,702 | 1,168,622 |
| (iv) Karnataka Coast | 18,732 | 3,910,490 |
| (v) Kerala | 38,863 | 29,032,828 |
| b. Eastern Coastal Plains | 204,141 | 77,519,837 |
| (i) Orissa Coast | 40,166 | 15,012,940 |
| (ii) Andhra Coast | 92,906 | 28,700,967 |
| (iii) Tamil Nadu Coast and Pondicherry | 71,069 | 33,805,930 |
| c. The Islands | 8,281 | 330,792 |
| (i) Lakshadweep | 32 | 51,681 |
| (ii) Andaman and Nicobar Islands | 8,249 | 279,111 |
| INDIA | 3,287,263 | 844,324,222 |

*Notes:* 1. The above regional scheme bears quite a similarity with that devised by the census of India (1988). It uses the district as the basic spatial unit for delineation of regions.

2. Population and area statistics in chapter 14-17 are based on the 1991 census of India unless stated otherwise.

People in the Western Himalayas find their racial roots largely in the Mediterranean region and Central Asia; those in the Eastern Himalayas are of the Indo-Burmese and Indo-Tibetan stock. For the last over one century, outmigration for recruitment in army and city jobs has been more typical of the Western Himalayas; and inmigration to tea plantations, agricultural wastelands and new construction sites is characteristic of the Eastern Himalayas. Paradoxically, the latter is economically less developed than the former. Much of its development was superimposed and confined to few pockets.

Administratively, the Western Himalayas are divided into Jammu & Kashmir, Himachal Pradesh and Uttarakhand, and the Eastern Himalayas into the Tista valley (Sikkim and the Duars), Arunachal Pradesh, Nagaland, Manipur, Mizoram, Tripura, Meghalaya and the Assam valley (Assam) (Figure 14.2). Meghalaya is structurally a detached portion of the peninsular India and the Assam valley an extension of the Ganga plain. But both have a strong historical, social and economic association with the Eastern Himalayas in general, and both have been treated along with the Himalayas.

## Western Himalayas

Known for the natural beauty of its majestic mountains, green forests, extensive meadows and fast flowing rivers, Jammu & Kashmir, the "Switzerland of India", is a crown on the nation's body. The metaphor goes a bit further—the crown makes the head uneasy if not properly cared for. The current militancy in the state could be avoided if a right kind of development strategy would have been adopted.

Jammu & Kashmir is the only Muslim (nearly two-thirds of total population of 74 million) majority state of India—a symbol of country's secularism. It opted to be a part of India rather than that of Pakistan. A special status under the Article 370 of the Indian Constitution entitles it to a greater degree of autonomy.

The state covers an area of 222,236 km$^2$ but more than half of it is under the occupation of Pakistan (78,114 km$^2$) and China (42,735 km$^2$). Though the legal and constitutional position is in favor of India, yet the territorial disputes across the border persist. For that reason, Jammu & Kashmir is a critical factor in the formulation of India's defense policy.

The region offers a great physiographic variety: the trans-Himalayan Ladakh plateau and Korakoram ranges (altitude 3,000 to more than 8,000 m), the Great Himalayas and Zanskar range (3,000 to more than 6,000 m), the valley of Kashmir (1,500 to 4,500 m), the Outer Himalayas in the form of Pir Panjal and Dhauladhars (1,000 to more than 6,000 m), and the sub-Himalayan Jammu (300 to 1,000 m).

**Figure 14.2**
*Western Himalayas*

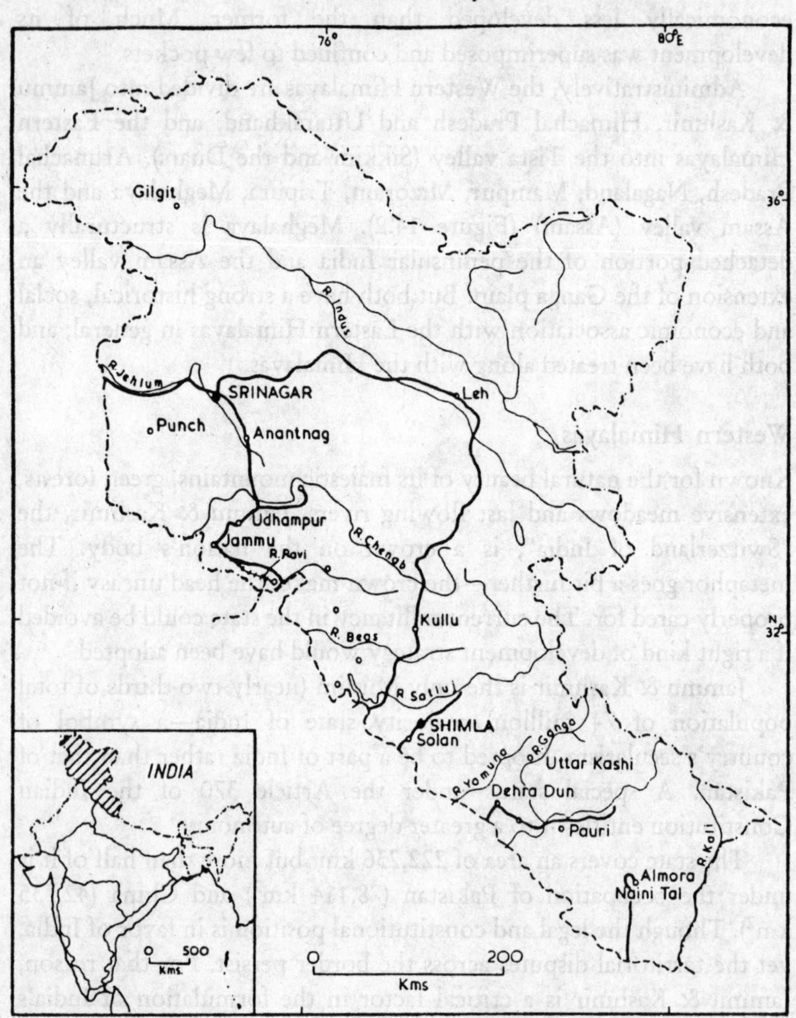


As much as 90 per cent of the area is mountainous and hilly. The range of relief is from about 300 to 8,611 m at $K^2$.

The Indus is the master stream which separates the trans- and the Great Himalayas from each other. Its tributary—the Jhelum—contains the picturesque Kashmir valley of a lacustrine origin along its course. The two other major tributaries—the Chenab and the

Ravi—traverse through the Jammu part of the region.

The imprint of relief on climate and vegetation is pervasive. The trans-Himalayas is a cold desert; Leh receives hardly 8 cm of precipitation in a year. Located at an altitude of 3514 m, it has January mean temperature of -7°C and June mean temperature of 18°C. The Great Himalayas are again very cold, with little plant life beyond 4000 m. The Kashmir valley has pleasant summers and snowy winters. Srinagar, at an elevation of 1587 m, records a mean temperature of 25°C in June and of 1°C in January. Its annual precipitation of 65 cm is almost equally divided between summer and winter. The Jammu part is sub-tropical, with hot summers and cool winters.

Vegetation ranges from tropical to alpine, conforming to altitude and aspect, sun facing or otherwise. Excluding the vast barren Ladakh, about one-half of the region is under forest. Extensive meadows form a conspicuous part of the landscape.

Agriculture, tourism and household industries define the major contours of economy. Only 5 per cent of the state's area is under agriculture with sharp regional variations: around 25 per cent in the Kashmir valley, 33 per cent in the Jammu sub-region, and less than one per cent in Ladakh. Only one crop a year is possible at altitudes above 2,000 m; beyond 4,000 m, agriculture is virtually absent. Nearly a half of the agricultural land gets irrigation by diversion of streams (*kuhls*) on the hill slopes and by canals in the valley and Jammu sub-region. Rice and maize in the valley, wheat, rice and maize in Jammu sub-region and barley in Ladakh describe the foodgrain basket of the state's different parts. The Kashmir valley is also famous for its orchards of apples, plums, apricots and walnuts. Raising of vegetables on artificially-raised floating islands in the Jhelum is a special feature. Saffron is the golden crop on the *karewas* (natural terraces) of the valley.

*Pashmina* shawls, woolen carpets, carved wooden items and embroidery products of Jammu & Kashmir are world famous. Most of these originate from the household industry which accounts for two-thirds of the production in industrial sector. Modern industry, producing goods like watches, machine tools, woolens and silk, is confined to few places like Srinagar and Baramula.

Natural beauty, combined with mild summers, has promoted a lucrative tourist industry. Tourists use Srinagar, the state capital, as the base for a visit to other attractions like Gulmarg, Pahalgam and Amarnath. The city, with a population of more than half a million, is the most populous hill station of India. It is famed for its terraced gardens, Dal lake and houseboats. Gulmarg, at 2653 m, has the highest golf course in the world. Pahalgam is a quiet retreat on Lidder river. Located nearby is the ancient shrine of Amarnath which is visited by thousands of Hindus on the occasion of *Raksha Bandhan* to pay obeisance to its Shivalingam made of ice. Ladakh, the mysterious land of *gompas* (monasteries), has been attracting an increasing number of adventurers and scholars since 1974 when free entry to this forbidden land was permitted. Leh and its closeby monasteries are the main attraction. The Vaishnodevi temple, near Jammu, is the destination of millions of Hindu devotees every year.

A less highlighted feature is the necessary presence of defense forces in a large number in this sensitive region bordering with Pakistan and China. This can be turned to a great advantage for the region's economy. A desired link can be established between the routine requirements of the army and their production at local level. This will go a long way in taking care of the discontent which does have rising unemployment and undiminished poverty as its roots. The region's per capita income is about 20 per cent lower than that of the national average.

Historically, Jammu & Kashmir is a conglomerate of the Buddhist Ladakh, Muslim Kashmir and Hindu Jammu. Each sub-region generates sub-regional sentiment, the most intense being in the Kashmir valley. The Jammu-Srinagar-Leh national highway is the lifeline of the economy. A balanced and interrelated development of the three sub-regions through decentralized planning will be in the state's interest.

## Himachal Pradesh

Formed initially in 1948 by an amalgamation of 30 princely states and later enlarged and consolidated by the merger of *Pahari* (a dialect of Hindi)-speaking parts of the former Punjab in 1966, Himachal Pradesh is among the most dynamic hill states of India. Its most notable accomplishments have been in the field of road transport, horticulture, education and tourism. Much still remains to be done as its per capita income has just approached the national average now.

The physiography of Himachal Pradesh is a part of the Himalayan system. There is a typical longitudinal arrangement of the various landforms: the trans-Himalayas, the Great Himalayas, the Dhauladhar-Shimla hills, and the Siwaliks, enclosing the faulted valleys (*duns*) of the Pinjore, Jaswan and Paonta. The relief ranges from 300 m to more than 6000 m. About one-half of the total area is higher than 3000 m.

Around two-fifths of the area (55,673 km$^2$) is under forest. Cultivated land is restricted to only one-tenth of the total, and is confined mainly to the Kangra valley, the *duns*, the valley of the Satluj, and the terraced hill slopes. Cropping pattern varies from wheat, maize and rice at lower elevations to barley and buckwheat at higher locations. Potatoes, peas and ginger-root are the main cash crops. Lahul valley is famous for quality potato seeds.

Remarkable expansion has taken place in pomiculture. Apples, plums, peaches and apricot are the chief products. The state is now described as the "apple belt" of India.

Tourism is being promoted in a big way in Himachal Pradesh. The Jammu & Kashmir problem has indirectly helped this process. The mild summers, scenic charm and tourist infrastructure have stimulated tourism at Shimla, Manali and Dalhousie. Shimla, situated on a ridge which separates the Ganga and Indus drainage systems, at an altitude of 2,206 m, was the summer capital of British India. It still exudes the nostalgia of the Raj through its Vicereal Lodge, Christ Church and the Mall. Dalhousie, another hill station at 2,036 m altitude, is built around five forest-clad hills and has many old English houses. Manali, a tourist resort, is located on a picturesque site along the Beas river. Manali's popularity is mostly a post-Independence phenomenon.

The major streams, including Satluj, Beas, Ravi and Chenab, as also their numerous tributaries, contain several rapids on their courses and are most suited for the generation of hydropower. This potential is being harnessed. Plans are afoot for generation of additional electricity for commercial supply to the adjoining states. Within the state, all the settlements, both urban and rural, are electrified.

Traditional household industry, such as spinning of wool and shawl making, predominates. This goes with raising of sheep and goats in large numbers. Some manufacturing industry in the form of breweries at Solan and Kasauli, tea processing at Jogindernagar and foundry works at Nahan had started before Independence. New industry, such as electronics, fruit processing and yarn spinning is set up at Parwanoo, Barotiwala and Baddi in Solan district.

Some sociological aspects of the state are noteworthy. As much as 96 per cent of the population is Hindu—the highest proportion for any Indian state. Practically, the same percentage of the agricultural workers is owner cultivator, representing an egalitarian agrarian structure. There is one government employee for every 25 persons or five households. All these ratios are among the highest for Indian areas.

Only 8.7 per cent of the state's population of 5.1 million is urban as compared with 25.7 per cent for India, but the literacy rate of 64 per cent excels the national average of 52 per cent. Among the urban places, the capital city of Shimla shows a high degree of primacy. Its population of 109,860 is more than four times of that of the next town. Notably, all the towns developed by the British are sited on ridges at an elevation of around 2000 m, and most of the towns of indigenous origin are sited along the river courses.

The physiographic diversity notwithstanding, the people of the state identify themselves as belonging to two sub-regional backgrounds: the old Himachal and the new. While the erstwhile princely states constitute the old, the new was earlier under the British as a part of Punjab. The former is less developed than the latter. Through government intervention, this disparity at the sub-regional level is being reduced with success.

Basic development problems of Himachal Pradesh relate to an optimal utilization of natural resources and repair of any ecological damage caused by deforestation or extension of cultivated land onto marginal areas or overgrazing of pasture lands. The state rightly complains that large chunks of its scarce cultivated land were lost under the major multipurpose irrigation and power projects, like the Bhakra, Pong and Beas-Satluj link which benefited mostly the neighboring states. Such a regional consciousness is impelling the state to conserve its resources. The state recently banned the export of timber.

## Uttarakhand

Covering the northern hilly and mountainous part of Uttar Pradesh, this region is composed of the upper basins of the Yamuna, Ganga, Ramganga and Sharda rivers. If the Kashmir Himalayas is a "paradise on earth", and the Himachal Himalayas the "apple belt", the Uttar Pradesh Himalayas is the "holy land". It is the source region of the sacred Ganga and its numerous tributaries. Thousands of pilgrims visit every year places like Gangotri, the source of Ganga, Yamunotri, the source of Yamuna, and Hardwar, the entry point of the Ganga into the plains.

The region has frequently asserted its identity by raising a demand for a separate hill state carved out of Uttar Pradesh. Accusations of neglect and discrimination in devolution of funds and in recruitment to government service are made. The feeling has visibly been moderated following the establishment of the Hill Development Board to look after its planning needs.

The region rises from the Siwalik hills to the glacier-capped Himalayas attaining an altitude of 7817 m at Nandadevi. The typical zonation in the Himalayas is represented as follows: the Greater Himalayas above 4000 m, the Inner Himalayas 2000-4000 m, and the Outer Himalayas, the Siwalik hills and the Dehradun at successively lower elevations. A large part of the region lies at an elevation of 2000-4000 m.

Climate and vegetation generally follow relief. Hot and humid conditions are typical of the Siwalik hills and Dehradun. On the Outer Himalayas, the winter mean is around 5° C at 2,000 m and freezing point at about 3,000 m. Summer means lie between 25° C and 30° C. Precipitation ranges from 150 to 250 cm, decreasing both toward north and south. On the upper reaches, climate is alpine. Winter snowfall is common at altitudes higher than 2000 m.

Broadly speaking, one-sixth of the region remains perpetually under snow. Nearly one-half is under forest. Cultivable land is barely one-eighth of the total area.

In land use, a typical valley has three vertical sections: (i) *katil* land—on higher elevations, with hoe culture raising mainly small grains, (ii) *upraon* land—heavily terraced having unirrigated maize and wheat cultivation, and (iii) *talaon* land—low lying irrigated tracts with intensive cultivation in rice, wheat and sugarcane. On the southern margins is the *terai* (marsh) land which was colonized by the Punjabi Sikhs after Independence. This erstwhile malarial/marshy land has now been converted into a prosperous agricultural belt.

The region has enormous hydroelectric potential for industry but the small local market is a constraint. Traditional weaving continues on subsistence lines. Some modern industry has come up at places like Dehradun, Rishikesh and Nainital where government has provided special incentives to both private and public sector entrepreneurs.

Urbanization level is low, i.e., only 20 per cent. Existing towns are primarily administrative centers (Dehradun), hill resorts (Nainital and Mussoorie), cantonments (Ranikhet and Landsdowne), and religious sites (Hardwar and Rishikesh). Dehradun (367,411) is a regional center. It is the main entry point for various places in the Uttarakhand. Important institutions like the Oil and Natural Gas Commission, Survey of India, Forest Research Institute and Indian Military Academy are located here. Retired elite find this place climatically mild and environmentally calm for a permanent abode. Located nearby is the hill resort of Mussoorie at 2,005 m, which houses the prestigious National Academy of Administration. At some distance is the more glamourous hill town of Nainital, built around a beautiful lake.

A population of nearly 6 million spread over an area of 51,125 km$^2$ gives a density of about 120. Agricultural density approaches 1,000, representing intense pressure on land under cultivation. This found an outlet in outmigration. Recruitment in army has been popular. This exposed the region to external influences, giving a stimulus to the spread of education in particular. Many educated persons had to outmigrate because local employment opportunities were few. Migration of the semi-educated or illiterate poor to cities like Bombay (now Mumbai), Delhi and Lucknow for labor was also considerable. All this resulted in a "money order economy" for thousands of families whose male heads had moved out for a living.

Uttarakhand, on the whole, is an underdeveloped region whose resources, both physical and human, got considerably transferred to other parts of India. Deforestation took place on a massive scale, rendering the ecology fragile. The *chipko* movement (cling to the trees to save them from the contractor's axe) has become quite popular. Likewise, a strong resistance to the Tehri Dam Project, which will submerge large pockets of forest and agricultural land, is not without reason. Additionally, outmigration of thousands of young adult males deprived the region of a substantial part of its productive workforce. An employment-oriented strategy of development is a dire need of the area.

Uttarakhand is generally divided, on physiographic lines, into three sub-regions: the Dehradun, an agriculturally fertile valley; the Kumaon, covering the outer and inner Himalayas, with an elaborately organized terraced agriculture; and the Bhotiya valleys, noted for pastoral-transhumance at higher elevations. It is, however, appropriate to divide the region into two parts, namely the Garhwal Himalayas in the west and the Kumaon Himalayas in the east. The two are distinguished by the dialects spoken locally. A large part of the former has been under the rule of the native princes while most of the latter part was under the British rule. The former is more backward than the latter, and experienced outmigration on a large scale.

## Eastern Himalayas

Spread over the state of Sikkim and the *duars* (districts Darjeeling, Jalpaiguri and Kooch Bihar) of West Bengal, this region coincides with the basin of the Tista—a tributary of the Ganga (Figure 14.3). It makes an international border with Nepal to the west, Bhutan to the east, and China (Tibet) to the north. By reason of location, it combines the impact of the Buddhist Bhutan and Tibet on Sikkim and of the Hindu Nepal on the *duars*. An urge to preserve the ethnic identity is strong in Sikkim, whereas a sub-regional sentiment in favor of Gorkhaland is vociferous in the *duars*.

Sikkim was earlier a protectorate of India under the rule of a local prince. Its Chumbi valley provides the shortest route between India and Tibet. Under considerations of security, it was included within the state structure of India in 1975.

Sikkim is entirely mountainous and the *duars* largely sub-mountainous. A bold variation in relief from less than 100 m to over 8,000 m is observed within a distance of around 100 km Kanchenjunga (8603 m) here is the third highest peak in the world. Likewise, extremes exist in climate from humid tropical to dry polar. Precipitation ranges from 300-500 cm over a large part and landslides on steep slopes are frequent. Altitudinal variations permit practically every botanical zone to be represented. One-third of the total area is under forest.

**Figure 14.3**
*Eastern Himalayas*

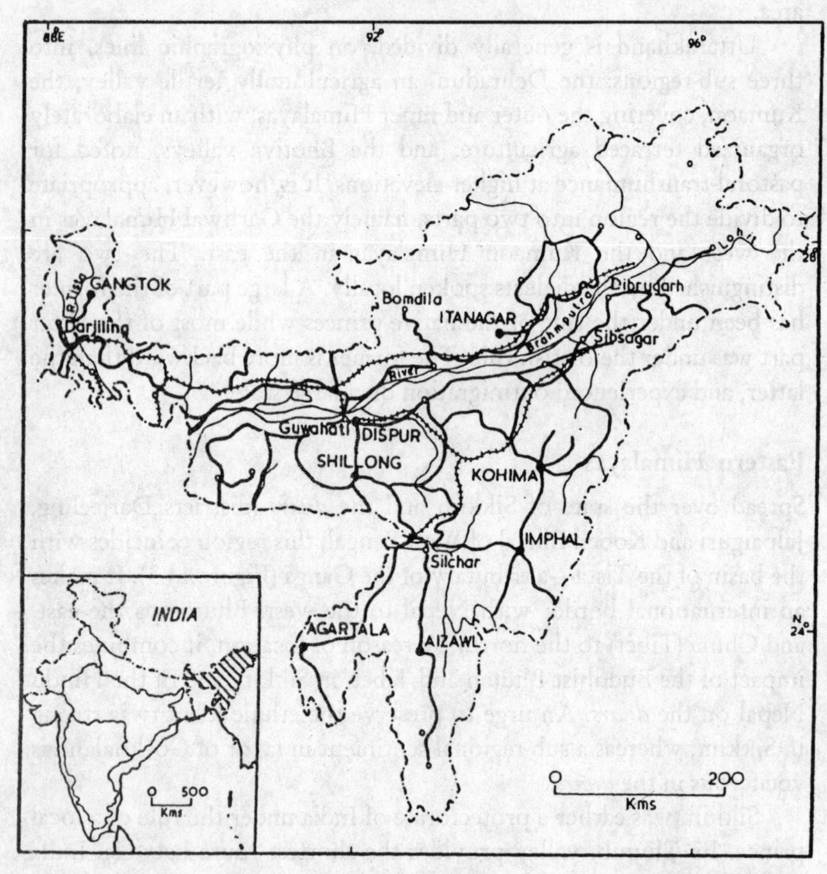


Agriculture is the staple economy. Most of the cultivated area is restricted to altitudes below 3,000 m. It accounts for about one-tenth of the total area. Terracing of fields is almost universal. Maize, rice, wheat and potato are the foodcrops. Cardamom is produced here in greater quantity than anywhere else in India. Orange is another lucrative cash crop. Sheep and yak are grazed at higher elevations. At 1,500 m or less, tea plantations are well developed in the Darjeeling-Siliguri tract. Much of the wasteland in the *duars*, originally a marshy tract, has been reclaimed for cultivation of rice and jute.

Industrial development is meager and concentrated in a few places. Entire Sikkim has been declared as industrially backward by the Government of India for special incentives. Some agro- and forest-based industry, such as fruit juice, jam, beer and matches, has come up. The Darjeeling area is famous for production and processing of high-quality tea.

Sikkim has a population of less than half-a-million on 7,096 $km^2$, and the *duars* of about 6 million on 12,763 $km^2$. This gives contrasting densities of around 50 and 500 in the two, respectively. Population in Sikkim lives in scattered hamlets; villages with monasteries make bigger clusters. Its capital town of Gangtok has a population of 24,971. By comparison, Darjeeling's population (73,088) is almost three times larger. It is a prominent hill resort as well as an educational center. At an altitude of 2134 m, this place offers a direct view of Mt. Everest (Sagarmatha) (8848 m) to the northwest and Kanchenjunga to the north. The Himalayan Mountaineering Institute is also located here.

Sikkim is traditionally the land of Lepchas and Bhutias who are Buddhist by religion. Monasteries form an important part of the cultural landscape. Nearly a century ago, the British invited the Nepalis in large numbers to construct roads and extend agriculture in this strategically located area, which is adjacent to Tibet. This led to a dramatic change in its population composition. In 1947, the Nepalis outnumbered the locals. Recent years have experienced considerable inflow of technical, professional and administrative personnel from other parts of India. This further reduced the share of the natives to only about 30 per cent. They, of course, remain in possession of a large part of the land. Nevertheless, an ethnic conflict between the natives and outsiders does exist.

Likewise, the *duars* received migrants from Nepal and Bangladesh to its tea plantations, agricultural wastelands and urban centers. Its demographic base was enlarged. This, however, did not pose any ethnic problem because the immigrants were from a similar background.

The region easily lends itself to a two-fold division: the Sikkim and the *duars*. The former is still largely ancient and sparsely populated, and the latter contemporary and crowded.

## Arunachal Pradesh

Popularly known as the "land of the rising sun" in India, this is demographically the least populated and politically the most pacific part of the country. The region is strategically important as it borders with China, Burma and Bhutan. It was brought into a limelight by the Chinese invasion in 1962. Then, an era of extensive road construction and other development activities followed. A wise policy on the part of the union government, whereby the local tribes were directly involved in the development process and their traditional institutions were duly protected, yielded positive results. This pre-empted any popular discontent. The region has remained exceptionally free from any insurgency, as is typical of many other parts of the Eastern Himalayas. It was granted statehood in 1987.

Arunachal Pradesh has a population of 0.9 million, and area over 83,743 km$^2$. This gives a density of nearly 10. Population is, however, diverse and segregated, representing some 30 tribes speaking 50 dialects. Important tribes include Sherdukpems, Mompas, Daflas, Apa Tanis, Miris, Abors, Mishmis and Wanchos. Over the years, some in-migration from other parts of India has also been taking place in response to the employment opportunities associated with government sponsored development programs. These migrants now constitute almost one-third of the total population.

Extremes of relief, from less than 100 m to almost 8,000 m, exist. Bomdi La range, Miri hills, Dafla hills, Abor hills and Mishmi hills get their names from the local tribes. The Brahmaputra river divides this mountainous region into two parts—the western and the eastern. Precipitation is high on both. It exceeds 1,000 cm at places, producing a thick covering of tropical rain forest at lower altitudes.

Forest covers 62 per cent of the land. Higher elevations are unfit for agriculture. Hardly 4 per cent of the total area is under cultivation. Shifting cultivation—slash and burn—is practiced on more than a half of the cultivated area. Rice is the principal crop; maize, pulses and oilseeds are the others. Orange and potato are also raised.

Forest, hydro and mineral (coal, oil and limestone) resources are abundant but the small size of population at the tribal mode of living is a constraint to their utilization. Traditional industries like weaving and basket-making are a part of the tribal economy. Some industry, related to the manufacturing of tea chests and plyboards and processing of fruits, is at an incipient stage. The Institute of Science and Technology located at Itanagar serves all the states in northeastern India.

This serene land of a few and far between settlements, and tranquil environment is not without some attractive places of tourist interest. Apart from the newly built state capital of Itanagar, is Tawang which has a historic Buddhist monastery nearby. A majority of Arunachal Pradesh's population adheres to tribal persuasions. The Hindus make less than one-third and the Buddhists one-seventh of the total population. The Christians number just a few. By comparison, other tribal areas in the northeast contain Christian majorities.

## Nagaland

Formed as a separate state in 1963, Nagaland has been in the forefront in pursuing separatism. Its distinctiveness lies in having a high level of political consciousness and unity of purpose despite internal tribal diversity. It has a population of 1.2 million on 16,579 km$^2$. Here 84 per cent of the population is tribal, 80 per cent Christian, and 61 per cent literate.

The Nagas do not make a single group. They belong to different tribes, such as Aos, Semas, Angmis, Konyaks, Tuesangs, Lothas and others, whose mutual rivalry to each other is manifest in the formation of various districts primarily on the basis of dominant tribes. Their dialects are not mutually intelligible. English is the link language. It enjoys the status of being the state language also.

The region is a long narrow strip of hills (300-3,000 m) with a northeast-southwest orientation. Saramati (3,826 m) in the Patkai range is the highest peak. Most of the Naga settlements are confined to elevations ranging from 1,000-2,000 m. These are typically sited on hill crests. Such locations were not only safe from a security viewpoint in a tribal ethos but are also free from malaria which was earlier endemic in the valleys. Kohima (53,122)—the state capital—is also located on a hilltop. The flower-decked war cemetery here is reminiscent of the Japanese invasion of India in 1942.

Agriculture absorbs around 70 per cent of the working force but occupies hardly 6 per cent of the total area. Three-fourths of the cultivated area is under shifting cultivation and the remaining one-fourth under sedentary terraced cultivation. Rice supplemented by meat makes the staple food of the Nagas. Dog's meat is a delicacy. Rice beer is a favorite drink.

Although modern industry is small—a sugar mill, television

assembly unit and brick plant at Dimapur, paper mill at Tuli, newsprint unit at Mokokchung, and plywood factory at Tizit—Nagaland can take a pride in the attractiveness of its colorful handlooms. Equally famous is the Naga dance. This land of scenic beauty, mild climate and easy access can be groomed for tourism.

### Manipur

The abode of Shiroy lily (the paradise flower), not found anywhere else in the world, Manipur is a stronghold of Hinduism amongst the tribal territories around. It was a principality before Independence; now it is one of the small states of India, with a population of 1.8 million over an area of 22,327 km$^2$.

Physically, the region has the Imphal valley enclosed within the fold of northsouth ridges around. The hills account for 92 per cent of the total area and range in elevation from 800m to over 2000m. Siruhi peak is the highest point at 2569 m. The valley is drained by the Manipur river which is a tributary of the Irawaddy in Myanmar (Burma). The river originates from the Loktak lake which is part of the valley itself.

Over 60 per cent of the population is Hindu and about 30 per cent Christian. There are some Muslims also. The Christians are tribal by background and concentrated mainly in the hills. On the other hand, both the Hindus and Muslims are confined mainly to the valley. They speak the Meitei language around which the ethnic sentiment is woven. The valley did have its own share of insurgency and political instability.

The Imphal valley, sharing 8 per cent of the area, accommodates two-thirds of the population. This disparity in population distribution is associated with the availability of agricultural land. The valley is heavily cultivated with 80 per cent of the area devoted to rice. It serves as the granary for the surrounding food-deficit areas. Other crops include sugarcane, potato, mustard and pulses. On the hills, rice remains the main crop. The Nagas in the northern part use terrace cultivation while the Kukis in the south practice shifting cultivation. On the whole, only one-tenth of the region's area is under cultivation. Forests cover two-thirds of the land.

Handlooms of Manipur enjoy a reputation for their quality and design and find a ready market in Calcutta, Guwahati, Shillong and other cities. An interesting feature of the cottage industry is its management which is carried out through cooperative marketing societies. In addition, some new industry, such as the manufacturing of television sets, bicycles, hydrogenated oil, dehydrated ginger and cement, is coming up in and around Imphal.

Imphal is the capital and the main tourist attraction of the region. Its Khwairamband bazaar is manned only by women—some 2,500 or more, selling clothes and vegetables. This city, with a population of 200,615, is a premier center of education in this distant part of India.

## Mizoram

If Nagaland is to be singled out as a forerunner in demanding tribal autonomy, Mizoram is to be noted for carrying on the longest spell of insurgency culminating in attainment of statehood in 1987. It was one of the districts of old Assam state till 1972. During 1972-87, it had the status of a union territory under the administration of the Government of India.

In its struggle for autonomy, the region was greatly favored by several factors. It is sandwiched between Myanmar (Burma) on the east and Bangladesh to the west. Necessary help from across the border was readily available. Secondly, it has a homogeneous demography: 94 per cent of the population is tribal, mostly belonging to the Lushai group, and 84 per cent is Christian. Because of remoteness from the mainland India, the Indian government could not keep strict vigil on it.

A special feature of the region is the recent large-scale reorganization of settlement structure. As a security measure and with a view to reducing the cost on provision of health, education, water supply, electricity and postal services, the dispersed villages were grouped together and relocated as "protected progressive villages" along the main roads. About two-thirds of the rural population was affected in the process. Migration from rural areas to the state capital of Aizawl was also encouraged, even by way of offering free urban land. The place has grown nearly five-fold during the last twenty years. Aizawl population of 154,343 is almost one-fourth of the state's total population of 0.7 million. Among all the states of India, Mizoram has the highest percentage of urban population. Its total area is 210,817 km$^2$ which gives a density of 33.

Mizoram, as its name denotes, is a hilly region. The ridges and valleys alternate each other in northsouth orientation. The average height of the hills is 1,000 m. Blue Mountain (Phawngpu) is the highest peak at 2,210 m altitude.

Three-fourths of the total area is under open forest. The remaining land is given to once-in-seven-year rotation for shifting cultivation. Therefore, not even 4 per cent of the area in a given year is under crops, which include mainly of rice and some oilseeds and maize; the last one is grown mainly as a feed for pigs. The agricultural productivity is low, and the state is deficit in foodgrains, milk and poultry, which are brought in from other parts of India. Efforts are now being made for sedentary cultivation. Sericulture and plantation of rubber, coffee and cardamom are also being encouraged.

Mizo *paun* (shawl for men and skirt for women), as a product of traditional industry, attracts markets even outside the state. Development of large-scale industry is likely to remain slow. It is constrained by the inner-line regulation which forbids the purchase of property or establishment of any industry and trade in the name of non-Mizos. This measure is considered necessary as a barrier against the influx of outsiders.

Although backward at present because of its remote location, low productivity in agriculture, and disturbed polity till 1987, Mizoram is poised for development in future. It is trying to replace shifting cultivation by a sedentary one, strengthen the physical infrastructure of power, transport and communication, and promote industrial development. In this task, the region is favored by the homogeneity of its people, a high literacy rate, a variety of resources, and liberal assistance from the central government. The grant of statehood in 1987 is expected to work as an additional stimulant.

## Tripura

Encircled by Bangladesh on three sides and lined with the rest of India only in the east, this erstwhile tribal principality is now one of the remote frontier states. It is the smallest in area size (10,486 km$^2$) of all the states in the northeast, but its population of 2.7 million is next only to that of Assam.

No other part of India has experienced such a demographic transformation as Tripura did as a result of large-scale migration from

Bangladesh across the border. Inflow of the Hindu Bengalis started on the eve of partition in 1947 and continued ever since. The native tribals, who were 90 per cent of the total population, have gradually been reduced to less than one-third because of this influx. Now 70 per cent of this 2.7 million population is Bengali-speaking, and the remaining 30 per cent, speaks tribal dialects.

The two populations are separated by their location. Non-tribal population is concentrated in the western plain, which accounts for 30 per cent of the total area, and the tribal population is concentrated in the eastern hilly tract, which occupies the remaining 70 per cent. A strong reaction from the tribals on their reduction to a conspicuous minority was natural. It found expression in insurgency. Notably they are largely Hindu. To mellow their feelings, the hilly tribal part of the state has been carved out as an autonomous district wherein the local people will look after their development activities.

Agriculture, which is the mainstay of economy, differs between the western plain and the eastern hilly tract. The former is noted for intensive cultivation of rice, jute, potato, sugarcane, oilseeds, and fruits; the latter has largely shifting cultivation in rice and oilseeds with a tendency toward sedentarization. Tea cultivation has made a big entry. On the whole, about one-fifth of the total area is cultivated. Nearly 60 per cent is under forests.

Handlooms and bamboo-cane products define the traditional industrial base of the region. Considerable small-scale industry, including aluminum utensils, steel furniture, fruit canning and saw milling, has come up during the recent years. Tea processing, jute textiles and fruit concentrates are other industries. A remote location and transport problems handicap the industrial growth. Discovery of natural gas is expected to give a boost to industry.

Only one-seventh of the total population is urban. Most of the towns are administrative centers and are confined to the western plain. Agartala—the state capital—is the biggest. Nucleated around a palace, the city is laid out open, is sprinkled with coconut groves, and is delightfully scenic. Its population of 157,636 is more than one-third of the state's urban population.

The state is politically volatile. Rivalry between the Communist Party and the Indian National Congress is bitter. In that sense, the state is a mini-transplantation of West Bengal.

## Meghalaya

Literally meaning the "abode of clouds", this is the wettest part of India. It bears the first frontal onslaught of the monsoons. Rainfall exceeds 1000 cm at places; it reaches the maximum of 1203 cm at Cherapunji.

Meghalaya was carved as a separate state in 1970 out of the state of Assam. This was a political solution to a persistent demand for statehood by the native tribals—Garos, Khasis and Jaintias—who constitute more than four-fifths of the total population. The state has a population of 1.8 million and area over 22,429 km$^2$.

In geographic literature, Meghalaya is referred to as the Shillong plateau. It is a single massif composed of the Garo, Khasi and Jaintia hills which get their names from the three main tribes who inhabit its different parts. The hills range in elevation from 600m. to more than 1800 m and have a radial pattern of drainage. Umiam-Barapani is the major stream. Its waters supplies hydroelectric power to both Meghalaya and Assam.

About 40 per cent of the total area is forested; 10 per cent is under cultivation, with the remaining falling largely in the category of uncultivable land characterized by rocky and rugged terrain. The insect eating "pitcher plant" is found only in this part of the world.

Shifting cultivation is more typical of the Garo hills and terraced cultivation of the Khasi and Jaintia hills. Rice occupies 60 per cent of the cropped area, followed by potato (15 per cent) and maize (12 per cent). Fruits like orange, pineapple and plum are grown extensively. Of special interest is the cultivation of arecanut which accounts for 6 per cent of the cropped area. Cultivation of tea, mushrooms and tomato has prospered recently.

Although rich in minerals like coal, sillimanite, limestone and glass sand, the region attracted recently modern industry. Shillong, the state capital town and previously the seat of power of the entire old Assam, has industry associated with manufacturing of electronics, plywood and watches. Cherapunji has a cement factory. The traditional household industries of weaving (including silken fabrics), blacksmithy (making of spade-hoes, spears, fishing hooks and axes), and basket/mat making are ubiquitous.

Shillong derives prominence from its status as a health resort, educational center, and historic stronghold of the Christian missionary activity. At an altitude of 1,500 m, its June temperatures average 21°C and January temperatures of 10° C. With a population of 222,273, it is the biggest hill station in the northeast.

A special sociological feature of the region is the general adherence to matriarchical system amongst the tribes. Family property is owned by mother and is inherited by one of the daughters, generally the youngest one. Traditionally animists by belief, now a majority of population is Christian.

## Assam Valley

Drained by the Brahmaputra river and its numerous tributaries, this region is the heartland of the entire northeastern India. Since 1826, it has been the focus of British expansion within and onto the surrounding tribal hill areas. After Independence, the peripheral tribal territories were gradually organized into four different states of Nagaland, Meghalaya, Mizoram and Arunachal Pradesh. The core emerged as a linguistically homogeneous state of Assam. Now it is reduced to an area of 78,438 km$^2$, only one-third of the original size.

The region has been a scene of intense political turmoil since 1979. Gripped by the fear of getting outnumbered by the illegal influx of aliens, mainly Bangladeshis and also Nepalis, the All Assam Students Union launched a mass movement demanding ouster of the foreigners. In 1985, it reached an agreement with the Government of India which stipulated that all foreigners who came to Assam on or after March 25, 1971 were to be detected and expelled and those who arrived between January 1, 1966 and March 24, 1971 were to be deleted from the voting rolls. This restored the political normalcy but it was short-lived. A demand for a separate Bodoland, within the valley, took a serious form. The Bodos are the plains' tribals. They are agitated over the transfer of their land to Assamese caste-Hindus, immigrant Muslims from Bangladesh, and migrants from other parts of India.

Roots of the problem are embedded in the massive inmigration of people with diverse ethnic backgrounds. Till early nineteenth century, the Assam valley was a land with sparse population—a demographic anomaly keeping in view its fertile soils, abundant water and, above all, proximity to the crowded Ganga delta. Any stable or prosperous settlement was prone to attack by the hill tribes. The valley remained

India's *lebensraum*. With the establishment of peaceful conditions under the British, tea plantations were laid which attracted labor from Bihar, Orissa and Uttar Pradesh. The migration of the Bengali Muslims onto the agricultural wastelands also started and assumed large proportions in the early decades of the present century. The Nepalese were also moving in, especially for cattle raising. The educated Hindu Bengalis found employment in government service and various other professions. The process continued after Independence. Things became alarming when infiltration from East Pakistan, now Bangladesh, remained unabated. The population of Assam multiplied 6.5 times from 3.4 million in 1901 to 22.3 million in 1991, whereas India's population grew by 3.5 times during the same period. The Assamese got worried over the prospect of getting reduced to a minority in their own home. They reacted sharply. It is worth stressing that political turmoil in the northeast has been typical not only of the Christian states like Nagaland, Mizoram and Meghalaya but also of the Hindu areas like Assam, Manipur valley and Tripura hills.

The valley is essentially an elongated structural trough, roughly 800 km by 80 km. It has a gentle gradient associated with a small fall in altitude from 130 m to 30 m. The river course is braided and builds several islands among which Majuli, with an area of 929 km$^2$, is the largest in the world. There are hillocks and monadnocks on both sides of the river give some diversity to relief.

Due to humid climate, Assam has nearly one-third of its total area under forest. The Assam teak is famous for its quality. Considerable area is occupied by water bodies, marshes and swamps. Cultivated area shrinks to around 40 per cent of the total.

Rice is the principal crop occupying more than 70 per cent of the cropped area. Tea plantations are frequent in the upper part of the valley; more than a half of India's tea comes from Assam. Jute is an important cash crop in the lower part of the valley. Tea and jute account for 7 per cent of the cropped area each. Oilseeds, pulses and sugarcane are the other notable crops.

Till the discovery of oil in the Gulf of Cambay in Gujarat and at the Bombay High in the Arabian Sea, Assam used to be its main supplier in India. Digboi, Moran and Naharkatiya are important among the oil fields. Now, only one-third of the oil drilled in the country comes from Assam.

Assamese women are expert at weaving, particularly in producing a special variety of cloth for social and religious occasions. A marriageable girl is expected to be expert in weaving. Every Assamese household must have a loom. The modern industry is largely agro-based and includes tea processing, jute textiles and food processing. Plywood industry is a handmaid of the tea processing industry. Petrochemicals industry has been facilitated by the availability of oil. A large part of the industry is concentrated around Dibrugarh in the upper part of the valley. Sibsagar in the middle, and Guwahati in the lower section. Guwahati, with its newly planned suburb of Dispur, is not only a capital city but also acts as a regional center for the entire northeast. It is a fast growing transport and educational center. It has a population of 577,591.

Though nature has been generous to Assam, the region remains one of the less developed parts of India. Per capita income here is less than that of the national average. An optimal use has not been made of the available resources—a reflection on the political situation prevailing in the state. Its peripheral location is another handicap. At the same time, the negative role of annual floods, on a disastrous scale, cannot be ignored. While for most parts of India, the scarcity of water is a serious problem, for Assam this problem is in reverse—how to get rid of the excess water.

## *References*

Baker, J.N.L., "Notes on Natural Regions of India", *Geography*, 14, 1928, pp. 447-55.

Cassen, R.H., *India: Population, Economy, Society*. London, 1980.

Census of India 1981, *Census Atlas: National Volume*. New Delhi, 1988.

Census of India 1991, *Provisional Population Tables, Series 1, Papers 1, 2 and 3 of 1991*. New Delhi, 1991.

Center for Science and Technology, *The State of India's Environment*. New Delhi, 1985.

Crane, R.I. (ed.), *Regions and Regionalism in South Asian Studies*. Duke University Program in Comparative Studies in Southern Asia, 1967.

Ginsburg, N.S. *et al*, *The Pattern of Asia*. London, 1958.

Government of India, *Report of the States Reorganization Commission*. New Delhi, 1955.
Government of India, *India: 1990, A Reference Annual*. New Delhi, 1991.
Muthiah, S. *et al* (eds.), *A Social and Economic Atlas of India*. New Delhi, 1987.
Schwartzberg, J.E. (ed.), *A Historical Atlas of South Asia*. Chicago, 1978.
Singh, B.P., *The Problem of Change: A Study of North-East India*. Bombay, 1987.
Singh, R.L. *et al*, *India: A Regional Geography*. Varanasi, 1971.
Spate, O.H.K., *India and Pakistan: A General and Regional Geography*. London, 1954.
Stamp, L.D., "The Natural Regions of India", *Geography*, 14, 1928, pp. 502-6.
Weiner, M., *Sons of the Soil: Migration and Ethnic Conflict in India*. Bombay, 1978.

# 15

# The Northern Plains

Spread over an area of 677,638 km$^2$ and inhabited by 314 million people, the region is among the most extensive and crowded plains of the world. Its population is larger than that of the United States; its density exceeds that of the Netherlands. It looks like a skewed crescent transverse to that of the Himalayas.

The Ganga, the sacred and ancient river of the Hindu scriptures, has been adored as a source of material prosperity and spiritual salvation by most Indians. This master stream, together with its tributaries, such as the Yamuna, Gomati, Ghaghara, Gandak, Kosi and Mahananda, built a vast alluvial plain between the Himalayas and the peninsula. To the west, the plain merges into another extensive alluvial tract laid by the Indus and its tributaries, namely, Satluj, Beas, Ravi, Chenab and Jhelum, part of which lies in Pakistan. The entire area is known as the Indo-Gangetic plain of the Indian subcontinent. The Northern plains are the larger and eastern segment of this twin plain.

The Northern plains have been an active theater of Indian history and represent India's "heartland". It received a variety of people in successive waves through the western frontiers, witnessed the rise and fall of several empires, gave birth or admitted major religions/sects, and synthesized all this in a composite Indian culture. The Grand Trunk Road, from Calcutta through Delhi to Amritsar, and beyond to Peshawar in Pakistan, has always been a major route to India. Now, the Northern plains are largely agricultural, rural, and thickly populated, with a transition toward industrialization and urbanization. Much of it is mineral-poor; parts of it are subject to occasional flooding or droughts.

The Northern plains are disposed to a two-fold broad division: the Ganga plains and the Trans-Ganga plains with the Yamuna as the divide. The former is the main component of the Ganga basin, and the latter a part of the Indus basin. The former is largely tropical sub-humid to humid; the latter arid to semi-arid. Paradoxically, the Trans-Ganga plains are agriculturally more advanced than the Ganga plains; thanks to its superior management of water through irrigation.

The Ganga plains are further divisible into three sub-regions: (i) Upper Ganga plain, (ii) Middle Ganga plain, and (iii) Lower Ganga plain. Roughly, the 150 m contour is the dividing line between the Upper and Middle Ganga plains, and the 50 m contour between the Middle and Lower Ganga plains. The three are distinguished by the language or dialect spoken, agricultural pattern, and political orientation. Though, both Upper and Middle Ganga plains are Hindi-speaking, but they differ in their dialects: *Khariboli* and *Brajbhasa* in the former and *Avadhi*, *Bhojpuri* and *Maithili* in the latter. The Upper Ganga plain has irrigation-based commercial agriculture in wheat and sugarcane; the Middle Ganga plain is largely subsistence in rice; and the Lower Ganga plain has an agro-industrial economy in rice and jute, focusing on Calcutta. The political sociology also offers a comparison. The Upper Ganga plain has a strong pro-farmer lobby; the Middle Ganga plain is sensitively divided between the forward and backward castes; and the Lower Ganga plain is politically oriented toward radical ideology.

Likewise, the Trans-Ganga plains are also divisible into three regions: (i) Punjab, (ii) Haryana, and (iii) Rajasthan desert. Punjab is famous for progressive agriculture; it is a Punjabi-speaking area. Haryana, earlier a part of Punjab, resembles the parent state in many ways but population here is Hindi-speaking. The Rajasthan desert derives its name primarily from a dry climate.

## Ganga Plains

### *Upper Ganga Plain*

Occupying the western part of Uttar Pradesh, the Upper Ganga plain

**Figure 15.1**
*Ganga Plains*

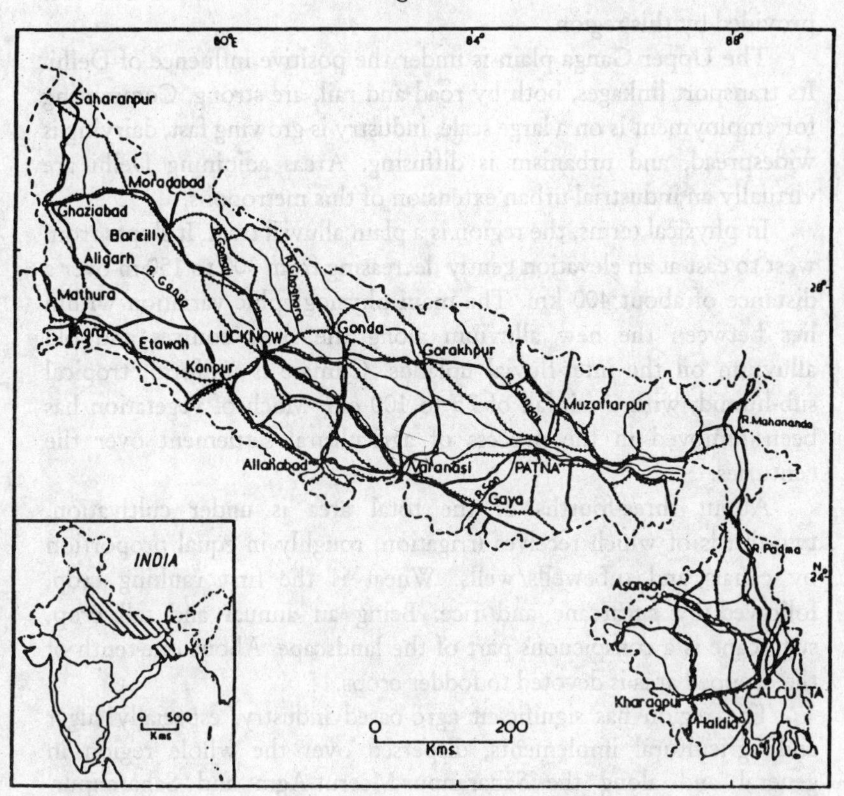


is agriculturally the most developed of the three segments of the Ganga plains (Figure 15.1). This is attributed to an early extension of canal irrigation during the later half of the nineteenth century and rapid expansion of tubewell irrigation after Independence. Railways, electrification and sugar industry had also arrived here by the beginning of the present century. Starting in mid-1960s, the Green Revolution, associated with the high yielding variety (HYV) of wheat, has made an impressive headway.

The ethos of commercial agriculture, coupled with the proximity to the national capital of Delhi, has given rise to an assertive farming community. It is highly mobilized for getting remunerative support prices fixed for their agricultural products, and is sensitive to any failure in supply of agricultural inputs, such as fertilizers or power for tubewells. A committed pro-farmer leadership at the national level is provided by this region.

The Upper Ganga plain is under the positive influence of Delhi. Its transport linkages, both by road and rail, are strong. Commuting for employment is on a large scale, industry is growing fast, dairying is widespread, and urbanism is diffusing. Areas adjoining Delhi are virtually an industrial-urban extension of this metropolis.

In physical terms, the region is a plain alluvial tract. It slopes from west to east at an elevation gently decreasing from 300 to 150 m over a distance of about 400 km. The main physiographic variation within lies between the new alluvium along the river courses and old alluvium on the inter-fluvial uplands. Climate is uniform: tropical sub-humid, with a rainfall of 75 to 100 cm. Much of vegetation has been removed in the process of agricultural settlement over the centuries.

About three-fourths of the total area is under cultivation, two-thirds of which receives irrigation, roughly in equal proportion by canals and tubewells/wells. Wheat is the first ranking crop, followed by sugarcane and rice. Being an annual and tall crop, sugarcane is a conspicuous part of the landscape. About one-tenth of the cropped area is devoted to fodder crops.

The region has significant agro-based industry, especially sugar and agricultural implements, dispersed over the whole region in general and along the Saharanpur-Meerut-Agra and Saharanpur-Moradabad-Bareilly rail routes in particular. An impressive concentration of modern industry, such as engineering goods, electronics, and chemicals has taken place at Ghaziabad in the neighborhood of Delhi. Mathura has a big oil refinery, Aligarh is famous for its locks, and Agra has an export-oriented footwear industry.

It is no surprise that the region with a long history of settlement, productive agriculture, and growing industry is densely populated: 52 million over an area of 89,876 km$^2$, giving a density of almost 600. About one-fourth of the total population is urban. Agra (955,694) is the city of the world famous Taj Mahal, a monument in marble to conjugal love. Meerut (846,954) is a cantonment city and focus of the Hindu Jat culture. Aligarh (479,978) is the seat of a Muslim University.

### *Middle Ganga Plain*

Including the eastern part of the Uttar Pradesh and northern section of Bihar, the Middle Ganga plain presents a contrast to the Upper Ganga plain in terms of socio-economic development. It is a thickly populated, overwhelmingly rural and agricultural, and poverty-stricken part of India. The per capita income here is around one-half of the national average, no less than 90 per cent of the population is rural, and nearly the same proportion of rural population is in agriculture. Almost two-thirds of the population is illiterate. All this fits into the description of this region as a "rural slum".

Such a scenario is a reversal of the historical eminence the region enjoyed as a "granary of India" before the colonial rule. A sequence of events happened which were lethal—economically and socially. The permanent land settlement, introduced by Cornwallis (1793-98), laid the foundation of an institutionalized landlordism under which the society got sharply divided between a small class of absentee landlords and a large mass of tenants. Conditions for progressive agriculture, based on personal initiative, disappeared. In addition, the region, which was a major source of recruitment to the British army, earned the wrath of the British rulers for participating in the uprising of 1857. As a penal measure, it was subjected to a gross neglect. At the time of Independence, it inherited not only a poor economy but also a society divided between the few forward-looking landowning class and the many backward landless castes. Today, the region, especially its Bihar component, is marked by an explosive caste division.

Physiographically, the Middle Ganga plain is a flat gently sloping area from 150 m in the west to 50 m in the east. It is drained by the Ganga and its numerous tributaries, among which Ghaghara, Gandak and Kosi are the major ones. The Mahananda river defines its eastern boundary.

The region is rich in river waters. Rainfall is also fairly high, ranging from 100 cm in the west to 150 cm in the east. Most of the rainfall is concentrated during the monsoon months of June to September. It causes frequent floods which are a bane of this area. Deforestation in the upper reaches of its various rivers has aggravated the situation.

About two-thirds of the total area is under cultivation. Nearly one-third of it is irrigated by canals and wells/tubewells. Intra-regional variations in irrigation are sharp. Canal irrigated areas are much better placed. Rice ranks as the first crop, accounting for about 40 per cent of the cropped area. Wheat, barley, maize and pulses are other major foodcrops. Sugarcane is the major cash crop. Jute is coming up in more humid parts. Vegetables are grown along the floodplains.

Industrial development is at a low level. Existing industries are based primarily on agricultural produce, such as sugarcane, rice and oilseeds. Kanpur has been an important center of cotton textiles and leather products since the British days. After Independence, it attracted significant public sector industry, such as aircrafts, chemicals and engineering goods. Establishment of public sector industry at some other places also deserves a mention: diesel locomotives at Varanasi, telephones, pumps and compressors at Naini near Allahabad, digital systems and scooters at Lucknow, oil refinery at Barauni, and railway wagons at Muzaffarpur. The Uttar Pradesh component of the region is more developed than its Bihar counterpart.

Problems of the region, having a population of 136 million on 206,851 km$^2$, are colossal. With population densities approaching 700, and most of the people living in rural areas and depending on agriculture, pressure on land is intense. Landholdings are pitiably small: just 0.87 hectare or around two acres on an average. This impelled massive outmigration. Migrants from here are found not only in the metropolitan cities like Calcutta, Mumbai and Delhi but also in far-flung tribal hill states of Mizoram, Nagaland, and Arunachal Pradesh. They are engaged mostly in factory, construction, or manual labor jobs. Their remittances back home are hardly enough to meet the basic needs of the families left behind. Productive investments at their native places are meager.

Ironically, this poverty-stricken region is rich in its historical and cultural heritage. Patna, the capital of Bihar, is an ancient city, spread along the southern bank of the Ganga. It is a gateway to the Buddhist centers of Vaishali, Rajgir and Bodhgaya. With a population of 1,098,572, it is the only metropolitan city of Bihar today. Among the cities in the eastern Uttar Pradesh, Varanasi (1,026,467) is a holy city situated on the bank of the Ganga. It is said to be the oldest *living city* in the world. Allahabad (858,213), on the confluence of the Ganga and Yamuna, is famous for its *Kumbh Mela* (bathing festival) which is celebrated every twelve years. It may cause a congregation of as many as 15 million pilgrims at this occasion. Lucknow (164,234), the capital of Uttar Pradesh, was the seat of power of the *Nawabs* (princes) of Avadh before it was taken over by the British in 1856. It is famed for sophistication of mannerism and conversation. Kanpur (2,111,284) is, of course, the biggest city of Uttar Pradesh. It originated as a cantonment, railroad center, and a commercial manufacturing town during colonial days. Now, it is the biggest industrial concentration between Delhi and Calcutta.

### *Lower Ganga Plain*

Coinciding largely with the Indian part of the Ganga delta, this region is a Bengali-speaking area. Calcutta is the focus of culture and economy. Its political distinctiveness lies in a sustained adherence to radicalism. The record of land reforms of the left-wing government, which has been in power for the last two decades, is creditable. The social distortion of a wide gap between the few big landlords and masses of peasantry, as typical of the Middle Ganga plain, has been corrected here to a large extent.

The path to radicalism was not easy. Initially it found an aggressive expression in the form of Nexalite movement which was led by educated, unemployed youth. Life was made difficult for the rich, including the big landlords and affluent businessmen. Industrialists, hailing from other parts of India, felt insecure in particular. The movement was successfully contained, and with political maturity taking shape, an era of stability and progress has firmly set in.

The region is a monotonous flat river plain and a delta, traversed by the Ganga (known here as Padma, branching into Bhagirathi and Hooghly lower down), and other streams, such as the Damodar, Rupnarayan and Subarnrekha, which originate from the adjacent Chota Nagpur plateau. Climate is humid maritime due to proximity of sea. Rainfall is high, between 150 and 200 cm, and the range of annual temperature is small, with summer mean temperature of 30° C and winter mean temperature of 25° C.

The original forest cover has been removed over large parts to make way for agriculture and industrial-urban establishments. The Sundarbans, the tidal forests along the mouth of the Ganga, are being preserved. The western plateau margins contain some forest.

The region is distinguished by an agro-industrial economy. The entire system focuses on Calcutta. This place has grown as a linear industrial conurbation along the Hooghly. A variety of agro-based, manufacturing such as jute textiles, cotton textiles, tea packing and rice processing; forest-based products, such as paper and plywood boxes; and livestock-based items, such as shoe making and leather tanning industries are concentrated here. Besides, a number of engineering, transport, heavy machinery, electricals and electronics industries are also located. Another industrial concentration, designed as a countermagnet to Calcutta, has come up along the Damodar river in the northwest, with Asansol, Chittaranjan and Durgapur as the major constituents. Iron and steel, locomotives and fertilizers are the main products. In addition, Haldia has been developed as India's biggest petrochemicals complex to the southwest of Calcutta. Public sector investment in industry is large in the Lower Ganga plain.

Nevertheless agriculture remains the livelihood of a majority of the population. Nearly two-thirds of the total area is under cultivation, and of it two-thirds is under rice cultivation. Jute is a prominent commercial crop. Vegetables and fruits are grown on a large scale to meet the demands of the big urban market.

The pressure of population on land is intense. A population of 62 million on 75,989 km$^2$ gives a density that exceeds 800—the highest for any part of India. About one-third of the population lives in towns and cities. The region is among the more urbanized parts of the country. Calcutta dominates. It had a colonial origin. By virtue of its status as the imperial capital of the British India till 1911, and the provincial/state capital thereafter, it has now grown into a megacity. Its population of 10,916,272 is almost the same as that of Beijing—the capital of China. The city is an agglomeration of as many as 129 municipal bodies. Located close to this city, and set in scenic surroundings, is Shantinekatan. This place houses the cultural university of Vishwabharti founded by the nobel laureate Rabindranath Tagore.

Most paradoxical is the continued inmigration to this crowded land. The region is sandwiched between two areas of acute distress—the Middle Ganga plain to the west and the Bangladesh to the east. A large part of the influx originates from here.

There has been sustained outmigration as well, both before and after Independence. This involved mainly the educated Bengalis. They moved to a number of north Indian cities, such as Guwahati, Shillong, Patna, Varanasi, Allahabad, Kanpur, Lucknow and Delhi, as teachers, engineers, doctors and government employees. An early exposure to the new educational system, introduced by the British, had produced surplus pool of Bengali intelligentsia, who migrated to the industrial towns of the Chota Nagpur plateau after Independence.

On the whole, the inflow has been bigger than the outflow. The region's population growth rate has been higher than the rate of natural increase. The specter of multiplying numbers is frightening. The federal government has come forward to provide help by way of establishing several public sector industries and rehabilitating refugees from Bangladesh in other parts of India. But that touches only a fringe of the problem.

## Trans-Ganga Plains

### *Punjab*

This traditional land of five (*Punj*) rivers (*ab*) has been subjected to several territorial reorganizations since Independence. It assumed the present form in 1966. In the process, it became demographically more homogeneous but reduced to one-seventh of its area size before Independence. It is the only Sikh majority state of India: 60 per cent Sikh, 37 per cent Hindu and remaining few adhering to other religious persuasions. About 85 per cent of the people are Punjabi-speaking. It has a population of 20 million on 50,362 km$^2$. Almost 30 per cent of the population is urban.

Although inland by location, the region has behaved as a coastal state. Emigration from here has been continuing for the last about hundred years. Punjabis are found in large numbers in the United Kingdom, United States and the East African countries. An outward orientation on their part is manifest also in a long tradition in army service. Their share in defense forces is five times of that in total population. They outmigrated in large numbers also to newly colonized agricultural lands in other parts of India. It is anomalous that Punjab, with among the highest per capita incomes in the country, should be an area of net outmigration. Outmigration from

**Figure 15.2**
*Trans-Ganga Plains*

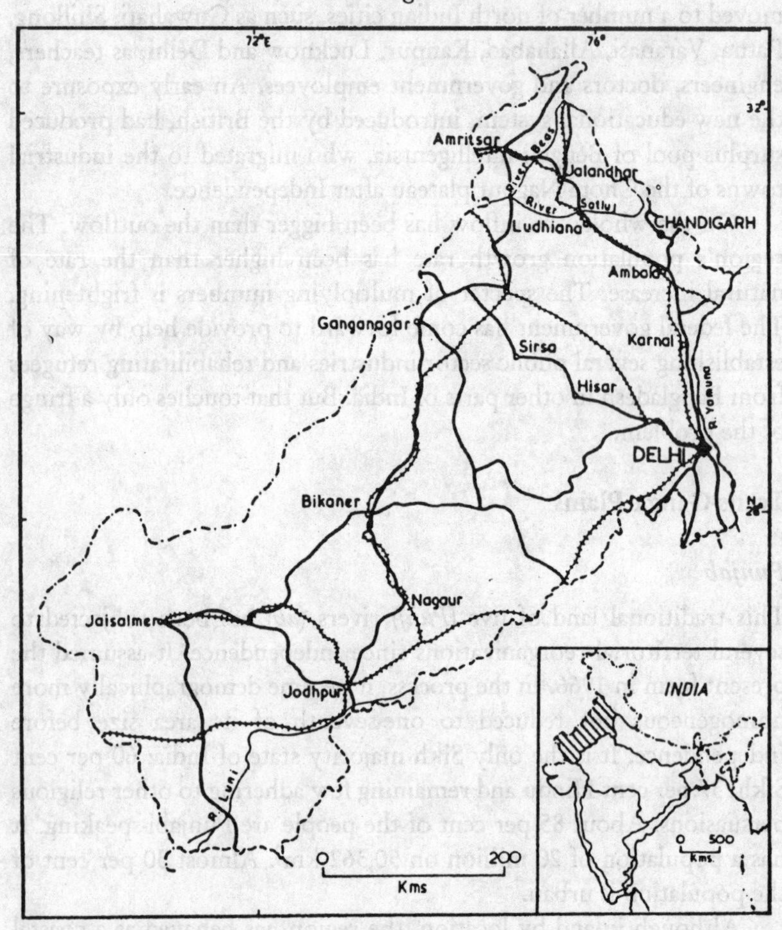


here is rooted not in distress but is stimulated by considerations of higher levels of prosperity.

Punjab is distinguished for the success it made in the Green Revolution. Agriculture is highly commercialized and prosperous. It is the largest contributor to the national pool of wheat and rice for the public distribution system. Its per capita income, next only to that of Goa among the Indian states, is nearly double of the national average.

During the last two decades, however, the accomplishments in agriculture were not matched by a parallel development in industry.

The economic growth rate gradually reached a plateau level. There was a limit to what could be accomplished on a sustained basis through agriculture. Concern about even maintaining the achieved level of agricultural productivity became serious. Meanwhile, unemployment among the educated rural youth was rising. All such economic distortions caused discontentment and degenerated the political situation. The economic issues gradually assumed religious complexion. The Sikhs attributed the evolving scene to discrimination against them on the part of the federal government including in matters of establishment of public sector industry. They sought greater autonomy, and the demand so became militant. Fortunately, the two main communities in Punjab—the Sikhs and the Hindus—at large do not bear any emotional divide, and the recent violent activity is restricted mainly to small extremist groups.

The region is mostly a flat, featureless plain. Nine-tenths of it is at an elevation ranging from 300 m in the northeast to 180 m in the southwest. The Siwalik hills, of 300 to 1,000 m in height, make a distinct appearance all along the northeastern border. Sand-dunes are sporadically distributed in the southwest.

The Ravi, the Beas and the Satluj are the perennial streams. Their waters are most critical to the Punjab agriculture. The Bhakra-Nangal project for irrigation and power on the Satluj has transformed the economy not only of this region but also of adjoining areas in the neighboring states. The Indira Gandhi canal in Rajasthan, one of the largest in the world, has been taken out from the Beas-Satluj confluence at Harike. Its water is reserved for irrigation in the Rajasthan desert.

The region has a continental semi-arid to sub-humid climate. Rainfall decreases from around 100 cm in the northeast to less than 25 cm in the southwest. Irrigation is indispensable for good agriculture. This challenge has been admirably met. About 89 per cent of the cultivated area is irrigated: the highest percentage for any Indian state. In Ludhiana and Amritsar districts, virtually the entire cultivated area receives irrigation. On the whole, tubewells take care of about 60 per cent of the irrigated area and canals of about 40 per cent.

Cultivated area is as much as 85 per cent of the total. Three-fourths of it is sown more than once in a year. As a result, the cropped area is 1.5 times of the region's total area. Wheat is the king crop sharing 40 per cent of the cropped area. Rice accounts for 25 per cent. Both are cash crops. Cotton is grown over 10 per cent of the cropped area. Vegetables, mainly potato and onion, occupy 5 per cent of the cropped area. Ecologically, rice is not a crop of this largely semi-arid region. It demands heavy irrigation under conditions of highly inadequate rainfall. Several rice producing areas depend upon tubewell irrigation. Excessive withdrawal of sub-surface water has caused an alarming decline in the water-table. Agricultural prosperity of the region carries a heavy ecological cost.

Punjab's performance in industry does not compare well with its success in agriculture. The region was a priority area for agriculture at the national level but not for industry. The pace of industrialization has been steady rather than rapid. Recent policies aim at a spurt to industry. Gains in industrial development would have been greater if the state had not faced the current political turmoil and associated feeling of insecurity.

Much of the industry in the region is agro-based, processing agricultural raw materials, or agro-oriented, producing agricultural inputs, or agro-income beneficiary, responding to a lucrative rural market. Food products, textiles, sugar, fertilizers, agricultural implements, rerolled steel, machine tools, sports goods, furniture, transport vehicles, and electronics rank high in the inventory of industrial products. Most of the important industrial centers are located on the Amritsar-Delhi railway line. The industrial city of Ludhiana is the biggest city, followed by Amritsar and Jalandhar. Mohali near Chandigarh is coming up as an important center for the manufacture of electronics.

The region is heading toward a sound, agro-industrial economy. Rural-urban linkages are strong. Practically, every village is electrified and connected with a paved road. Several central villages have acquired urban functions of an agricultural market center, a health center and a college. Towns and cities are closely spaced and rural-urban commuting is common. In contrast to the general pattern in India, the average assets of a rural household are two times of those of its urban counterpart, and the incidence of rural poverty is less than that of urban poverty.

Punjab needs to be visited. It is a model for what an Indian state could achieve through its agricultural transformation, rural-urban integration, and enterprise of the people. Amritsar is the most holy city for the Sikhs. Its golden temple is visited by devotees from all over the world. With a population of 709,456, it is now Punjab's second largest city. Ludhiana, with a population of 1,012,062, is the only metropolitan city in this part of India. The Punjab Agricultural University located here has made a significant contribution to agricultural advancement of the region. Patiala, a city of a quarter-million, still retains some of the glory of the past when it was a seat of a princely state.

One of the by-products of the Punjab's reorganization on linguistic lines in 1966 was the formation of the union territory of Chandigarh. The territory is virtually conterminous with the city. It was the capital of the joint Punjab; now it functions as a triple capital: of Punjab, Haryana and the union territory. Chandigarh is a planned city. This horizontally laid city was designed by the same grand master, Le Corbusier, who had influenced the planning of the vertical city of Brasilia. It is a great experiment in city design; a Mecca for every student of architecture. The city is visualized as a living organism. It is divided into rectangular sectors, generally 1,200 by 800 m, which are self-contained neighborhood units. All the sectors are integrated amongst themselves and within through a well-conceived road system.

Chandigarh, the "city beautiful" is an international attraction. Its capitol complex, consisting of the secretariat, the legislative assembly, and high court, represents the three wings of the government. The Punjab University campus is a beautifully landscaped area. The "Rock Garden" is an art fantasy. Here garbage, consisting of broken pieces of chinaware, fused tubes, bottle tops and all such waste material, is shaped into dolls, animals, birds and what not, creating an altogether delightful scene. Independent India had some other planned state capitals also; Bhubaneswar and Gandhinagar to name two. But these do not match Chandigarh in elegance and quality of life. Part of the credit for this goes to its meticulous planning, and part to a liberal financial support from the union government. The city, founded in 1952, is a vibrant city of 574,646 (in 1991).

### *Haryana*

This state was carved out of the former Punjab in 1966. This was in recognition of its language-based cultural identity. It was the Hindi-speaking part of old Punjab. Haryana has played a crucial role in Indian history. The region witnessed the evolution of early Aryan civilization, served as an arena for the three battles of Panipat which decided the fate of several empires at Delhi, and played an active role in the 1857 uprising against the colonial rule. The Haryana territory was made a peripheral part of the Punjab and was subjected to neglect.

The region today is, however, the most progressive among the Hindi-speaking states of India. Its per capita income is next to that of Goa and Punjab. Several factors favored this. The extension of canal irrigation in the northwestern and eastern parts laid a foundation for the Green Revolution. Proximity to the national capital of Delhi provided a thrust to the industrial and urban development. A speedy provision of physical infrastructure, subsequent to attainment of statehood, transformed its life and economy. By 1971, the state had electrified all its villages; by 1975, every village had been linked by a metalled road; and by 1985, 85 per cent of the problem villages had been provided with safe drinking water.

Lying between the Yamuna to the east and the Ghaggar rivers in the north, the region is essentially a water-divide between the Ganga and Indus river systems. It is mainly a plain tract at an elevation of 200 to 300 m. The Siwalik hills in the northeast, the Aravalli outliers in the south, and sand-dune sprinkled topography in the southwest render some variety to its physical layout.

Inland location and proximity to the Rajasthan desert give the region a continental semi-arid climate. Rainfall ranges from 25 to 50 cm in the western half, and 50 to 100 cm in the eastern counterpart. This is far too inadequate under tropical conditions. The scarcity of water is accentuated by a lack of surface drainage. The only perennial stream of the Yamuna marks the eastern boundary; the Ghaggar in the north is a misfit stream with hardly any water during summer. Nor is the underground water much obliging. The sub-soil water is brackish or saline in the central and western parts which account for almost two-thirds of the total area. This discourages tubewell irrigation. Scarcity of water is the region's foremost problem.

Despite these physical constraints, the region has made tremendous progress in agriculture with the help of canal and tubewell irrigation. As much as 86 per cent of the total area is cultivated, of which 60 per cent is irrigated. Canals and tubewells share the irrigated area in almost equal proportion.

Four-fifths of the cropped area is devoted to agriculture. Wheat is the first ranking crop over a large part of the region. Gram predominates in the west and bulrush millet in the southwest. Rice, like wheat, is an important commercial crop. Its cultivation is concentrated mainly in the tubewell irrigated northeastern part of the state.

Haryana is famous for the quality breed of livestock. It is an important exporter of cattle to other parts of India. Dairying has developed on a large scale. Production of milk has been stimulated by an expanding urban market.

The region occupies an important position on the industrial map of India. The proximity of Delhi is most helpful. Equally creditable are the efforts of the Haryana government in offering several incentives to new industry. Industrial peace has also been instrumental in attracting considerable industry from other states, particularly the neighboring Punjab.

Faridabad is an industrial hub, producing a wide range of goods—tractors, motor cycles, electronics, scientific instruments, and beverages. Other places near Delhi, such as Gurgaon, Sonipat and Panipat have also grown into important industrial centers. The Maruti Udyog, in collaboration with Japan's Suzuki, at Gurgaon has brought a kind of car revolution in Indian cities. Sonipat manufactures one-fifth of the bicycles in the country. Panipat supplies 75 per cent of the Indian army's requirements for woolen blankets. Located in the region's northeast, Yamunanagar meets 60 per cent of the demand for ammunition boxes. Ambala accounts for one-third of the nation's exports of scientific instruments.

Reclamation of agricultural land with the help of the Bhakra canal irrigation system, and rapid industrialization have induced sizable migration to the region since Independence. In the process, it was transformed from a relatively low to a high density area. Its population of 16 million in 1991 over an area of 44,212 km$^2$ gives a density of 369, significantly higher than India's. About one-fourth of the total population is urban. Faridabad, with a population of 613,828, is the biggest city.

An outstanding achievement of Haryana lies in the field of

tourism despite the virtual absence of any scenic landscape. Tourist resorts have been developed by way of reviving the glory of historic or religious sites and by creating artificial lakes close to the big towns located on the national highways. Pinjore Gardens (Chandigarh), Surajkund (Delhi), Sohna (Gurgaon) and Karna Lake (Karnal) can be listed among the major tourist attractions. In addition, Kurukshetra is an ancient religious place of historic importance.

To the southeast of Haryana is located the union territory of Delhi which encloses the national capital Delhi. Covering an area of 1,483 km$^2$, it has a population of 9.4 million. Delhi is a huge fast expanding metropolis in the union territory.

Seemingly eccentric in location in relation to the physical disposition of India, Delhi held a strategic position in the historic context. It was a natural capital site in the heart of the Indo-Gangetic plain, located close to the Himalayas, the peninsular India and the Rajasthan desert. It remained the capital of India for centuries. The British shifted their capital from Calcutta to this place in 1911 and built a planned city of New Delhi as an adjunct to old Delhi. The post-Independence India found this place as the most suitable for national administration.

Delhi has grown explosively over the recent decades. Its population increased from 1.7 million in 1951 to 8.4 million in 1991, nearly three times the population of Washington, D.C. There has been phenomenal proliferation of administrative, industrial and commercial functions. The number of government employees exceeds half a million; industrial employment approaches three-quarters of a million. Important industries include machine tools, autoparts, bicycles, textiles, plastic items, rubber manufactures, and a variety of daily consumption goods.

An unabated growth of Delhi has become a matter of great worry. Its built-up area is now almost four times of that in 1947. Rate of population growth has been very high (8-9 per cent a year) causing a heaving strain on housing, transport, water supply, electricity and other services. The "National Capital Region" was constituted in 1985 to consider ways to deal with the city's explosive problems. It covers the union territory of Delhi as a whole and its contiguous area in Haryana, Uttar Pradesh and Rajasthan (to a distance of around 100 km). Implicit in this spatial strategy was dispersing the urban growth from the core to the periphery. This experiment in inter-state cooperation seeks to achieve a manageable Delhi on the one hand, and a harmonized and balanced development of its region on the other.

### *Rajasthan Desert*

The region is unique among various parts of India being a hot desert where summers are scorching (noon time maxim approaching 50°C), rain scarce (10 to 26 cm), and sand-dunes pervasive (over 95 per cent of the land). The entire area was under the rule of princes before Independence, leaving a heritage of magnificent palaces, fortified cities, and splendid temples, as also of social conservatism and economic backwardness. It has an international border with Pakistan for about 1,200 km. This has made it necessary to station defense forces in large numbers, construct roads, and facilitate free movement of supplies.

The desert is a plain overlain by sand-dunes of various types, longitudinal, transverse and crescent. Relief has a range of 150 m in the west to 300 m in the east. There is evidence to suggest that the region had a thick cover of forest at the beginning of the Christian era. The subsequent process of erosion caused by disruptions in the drainage system and gradual deforestation led to its degeneration into a desert.

The surface water is virtually absent, except in the valleys of Ghaggar, Luni and Sukhani streams. Water-table is generally deeper than 100 m. Rainfall is highly variable. Any failure of rains is followed a triple famine of water, fodder and food.

On an area of 208,751 km$^2$, the desert has a population of 17 million. In area it is two-thirds of Arizona but carries a population which is five times as large. Its density of 80 is higher than that of the United States. This is ascribed partly to a long history of settlement and partly the recent settlement of newly reclaimed agricultural lands.

The introduction of irrigation from the Gang canal in the upper parts of the region during the 1930s and the post-Independence construction of the Indira Gandhi canal to irrigate land further southward permitted agricultural colonization, in addition to improving the existing agriculture. Percentage of cultivated land now ranges from 7 in the southern Jaisalmer district to 70 in the northern Ganganagar district. The proportion of irrigated area ranges from zero to 50 per cent of the cultivated area. On the whole, about a third of this desert region is under cultivation, of which one-fourth is irrigated.

Crop cultivation, combined with pastoral activities in the form of raising sheep, goat and camel, describes the economy. Bulrush millet predominates. Some pulses are also grown. In the canal irrigated Ganganagar district, gram (chick peas), wheat, and cotton are the major crops in addition to bulrush millet. Camel, the mount and pack animal of the desert, is raised additionally for the camelry section of the defense services. The Suratgarh farm was established by the central government in 1956. Spread over an area of 125 $km^2$, it is a research-cum-demonstration institution for raising crop productivity and improving livestock quality under desert conditions.

Mining activity is connected with gypsum, inland salt and building stones, which are found in abundance. The white marble of the Taj Mahal at Agra was brought from here. Some industry based on local raw materials has also grown. Bikaner is famous for wool baling and processing, Ganganagar for cotton ginning and textiles, and Makrana for marble quarrying.

About one-fourth of the total population is urban. While rural population densities are low, there are several large cities in the region. Jodhpur (648,621), Bikaner (415,355) and Jaisalmer (38,813) are the major desert cities. These were earlier the capitals of big princely states, and were located on major caravan routes. Their tourist attraction is associated with the history and architecture of magnificent palaces and forts.

# 16

# The Peninsula

More than two billion years in age, the Indian peninsula is earth's one of the most ancient shields or crustal blocks. Its northward drift, synchronizing with the southward movement of the Siberian shield, was responsible for uplifting the material deposited in the intervening Sea of Tethys into the fold mountains of Himalayas some 60-65 million years ago. The peninsula has not been geologically as quiescent as is commonly believed. It experienced periodic upheavals, major among these being the folding of the Aravallis, fissure eruptions/giving rise to the Deccan lava country, and subsidence of the land to the west of the Western Ghats.

Physically, the peninsula (1,504,580 km$^2$) equals Alaska but its population (301 million) is more than 500 times. By Indian comparisons, however, the region is not densely populated. On nearly a half of the country's area, it has just over one-third of its population.

The peninsula is divisible into two secondary regions: (a) Central Highlands and Plateaus, and (b) Deccan Plateaus. The former acted as the foreland and the latter as the mainland at the time of the Himalayan orogeny. The topography of the former is far more complex than that of the latter. The linguistic patterns of the two also differ. The former is predominantly a Hindi-speaking zone, and the latter a mosaic of Marathi, Telugu, Kannada and Tamil-speaking areas.

Central Highlands and Plateaus constitute a shatter zone in consequence of its role as a foreland. The region has been subjected to intense earth movements causing, among other things, a repetitive rejuvenation of the Aravallis, faulting of the Narmada and Tapti valleys, and fissure eruption in the Malwa plateau. The overall pattern of drainage of the peninsula is radial: with Mahanadi flowing to the east, Narmada to the west, tributaries of the Yamuna to the north and those of the Godavari to the south. The fluvial and aeolian processes played their own role in changing the face of the land.

Though fairly rich in mineral and forest resources, the region is comparatively backward in development. On one-fourth of India's area, it has one-sixth of the population. This is explained largely by constraints on agriculture imposed by topography and lack of irrigation and by concentration of tribal population in several areas. Moreover, most of this region was earlier fragmented into a number of princely states inhibiting an integrated development.

On the basis of topography and associated landscape, economy and and demography, the region can be further divided into the following regions: (i) Aravalli Hills and Rajasthan Upland, (ii) Bundelkhand, (iii) Northern Madhya Pradesh Upland, (iv) Central Madhya Pradesh Plateau, (v) Southern Madhya Pradesh Upland, (vi) Northeastern Peninsula, and (vii) Chota Nagpur Plateau.

Deccan Plateaus present a similar topography. These are a series of tablelands at varying elevations. Their drainage system is mature. Highly graded beds are typical of the Godavari, Krishna and Kaveri rivers which have carved out a senile topography. Residual crests separating the river basins, mesas and buttes in the lava country, and tors and ridges developed on the archean rocks do lend some variety to the relief.

Before Independence, a part of the region was under the British and a part divided into a number of princely states, of which Hyderabad and Mysore were the bigger ones. The British-administered areas were generally more developed. Mysore, as a princely state, was a notable exception. It registered a remarkable progress in canal irrigation, road construction, electrification and modern industry under the aegis of a benevolent ruler during the 1930s.

The Deccan Plateaus may be further organized into four regions on the basis of the spoken language. These are: (i) Inland Maharashtra, (ii) Inland Andhra Pradesh, (ii) Inland Karnataka, and (iv) Inland

Tamil Nadu. The first is Marathi-speaking, the second Telugu, the third Kannada, and the last Tamil. On a little over one-fifth of the country's area, these together have slightly less than one-fifth of the population.

## Central Highlands and Plateaus

### *Aravalli Hills and Rajasthan Upland*

Contrary to the popular image of Rajasthan as a desert, this eastern part of Rajasthan is a productive agricultural upland, partly because it is hilly and forested. On 133,488 km$^2$, it has a population of 26 million, giving a density of about 200. Some pockets are inhabited by tribes, such as Minas around Jaipur and Bhils in Dungarpur, Banswara and Sirohi districts.

Like the Appalachians, the Aravallis are among the oldest mountains in the world, over 600 million years in age in contrast to the Himalayas which are the youngest. These are broad and high in the southwest with a fragmented linear orientation toward the northeast for a total length of 692 km. Their general elevation ranges from 500 to 1,000 m Mt. Abu (1,722 m) is the highest peak.

To the east of the Aravallis is located the Rajasthan upland. It is the drainage basin of the Chambal, a tributary of the Yamuna to the north. Most of it lies between 300 and 600 m.

The Aravallis are virtually a "mineral museum" containing a wide variety of minerals. Mining of non-ferrous items like lead, zinc, silver and copper is particularly important. The economy here exists of subsistence cultivation of sorghum and bulrush millet and livestock grazing.

The upland, by comparison, is relatively developed in agriculture. Over one-half of the total area is cultivated, of which nearly one-third is irrigated, mostly by wells and partly by canals. Wheat is an important crop, followed by sorghum and bulrush millet. Pulses and oilseeds are other crops grown.

For long, the region was under the rule of princes. Such a politico-economic set-up was not conducive to development, but considerable industry was started after Independence. The region is benefiting from investment by the Marwari industrialists who shifted some of their industry from the troubled parts of India to their native

**Figure 16.1**
*Central Highlands and Plateaus*

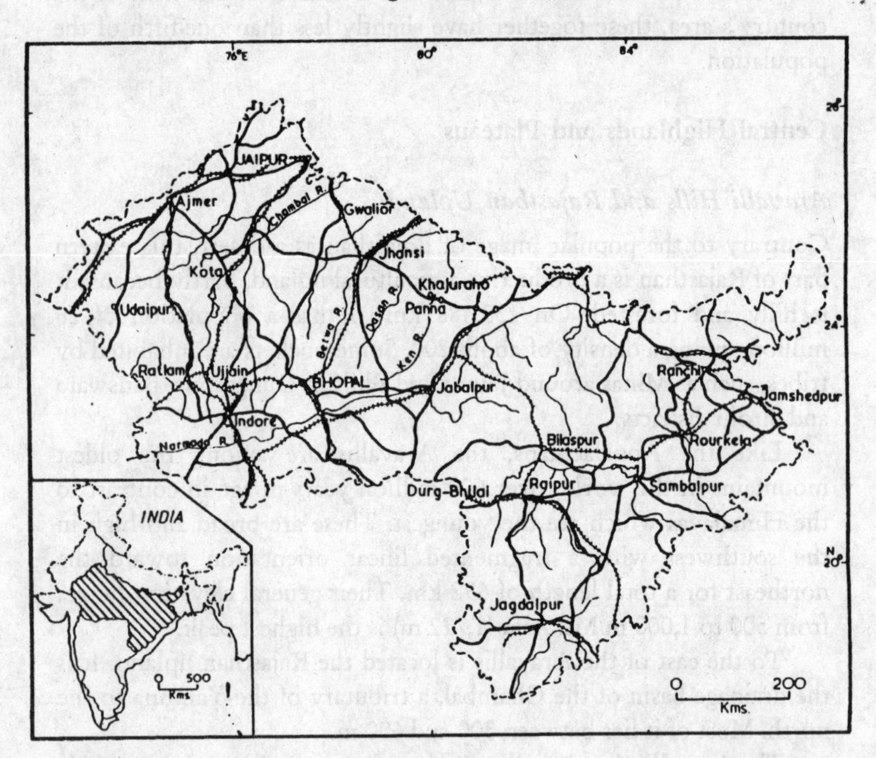


area. Most of the industry is located along the Jaipur-Udaipur and Delhi-Kota railroad routes. Important industrial centers include Jaipur (electricals and electronics), Ajmer (textiles and machine tools), Bhilwara (textiles), Udaipur (zinc smelting) and Kota (textiles and electronics).

Jaipur, the state capital and a metropolis with a population of 1,514,425, is famous as the "Pink City". Founded in 1728, the city was designed and constructed on planned lines by Maharaja Jai Singh. Rose pink stone was used as the basic material. It is an internationally known attraction. Udaipur (307,682), known as a "City of Lakes, Palaces and Fountains", is another princely city nestling in the lap of forest-clad Aravalli range. Ajmer (401,930) became important as a religious center and a place of pilgrimage for both Hindus and Muslims.

### *Bundelkhand*

This land of Bundelas, a warrior clan inhabiting the trans-Yamuna tract of Uttar Pradesh, is distinctive on linguistic and historical lines. It is a Bundeli (a dialect of Hindi)-speaking area. Its location at the junction of the rival Mughal, Maratha and the British powers during the eighteenth century was highly strategic at a critical phase in the Indian history. The process saw a culmination in the challenge posed by a local princess to the mighty British power, heroically though unsuccessfully, during the 1857 uprising. For reasons of a turbulent history, the region could not have a high population density like the adjoining Ganga plain. Its population of nearly 7 million on an area of 29,417 km$^2$ gives a density of less than 250.

Bundelkhand is an upland plain, with general elevation between 200 and 300 m. Thin alluvium covers its archean basal rocks. Deforestation has led to gully erosion and ravine formation at many places. The ravine-ridden topography, particularly near the administratively soft trijunction of Uttar Pardesh, Madhya Pradesh and Rajasthan, has exposed the region to the menace of decoity which was controlled recently.

About two-thirds of the total area is under cultivation. Because of a relatively low population density, landholdings are comparatively large, two to three times of those in the Ganga plain immediately to the north. About one-fourth of the cropped land is irrigated, mainly by canals and by wells. Construction of canals during the colonial rule and completion of irrigation projects after Independence have helped agriculture. There is a surplus production of food in the area unlike most parts of the country. Pulses, wheat and sorghum are the principal crops. Pulses share more than one-third of the cropped area, wheat around one-third and sorghum about one-fifth.

The picture is not so encouraging on the industrial front. Jhansi (368,580) is the only large industrial center. It is known for a big railroad workshop. It is also famous for its fort.

### *Northern Madhya Pradesh Upland*

This region is an eastern continuation of the Rajasthan upland with a difference; its topography is much more varied consisting of ravines, river bluffs, and hill ranges superimposed upon the base of an upland. Like Bundelkhand, it is drained by Chambal, Betwa and Ken rivers, which are tributaries of the Yamuna. Ecologically, it is a degraded area due to extensive deforestation, and for long it remained decoit-infested. On 10,388 km$^2$, it carries a population of 16 million, giving a density of abut 160.

It is not without reason that some geographers consider the region as simply a southern extension of Bundelkhand. This may not be accepted in the light of some striking differences between the two. Bundelkhand was under the colonial rule; northern Madhya Pradesh upland was fragmented into a number of princely states. The former saw the development of canal irrigation, whereas the latter is mostly unirrigated. Above all, the topography of the former is simpler than that of the latter.

About one-half of the area is under cultivation, the remaining being rugged or forested. At the district level, the cultivated area varies from 25 per cent in Panna to 75 per cent in Bhind. Contrasts in the proportion of irrigated area also range from less than 3 per cent in Sidhi to more than 30 per cent in Morena. Irrigation is highly inadequate for any good agriculture in this region with a variable rainfall of around 100 cm. Among the crops, wheat, pulses and oilseeds are prominent. Pulses occupy as much as one-fifth to more than one-third of the cropped area. In irrigated pockets, rice emerges as important.

Panna is famous for its diamond mines. From industrial point, only Gwalior (720,068) is the only industrial center of note, containing cotton textiles, woolen textiles, engineering goods and pottery industries. This former capital of the Scindia rulers is a place of great historical importance, best known for its fort. Another tourist attraction is Khajuraho whose sculptures are famous for their erotic expression. These form a part of a temple complex which was built around 1000 A.D.

### *Central Madhya Pradesh Plateau*

Comprised of the three plateaus of Sagar, Bhopal and Ratlam, flanking the Vindhyachal range, this region coincides with the traditional Malwa. Malwa is important for black soils, productive agriculture and agro-based industry. Population of the region is 18 million and the area 101,691 km$^2$.

The region is at a general elevation of 115 to 600 m. Rainfall ranges from 100 to 150 cm. Cultivated area makes one-half to two-thirds of the total. Lack of irrigation is the major problem; hardly 5 per cent of the cultivated area is irrigated, mostly by wells. Landholdings, in general, are comparatively large, and agriculture is commercialized.

Wheat and sorghum are the main foodgrain crops, whereas pulses, cotton and oilseeds are the main cash crops. Significant industry based on cotton is located at Indore, Ujjain and Ratlam. Bhopal is the seat of a public sector heavy electrical undertaking—BHEL (Bharat Heavy Electrical Limited). About one-third of the population is urban. Bhopal (1,063,662) is the state capital; its Taj-ul-Masjid is the biggest mosque in India. Indore (1,104,065) is one of the biggest cotton textiles manufacturing and trading centers in India. Ujjain (367,154), the capital of the ancient state of Avanti, was the prime meridian of the Indian astrologers. All the cities mentioned above, were the capitals of relatively prosperous princely states before Independence.

### *Southern Madhya Pradesh Upland*

Coinciding largely with the catchment area of the Narmada, this region is characterized by complex topography consisting of the Narmada and Tapti troughs, Satpura, Mahadeo and Maikal relict hills and several small dissected plateaus. It is marked by higher elevations of 600 to 900 m in the southeast and lower levels of 150 to 300 m in the west.

The Narmada trough, with a northeast to southwest orientation, is of great physical and human interest. This elongated rift valley is the most cultivated and populated part of the region. The Narmada Valley Project is India's largest single river development programme. It involves the completion of 30 major, 135 medium and 3000 minor irrigation schemes at an estimated cost of five billion dollars.

About 40 per cent of the total area is under cultivation, of which nearly 10 per cent is irrigated. A rainfall of 100 to 200 cm and fertile black soils are very useful for agriculture. Wheat, pulses and oilseeds mainly represent the food system. However, there is some irrigated rice. Cotton is the only cash crop.

Around one-third of the total area (102,673 km$^2$) is forested. Tribal groups form nearly 15 million. These groups include the Gonds in the southeast and the Bhils on the western margins.

Jabalpur, located on the Narmada near its source, is a prominent industrial center for defense production, engineering goods and saw-milling. The town (887,188), which was a cantonment area 1819, later grew as an important transport, manufacturing and educational center.

Virtually, the entire region was under the colonial rule before Independence. Its rail-road links with Mumbai (then Bombay) were established at an early stage, supplying it with cotton and labor for its textile industry.

### *Northeastern Peninsula*

Largely forming the upper basin of the Mahanadi, this vast region (262,045 km$^2$) is spread over the eastern part of Madhya Pradesh, non-coastal segment of Orissa, and southeastern tract of Uttar Pradesh. It can be visualized as consisting of the Chattisgarh plain (elevation of 250 to 350 m) at the core with a wide rim of plateau (550 to 750 m) and hills (up to 900 m) around. The region abounds in forests, numerous tribes and has some mining and industrial centers. Before Independence, the Chattisgarh plain was under the colonial rule and its surrounding hills and plateaus were split into a number of tribal states. Major tribes include Gonds, Kawar, Kol, Khond, Saora, Munda and Oraon. Tribal population makes one-third of the total population of 37 million.

The region is rich in its water, mineral and forest resources and includes a number of irrigation and power projects, namely, Hirakud, Rihand, and Mahanadi. Several mining towns, such as Bhiladilla for iron ore and Korba for coal, have been recently developed. Besides, Bhilai and Rourkela have been developed as "steel cities". Dandakaranya Colonization Project, covering the Bastar district of Madhya Pradesh and Koraput and Kalahandi districts of Orissa, was started in 1957 to rehabilitate the displaced persons from Bangladesh, in this region.

Agriculturally, the Chattisgarh plain is the most productive part of the region. It is a surplus producer of rice, pulses and vegetables and is known as the "rice bowl" of Madhya Pradesh. Nearly a half of its total area is under cultivation and around one-fourth of the cropped area is irrigated. Rice, pulses and oilseeds are the main crops of the hills and plateaus surrounding Chattisgarh plain, but the productivity is much less. The proportion of cultivated area drops to one-third and of irrigated area to less than one-tenth of the total land. In Bastar, for example, 99 per cent of the cropped area has to do without irrigation.

Recent mining and industrial activity has promoted urbanization. At the time of Independence, the region was overwhelmingly rural; now about one-sixth of the total population is urban. Some places have grown into big cities. The three industrial cities of Durg-Bhilai (688,670), Raipur (461,851) and Rourkela (398,692) were all developed after Independence. All such important places in the region are located on the Bombay-Calcutta railroad route.

### *Chota Nagpur Plateau*

Described as the "Ruhr of India", this region occupies the southern part of Bihar. Large quantities of several minerals, in association with different rock formations, are found in the region: iron, manganese and copper in Singhbhum district, mica in Hazaribagh district, coal in the Damodar valley and building stones in the Son valley. About 40 per cent of the estimated mineral wealth of the country is located here.

Essentially a plateau at a general elevation of 600 to 1,000 m, the region has considerable physiographic diversity: of the Hazaribagh, Ranchi and Kaimur plateaus, faulted valleys of the Damodar and Son, and Rajmahal hills on the eastern margins. Parasnath, the holy place for the Jains, is the highest point at 1365 m. The climate can be described as tropical sub-humid. Rainfall ranges from 125 to 150 cm.

Cultivated area accounts for one-third of the total, forests another one-third, and other land uses the remaining one-third. Irrigated land accounts for less than one-tenth of the cropped area. The agricultural landscape follows a typical pattern: irrigated rice in the valley bottoms, unirrigated rice on lower terraced slopes, and maize, pulses and oilseeds on the higher slopes, with the hilltops lying as wastelands. Agriculture varies vertically more than horizontally.

Crop yields are generally low due to soil erosion caused by extensive deforestation. The damaged ecology calls for restoration. The inter-state Damodar Valley Project has been patterned after Tennessee Valley Authority.

For a number of reasons, Chota Nagpur plateau witnessed significant development of industry. Availability of mutually complementary minerals like coal, iron and manganese, proximity to Calcutta, and large-scale investment by the central government most favorable conditions for industrialization. The beginning was, however, made with the private sector Tata Iron and Steel Company (TISCO) at Jamshedpur in early years of the present century. The place grew into a nucleus for several steel-based engineering and transport industries. The post-Independence era saw the establishment of several public sector industrial units: heavy engineering at Ranchi, fertilizers at Sindri, steel at Bokaro, aluminum at Muri and copper smelting at Ghatsila. In addition, coal industry in the Damodar valley, and cement, paper and sugar industries are concentrated in the Son valley.

The economic dynamism of the region generated sizable inflow of migrants from the adjoining Bihar plain, Orissa and West Bengal. Its present population of about 21 million on 97,714 km$^2$ gives a density about 220. More than one-fourth of the total population is urban. Major urban concentrations include Jamshedpur, Dhanbad, Ranchi and Bokaro. Jamshedpur (834,535) started as a planned steel city in 1907. Today, it has grown to be the "Pittsburgh of India". Dhanbad (817,549) has its economy centered on coal. Ranchi (614,454), at an altitude of 648 m, is a hill resort and known for its mild climate. Bokaro (415,686) has India's biggest iron and steel plant in the public sector.

The tribal population of the Chota Nagpur plateau has been asserting their regional identity by way of demanding a separate state of Jharkhand (the forest region). The movement has its roots in grouses relating to land and forest alienation, reduction of tribals to a minority status as a result of inmigration of the outsiders, and a loss of cultural identity of the tribals. The demand did not meet political approval primarily because tribals here form only one-third of total population and different tribes, such as Mundas, Santhals, Hos and Oraons, do not have any link language. Nevertheless, the event has been successful in highlighting the distinction between the Chota Nagpur plateau and the North Bihar plain as two culturally distinct sections of the state of Bihar.

## The Deccan Plateaus

### *Maharashtra Plateau*

Distinguished by a predominance of the Maratha people, this region has a distinct historic identity. The Maratha chiefs, such as the Peshwas of Poona, Scindias of Gwalior, Holkars of Indore, Gaekwads of Baroda and Bhonslas of Nagpur, founded a number of principalities in different parts of India, and extended their political and cultural influence extensively. Basically peasant-warriors by nature, the Marathas had a long religious tradition manifest in a great reverence for the priestly class of Brahmins. All this finds an expression in several forts and temples in areas of their domain.

Inland Maharashtra has a population of 59 million residing on an area of 276,985 km$^2$. Both in population and area, it is a little bigger than the United Kingdom. Linguistically, it is a Marathi-speaking area. Economically, it is the hinterland of Mumbai. Geologically, it is the Deccan trap country made of lava rocks. In topography, it is a plateau, 600 to 1,000 m, with the Western Ghats (600 to 1,500 m) on the western and the Wainganga basin (less than 300 m) on the eastern margins. Following the slope of the land, the Godavari and Krishna traverse eastward to the Bay of Bengal. The Tapti river in the north, however, flows from east to west in a faulted valley to the Arabian Sea.

Most of the region lies in the rainshadow of the Western Ghats. It receives a highly variable rainfall of 50 to 100 cm. Periodic droughts occur. On the Western Ghats, rainfall rises to 600 cm. This causes soil erosion under any deforestation. Soils are more fertile along the river courses, but less on the intervening uplands. Regional variations in agricultural productivity and population distribution are associated with variations in soil and rainfall distributions.

About 60 per cent of the area is under cultivation; of which nearly one-tenth receives irrigation. Wells account for 60 per cent of the irrigated area, canals for 20 per cent, and tanks for the remaining 20 per cent. Canal irrigation is largely confined to the foothill zone of the Western Ghats where topography favors construction of large reservoirs. Godavari and Parvara canal systems, completed in the early years of the 20th century, and the Jayakwadi project, recently constructed on the Godavari, are of special importance. Irrigated area

**Figure 16.2**
*Deccan Plateaus*

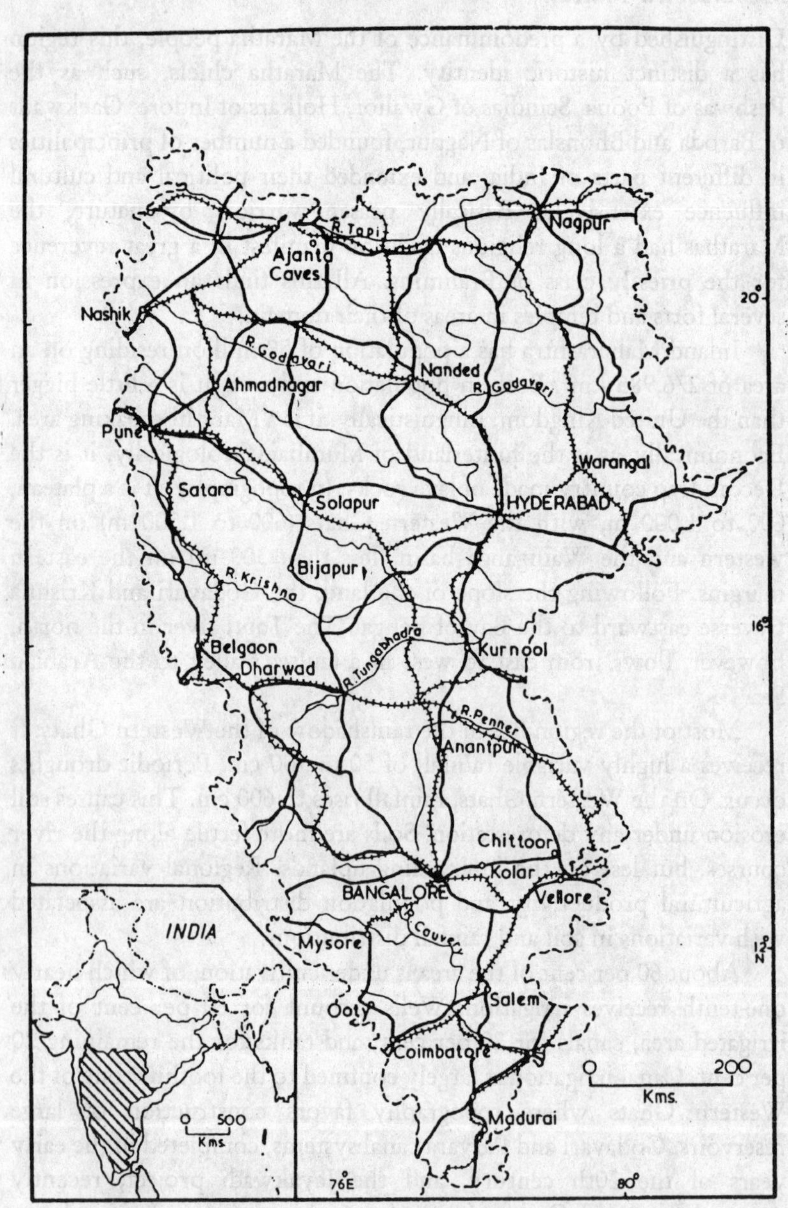

has increased two-fold since Independence.

Sorghum occupies one-third of the cropped area, bulrush millet one-tenth and pulses one-sixth. Cotton is the major cash crop on one-eighth of the cropped area. Sugarcane, linked with sugar manufacturing in the cooperative sector, is an important crop in the Maval (foothill) zone.

The region is one of the relatively industrialized parts of India. Engineering goods, chemicals, electricals, cotton textiles and sugar rank high among the industrial products. Industry is spatially dispersed.

Nearly one-third of the region's population is urban. Pune is a fast growing metropolis with 18 satellite towns. Its population of 2,485,013 approximate to that of Singapore. It is the historic seat of power and culture of the Marathas. Now it draws its eminence as an educational and industrial center. Engineering goods, electricals and antibiotics are the major industries. The Rajneesh Ashram of the adherents to the cult of uninhibited love is also located here. Nagpur (1,661,409) is another historic city—firstly as the seat of power of the Bhonslas, and subsequently the capital of the Central Provinces. It is the cultural heart of the Vidarbha sub-region of Maharashtra. Defense equipment, cotton textiles and chemicals industries are located here. Aurangabad (592,052) is a large cantonment. In its close proximity are located the famous Ajanta and Ellora caves known for their carved pillars, rock statues and wall panels belonging to the Buddhist, Jain and Hindu periods of Indian history.

Historically, the Inland Maharashtra is divided into three sub-regions: the Desh in the west, which was a part of the Bombay province; the Vidarbha in the northeast, which formed a part of the Central Provinces; and the Marathwada in the southeast, which was a part of the Hyderabad princely state before Independence. The three sub-regions are discussed in term of their socio-economic development, on historical background and geographical location.

### *Andhra Plateau*

This is the Telugu-speaking part of the peninsular interior. Spread over an area of 182,139 km², it has a population of 38 million. A density of around 200 is relatively low. With a per capita income less by one-fourth of the national average, the region is economically backward. This is explained by not only its physical constraints of a rugged topography, low and variable rainfall and infertile soils but also by a historical legacy of the princely rule of the Hyderabad Nizams over a large part. The prevalence of landlordism kept the vast masses of landless peasantry under perpetual poverty. There are three agricultural laborers for every two cultivators here in comparison to a ratio of 2:3 in India. The administrative separation of the region from the more developed coastal area during the colonial time has also been responsible for the stagnation.

The region is noted for frequent bouts of agrarian unrest. The ideological conflict between the landed aristocracy and landless labor is not free from a violent expression. Tribals in the north are more prone to the Naxalite (anti-landlord) movement.

Inland Andhra is largely a peneplain developed on archean rocks. Most of it lies at an elevation of 300 to 600 m with topography containing monadnocks and depressions. In the east are located the Eastern Ghats which are relict hills. Major rivers, namely, Godavari, Krishna and Penner have attained their base level.

A rainfall of 75 to 100 cm increases from southwest to northeast. It is highly variable and famines were not uncommon in the past when irrigation was grossly inadequate and the public distribution system virtually absent. Irrigation was mainly by tanks (man-made small lakes) which, being fair weather friends, were unobliging during a drought. Things have improved since Independence. The Tungbhadra, Nagarjunsagar, Telugu Ganga and other projects have extended the land under canal irrigation. About one-sixth of the cultivated area is now irrigated.

Sorghum, pulses and oilseeds are the major subsistence crops. Rice around the tanks, sugarcane in the canal irrigated areas, cotton along the river valleys and vegetables in the Anantpur-Chittoor basin are the other notable crops.

The region is endowed with a variety of minerals in association with its geological formations. It is a major source of asbestos and practically the entire barytes in India is found here. Both the minerals come from the Cuddaph-Kurnool tract. The Singhreni belt along the Godavari supplies coal to many parts of south India.

Manufacturing is picking up but almost two-thirds of it is concentrated in and around Hyderabad. Main industries include defense equipment, heavy electricals, aeronautics, machine tools, tobacco processing, cotton textiles and asbestos cement. Several big manufacturing units are also in the public sector.

Hyderabad dominates the urban scene. This former seat of princely Nizams is now the capital of Andhra Pradesh. Its population is 4,280,261. Another place worthy of mention is Tirupati (189,030), whose Tirumalai temple is rated as the most sacred and richest in south India.

Inland Andhra may be divided into two segments on historical grounds: Telengana and Rayalaseema. The former was the main component of the Hyderabad princely state and the latter a part of the Madras presidency before Independence. Sub-regional sentiment is strong in both. The separation of Telengana as a separate state was an important political issue during the 1950s but has died since.

### *Karnataka Plateau*

The region is a conglomerate with the old princely state of Mysore at the core. Its northwestern part was the Bombay Karnataka, northeastern part the Hyderabad Karnataka, and the east-central part the Madras Karnataka before reorganization of the Indian states in 1956. This historical-political disjunction is now cemented through the common language of Kannada.

Inland Karnataka covers an area of 173,059 km$^2$ and has a population of 41 million. This gives a relatively low density of 236. Population pressure, as typical of many parts of India, does not afflict this region which is characterized by a progressive economy in association with a commercial agriculture, widespread mining activity and high degree of industrialization. It has attracted considerable inmigration from the adjoining states of Andhra Pradesh, Tamil Nadu, Kerala, and even from north India.

The region, for that matter the whole of Karnataka, is now in limelight for its thrust on decentralized planning. Special administrative arrangements have been made to devolve the planning process from the state to the district and local levels. This was deemed imperative for ensuring the participation of people in planning, and for reducing inequality in development.

Topographically, the region is a series of plateaus which decrease in elevation from 800 to 1000 m in the south, from 600 to 800 m in the middle, and from 400 to 600 m in the north. The Western Ghats, locally known as Malnad, form a hilly tract along the western margins. The lithological arrangements divide the region into two segments: the southern with granitic/gneissic rocks and the northern with basaltic/lava rocks. These have given rise to red and black soil areas, respectively. The former is known as the Southern Maidan and the latter as the Northern Maidan. Broadly speaking, the former was under the princely rule and the latter under the British before Independence.

Rainfall pattern of the region is transverse to its topographic gradient. It shows a decrease from more than 100 cm in the west to less than 50 cm in the east. Cultivated area is about one-half of the total, of which about 10 per cent is irrigated. Tank and canal irrigation is more common in the southern part and well and canal irrigation in its northern counterpart. Canal irrigation in the south is through the Kaveri system laid before Independence, and that in the north through the Upper Krishna system developed during the post-Independence era.

There is a marked regional differentiation in crops. The Southern Maidan has finger millet as the main cereal with rice around the tanks. In canal irrigated areas, sugarcane, tobacco and groundnut become relatively more important. The Northern Maidan, on the other hand, has sorghum as the major cereal and cotton as the cash crop. Some wheat is also grown. On the forested Malnad, traditional rice and spices are also grown. Alongside, there are modern coffee plantations, especially in the Coorg area. In other parts of the region, finger millet, sorghum, spices and horticultural products are more conspicuous.

The region is rich in iron ore and manganese. It has virtually a monopoly in gold mining at Kolar. Gold mining gave an impetus to an early electrification here around the beginning of the present century. The diversity of resources helped industrialization. The capital city of Bangalore (4,086,548) combines the functions of industry, education, recreation, and of course, administration. A large variety of industries, such as engineering goods, machine tools, aircrafts, telephones, watches, cotton textiles and fruit canning, are concentrated in and around this city. Many units are in public sector financed by the central government. Bangalore's central location in south India, a mild climate, and above all, a cosmopolitan culture have turned this city into a great tourist attraction. Mysore (652,246) is famous for silk textiles and Brindavan Garden. It maintains its historic glory as the former capital of a prosperous princely state. Hubli-Dharwar (647,640) is a twin city, with the former as the industrial-commercial wing and the latter as the educational-cultural center.

Inland Karnataka may be divided into three subregions: (i) Northern Maidan, which is a black soil country with commercial agriculture in cotton; (ii) Southern Maidan, which coincides with the former princely state of Mysore and is the most developed part of the state, and (iii) Malnad, which is forested and hilly, with traditional subsistence agriculture coexisting with modern coffee plantations. The region is exceptional in some sense: its former princely administered part is more developed than areas formerly administered by the British prior to Independence.

### *Tamil Nadu Plateau*

Tamil Nadu plateau is composed of a series of plateaus and hills: (i) Coimbatore plateau, (ii) Salem plateau, (iii) western Tamil Nadu hills, and (iv) Nilgiri hills. This physiographic diversity notwithstanding, the region finds a unity through Tamil as the *lingua franca*. In comparison to the three regions described above, each of which was partly under the rule of princes and partly of the British, this region was almost entirely under the colonial administration.

It has a population of 23 million on 59,481 km$^2$. This gives a density of about 400. It is rather high for a plateau—hill region, particularly when only one-half of the area is under cultivation and only one-third of the cultivated area is irrigated. Pressure on agricultural land is high. Not only has it resulted in relatively low per capita income but also impelled sizable outmigration.

Part of this pressure is taken care by intensive cultivation even under not too favorable conditions and in part by diversification of economy. Sorghum and bulrush millet are the main cereals. Cotton, groundnut and vegetables are the major cash crops. On the Nilgiris, tea and coffee plantations are extensive. The Coimbatore plateau is especially noted for garden cultivation in onion, sweet potato, turmeric, chilies, and a variety of other vegetables grown with the help of well/tubewell irrigation.

Some significant mining and decentralized manufacturing have diversified economy. About 40 per cent of the working force is outside agriculture. Salem area is noted for iron ore, lignite, bauxite and limestone. Industry is highly dispersed. This dispersal is related to the use of electricity as the main source of power. Associated with this is the dispersed pattern of the urbanization. Temple, as the nucleus of many urban centers, reinforces this tendency. *Gopurams*, intricately carved out temple towers, are a bold feature of the landscape. On the whole, over one-third of the total population is urban.

Coimbatore (1,135,549) is the biggest city. Primarily an industrial center, it links Tamil Nadu with Kerala through Palghat. Cotton textiles, handloom products, motor pumpsets and engineering goods are the major industries here. The city finds a rival in Madurai (1,093,702), which again is a big industrial center for cotton textiles and engineering goods. The latter, however, is more famous for its ancient Meenakshi temple, which is a legend in architecture in honor of Lord Shiva's consort. Salem (573,685) is still another industrial center, known for iron and steel. Salem is also famous for handloom silk scarves. Among hill stations, Ooty (81,726) is the most populous and popular. It is set amidst sprawling tea plantations at an altitude of 2,286 m about 100 km northwest of Coimbatore. Now it is called as Uthagamandalam as a part of indegenization of place names. A film photo unit in public sector is located here. Kodaikanal (27,461) at 2,133 m, some 120 km northwest of Madurai, is developed around a picturesque lake. It has prestigious residential schools which draw students from many countries of the world.

The region gives a mixed feeling. Economy is diversifying but population pressure remains heavy. A vigorous programme in favor of family planning has rightly been adopted. Political and administrative commitment is strong. Results are already visible. The region has one of the lowest birth rates in the country, i.e., 22 per thousand against 30 in India.

# 17

# The Coasts and the Islands

Maritime India includes the Western Coastal Lowlands, the Eastern Coastal Plains and the Islands. The tripartite division is not merely a matter of location but also reflects the physical and cultural contrasts between the three regions. While the former western coastal region is a product largely of submergence of the land, the eastern coastal region is emergent in origin wherein the peninsular rivers extended land onto the sea by deposition of material brought along. As a result, the former is narrow, the latter wide; the former has virtually no delta, the latter has several; the former is a lowland, the latter a flat plain. Comparisons go farther. The western coastal region is more humid than its eastern counterpart. The former is rich in minerals like petroleum, iron ore and uranium, the latter is deficient in minerals. The former is linked more through rail with other parts of India; the latter is better connected within. Above all, the former has historically been more exposed to Southwest Asia, East Africa and Western Europe, the latter had links with the Southeast and East Asia.

The oceanic islands form a category in themselves. Located at a distance from the mainland, these could become active partners in the national life only through passage of time. Their resource potential could be harnessed mainly after Independence.

The three regions identified above can be further sub-divided. The western coastal region has Gujarat, Maharashtra Konkan, Goa, Karnataka coast and Kerala as its major components. This division is primarily on the basis of linguistic affinity and historical experience of the five areas. The eastern coastal region is divisible into the Orissa coast, Andhra coast and Tamil Nadu coast. Again, these regional identities are distinguished on the basis of language.

The islands form two groups, one in the Arabian Sea and the other in the Bay of Bengal. The two, namely, Lakshadweep and Andaman & Nicobar Islands, are administered as two separate union territories.

## Western Coastal Lowlands

### *Gujarat*

Though a specific cultural area for centuries, Gujarat in its present form came into being only in 1960 as a result of the bifurcation of the bilingual state of Bombay. Before Independence, three-fourths of it was parceled into 362 princely states/estates. Districts encircling the Gulf of Cambay were a part of the Bombay presidency.

Gujarat distinguishes itself in several ways. It is inhabited by enterprising Gujaratis who have a special aptitude for commerce and industry. They are habitual emigrants and have moved to different locations primarily for business in South Africa, East Africa, United Kingdom, United States, and other parts of the world. They have excelled in hotel industry. Above all, they have been very successful in cooperative movement in dairy farming, and manufacturing enterprises within their own homeland. Mahatma Gandhi, the father of the nation, was born in Gujarat.

The region has a population of 41 million on 196,627 km$^2$. This gives a density of 210. Pressure of population is not a problem. Emigration and outmigration were induced more by a desire for a better living rather than by any economic distress. A coastal location and proximity to Bombay (now Mumbai) facilitated this process.

The region may be divided into three sub-divisions: (i) Gujarat plain, the "heartland", which is at a relatively high level of agricultural and industrial development; (ii) Saurashtra, essentially a lava plateau, was a patchwork of a number of principalities before Independence; and (iii) Kachchh, though barren mudflats, is strategically important for having a border with Pakistan. Within the Gujarat plain, the Gandhinagar-Ahmedabad-Vadodara-Bharauch-Surat-Valsad-Vapi belt is agriculturally the most productive, industrially the most developed,

**Figure 17.1**
*Western Coastal Lowlands*

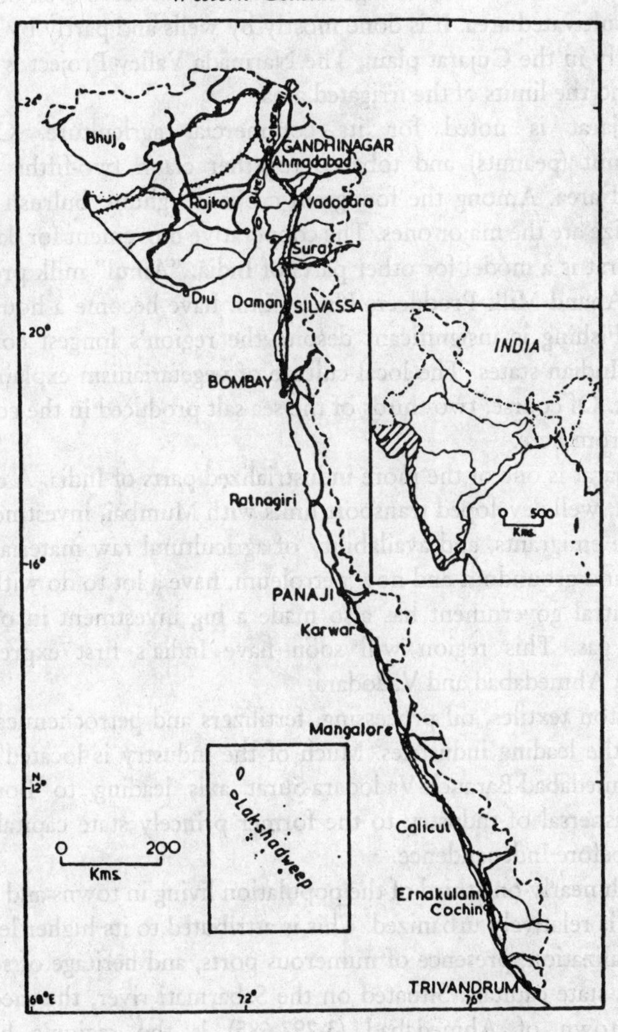


and the most urbanized. Areas surrounding to its west (Saurashtra and Kachchh), north (Banaskanth-Sabarkanth tract), and east (tribal zone) form the less developed periphery.

Practically, the whole of the region is drought-prone. Rainfall is less than 75 cm, except in the southern coastal strip. Much of the vegetation was removed in the process of settlement; only the eastern margins, inhabited by the tribals, have some forest cover.

Irrigation, necessary for agriculture, does not serve even one-fifth of the cultivated area. It is done mostly by wells and partly by canals, especially in the Gujarat plain. The Narmada Valley Project is meant to extend the limits of the irrigated area.

Gujarat is noted for its commercial agriculture. Cotton, groundnut (peanuts) and tobacco together claim two-fifths of the cropped area. Among the foodgrain crops, sorghum, bulrush millet and maize are the major ones. The cooperative movement for dairying in Gujarat is a model for other parts of India. "Amul" milk products of the Anand Milk Producers Union Ltd. have become a household name. Fishing is insignificant despite the region's longest coastline among Indian states. The local culture of vegetarianism explains this paradox. Of course, two-thirds of the sea salt produced in the country comes from here.

Gujarat is one of the more industrialized parts of India. A coastal location, well-developed transport links with Mumbai, investment by returnee emigrants, and availability of agricultural raw materials like cotton and groundnut and now petroleum, have a lot to do with this. The central government has also made a big investment in oil and natural gas. This region will soon have India's first expressway between Ahmedabad and Vadodara.

Cotton textiles, oil processing, fertilizers and petrochemicals are among the leading industries. Much of the industry is located along the Ahmedabad-Barauch-Vadodara-Surat axis leading to Bombay. Some dispersal of industry to the former princely state capitals had started before Independence.

With nearly one-third of the population living in towns and cities, Gujarat is relatively urbanized. This is attributed to its higher level of industrialization, presence of numerous ports, and heritage of several princely state capitals. Situated on the Sabarmati river, the medieval capital town of Ahmedabad (3,297,685) is the region's biggest industrial concentration. Cotton textiles, chemicals and engineering goods are the major industries. Its historic mosques, temples and palaces are a fine blend of the Muslim and Hindu architecture. Some 25 km to the north of it is located the planned city of Gandhinagar (121,746), which is the new capital of Gujarat. Surat (1,517,076), on the mouth of the Tapti river, was the site of the first British factory in

India in 1600. Now it is reputed for diamond cutting, silk textiles and silk thread embroidery. Vadodara (1,115,265), on the Mahi river, has a big petrochemical complex. The Royal Museum, Nazar Bagh Palace and Art Gallery are reminiscent of its glorious past.

Noteworthy among the tourist attractions of the regions are the game sanctuary (lion safari) at Gir, the ancient temples of Dwarka and Somnath, the Jain pilgrimage place of Palipatan, and the archaeological site at Lothal near Ahmedabad. There is evidently a fine combination of tradition with modernity in the overall spatial organization of Gujarat.

On the border of Gujarat and Maharashtra is located the union territory of Dadra & Nagar Haveli which was under the Portuguese rule till 1954. Covering an area of 491 km$^2$, it has a population of only 138,542. It is largely rural and tribal but its proximity to Mumbai is rapidly transforming it. Silvasa is its capital.

Daman & Diu, another union territory, is an enclave within the state of Gujarat. Along with Goa (which was formed as a separate state in 1987), it was also under the rule of Portuguese till 1961. It has a population of 101,439, over an area of 112 km$^2$. Daman is the capital.

### *Maharashtra Konkan*

Stretching from the north to south as a narrow belt of 60 km wide and over 500 km in length, the coastal districts of Maharashtra are traditionally known as the Konkan, where a Konkani (a dialect of Marathi) is spoken. About 19 million people live on 30,728 km$^2$. Over 70 per cent of the population is urban, mainly because of its proximity to Mumbai, on which the region focuses. For centuries, the region has been a gateway to India through the Arabian Sea. Its tradition in sea navigation is long. A sizable section of the Indian Navy is drawn from here.

This coastal region is a lowland rather than a flat plain. Ridges and spurs project out from the Western Ghats. Small youthful streams in the area are used for generation of hydropower but these are not suited to navigation or irrigation.

Land use reflects topographical variations: valley bottoms are used for intensive rice cultivation; spurs and ridges have forests, with millets and pulses as the main crops in the clearings. There is a large-scale market gardening in fruits and vegetables primarily for consumption in metropolitan Mumbai.

With a population which equals that of Los Angeles in the U.S., Mumbai (12,571,720) is unquestionably a giant metropolis. This colonial port city is the largest urban and industrial concentration in India today. Its backwash effects on the Konkan region cannot be overestimated. Not only it wrested youthful labor, quality professionals and investible capital but also caused extensive deforestation to meet the wood requirements of a regular massive ongoing construction work in the city. Its presence had a banyan tree effect on the growth of industry, commerce and towns in other parts of the region.

It is said that while Calcutta is a city of the past and Delhi that of future, Mumbai is the city of India's present. A cluster of seven islands in the Arabian Sea was a Portuguese emperor's gift to Charles II of England in 1661 as a part of his daughter Catherine's dowry. The real growth of the city had to wait till the middle of the nineteenth century when cotton textiles industry had started concentrating here, and it was the first to be linked by railroad with inland. Engineering and chemical industries followed. The establishment of Trombay Refinery nearby promoted petrochemicals and plastic industry. A further boost to these industries came from the discovery of oil from the Bombay High in the Arabian Sea.

By its eminence in film industry, Mumbai is known as the Bollywood of India. Ironic it may seem, over one-third of the city's population lives in slums. The place has the dubious distinction of having Asia's largest slum of Dharavi with a population of half-a-million. The city is the capital of Maharashtra. Financially, it is country's biggest stock exchange market and culturally, it is "mini-India" with residential localities displaying segregation among migrants from virtually all parts of the country. The city is rated as one of the best managed despite its colossal size and remarkable diversity. Part of the credit for this goes to the civic sense of the people who have learnt to live as a part of a multitudinous crowd.

The Konkan region is divisible into three parts, viz., north, middle and south. The north contains the megacity of Mumbai and is dominantly urban, industrial and commercial. The middle, represented by Raigad district, is noted for more subdued topography permitting good cultivation of rice, coconut, garden crops, betelnut and arecanut. The south, comprising Ratnagiri and Sindhudurg districts, is lateritic, much less fertile and characterized by subsistence agricultural economy. It has acted as a funnel for the labor force, mostly male selective, destined for Mumbai. Remittances from the migrants are critical to the household economy back home.

### *Goa*

Under the Portuguese rule for exactly 450 years, Goa was liberated by India in 1961. It is a special part of India with a deep colonial imprint on the landscape and lifestyle of the people. The region is now best known for tourism stimulated by its golden beaches, lush green landscape, proximity to Mumbai, and above all, a friendly culture. It offers something for everybody—food and liquor for the connoisseur, scenery and beaches for the naturophile, and churches, temples and mosques for the religious. It is a Konkani-speaking area, with some continuing usage of the Portuguese as a legacy of the past.

On an area of 3,702 km$^2$, the region has 1.2 million people. About 40 per cent of the population is urban. Towns are closely spaced at an average distance of 12 km. Urban-rural interaction is strong. About three-fourths of the working population is non-agricultural and employed in mining, manufacturing and a variety of services.

The sea-front of the region is occupied by a tiny delta made by the Mandovi river. At its back are the Western Ghats from where the Mandovi and other streams originate. The foothill zone is rich in minerals, particularly iron ore and manganese, which are exported.

Rice is the main crop, followed by coconut. Coconuts are an important ingredient in the Goanese food, especially for cooking fish. Cashewnut, sugarcane and fruits, such as pineapple, mango and banana are other cash crops.

Industry is making a headway. An electronics unit has come up in the proximity of the capital city of Panaji. A great scope exists for establishing an iron and steel plant at Marmagoa which is a fine port.

Goa for long has been a land of emigration to Portugal and its colonies. Some Indian cities like Mumbai were also a destination for migrants. This outward-looking tendency had a modernizing influence on the region's society. The Portuguese rule also left behind several Christian institutions. The Christians account for nearly one-third of the total population and the Hindus for two-thirds but churches are far more numerous and widespread than the temples.

Saint Francis Xavier Church, built in 1510, is a most impressive example of Goanese religious architecture.

### *Karnataka Coast*

This Kannada-speaking coastal region is a conglomerate of two segments: Uttar (North) Karnataka, which was a peripheral district of the Bombay province, and the Dakshin (South) Karnataka, which was a peripheral district of the Madras province. Both find linguistic unity in being Kannada speaking.

Physically, this coastal region is different than the Konkan region: it is more hilly than plain. Uttar Kannada is virtually an extension of the Western Ghats and South Kannada has a topography consisting of dissected hills, open valleys and coastal lowlands. Its population of 3.9 million on 18,732 km$^2$ gives a density of 210, a figure which is lower than for any other coastal region of India. About one-fourth of the total population is urban.

Rice cultivation and fishing are the main activities. Coconut is ubiquitous. Cashewnut and a variety of fruits are grown. Rubber plantations also exist. Forestry is important. Industry is concentrated mainly in the southern part. Mangalore (425,785) has a number of cashewnut and coffee processing industries. Its tiles are in demand throughout south India. The city is one of the major ports, especially for the export of coffee. Some 275 km to the north, Karwar port is being developed for export of iron ore and manganese.

The region has made a notable contribution to the spread of education, banking and hotel industry. The Christians from Mangalore have established prestigious educational institutions in several parts of India. Manipal is known for promoting banking and Manipal Medical College. Udipi hotels are popular eating places even in north Indian cities.

### *Kerala*

Formed by amalgamating the erstwhile princely states of Cochin and Travancore and the Madras presidency district of Malabar, this region assumed its present form in 1956. It is India's most homogeneous state linguistically; 96 per cent of the people speak Malayalam.

Although solidly homogeneous in language, the Kerala society is quite diversified by religion: Hindus number three-fifths of the population and the Muslims and Christians, one-fifth each. The Hindus are further divided amongst a number of castes: Namboodris, Nairs and Ezhavas to name the important ones. Before Independence, the region was described as the mad house of castes when some higher castes practiced extreme untouchability in their relations with low castes. This is markedly changed now. The new society is egalitarian; thanks to the political awakening, land reforms, high literacy, and not the least, the Christian missionary activity.

In 1991, Kerala recorded a population of 29 million on an area of 38,863 km$^2$, making exceptionally high population density of 747. That is three times the national average. However, the active family planning programmes have earned Kerala the name as a forerunner in fertility control. Kerala has also the distinction of containing literacy levels among the highest in India. Birth rate is only 16, death rate 6, and infant mortality rate 17 per thousand. Thus, demographically, Kerala has become a model for other Indian states. All this has been achieved when its per capita income is below the national average.

Politically, it was probably the first area in the world to vote Communist government into power in 1957. This was an outcome of its education-cum-poverty syndrome. Unemployment rate here is staggering—25 per cent. Wages have, however, not fallen because the working class is highly organized with an occasional tendency toward militancy.

"God who made Kerala had a green thumb", so goes a Malayali saying. Landscape is exceptionally beautiful: a handiwork of man and nature in unison. The sea, the coastal land, and the mountain backdrop are juxtaposed. The treelined backwaters, parallel to the coast, enhance the scenic effect.

Physiographically, Kerala's coastline is "emergent", contrary to the rest of the western coast of India, which mostly experienced submergence resulting in several coastal villages as "islands" surrounded by sea water. The threefold longitudinal division by physiography is typical: alluvial coastland up to 75 m in the west, low lateritic plateaus and hills (75 to 300 m) in the middle, and gnessic highlands (above 300 m) to the east. Anamudi peak (2694 m) is the highest point.

The region's cultural landscape is of less unique. Although it is one of the most thickly populated parts of the country, the crowding of people is not visible. Settlements are dispersed in the form of scattered homesteads, camouflaged by the shade of coconut and fruit trees. This is explained not merely by availability of water all around but also by the history of land settlement in which the landed aristocracy divided their estates among a large number of tenants, each having a separate homestead. Love for the native place is strong and people prefer to commute rather than migrate from rural to urban places. The presence of necessary service facilities in the countryside and availability of an efficient bus service have helped this process.

In response to the diversity of physical resources, a well-developed infrastructural base, and very special conditions of high density and universal literacy, Kerala's economy is significantly diversified. Nearly two-thirds of the workforce is outside agriculture. But agriculture remains at the foundation. Its main features may be listed briefly. Landholdings are exceptionally small, the average being 0.36 hectare as compared with 1.68 hectares in India. Rice and tapioca are the main foodcrops but meet only one-half of the food needs. There is an emphasis on raising of cash crops, including coconut, coffee, cashewnut, rubber, pepper, cardamom and arecanut. Coconut occupies one-fourth of the cultivated area, which occupies one-half of the total area. Coffee and cashewnut are big foreign exchange earners. The same is true of fishing; one-fourth of sea fish produced in the country is contributed by this region.

Household industries based on coconut are ubiquitous. Modern industry has also made a significant progress: cashewnut factories around Quilon, coir factories along the Alleppey coast, aluminum and fertilizer industries at Alwaye, and shipyard and refinery at Cochin. Many of these units are in the public sector.

Research organizations, such as Plantation Crops Research Institute at Kasargod, Tuber Crops Research Institute at Trivandrum, Institute of Fishing Technology at Ernakulam, and Space Research Organization near Trivandrum, are also worthy of note.

Over one-fourth of Kerala's population lives in towns and cities. Trivandrum (825,682) is the capital city. It is famous for the ancient temple of Padmanabha Swami. The port and industrial city of Cochin (1,139,543) is, however, bigger. It is a natural harbor. Its fort, built in 1500, is said to be the oldest European settlement in India. Calicut (800,913) has a long marine history. Vasco da Gama landed here in 1498. This harbor has several timber yards. The city is famous for handloom products.

The pressure of population on land is intense and unemployment acute. This has impelled migration not only to other parts of India but also overseas. More than half a million Malayalis have emigrated since the early 1970s particularly to the Gulf Region and African countries. Kerala experienced a construction boom; almost one-fifth of the housing stock here was raised during the 1970s alone partly subsidized by large remittances from abroad. There is outmigration to other parts of India as well, with males working as clerks and accountants in the offices and females as nurses in hospitals. Emigration and outmigration may give some partial and temporary relief but cannot be a lasting solution to the unemployment problem. Employment opportunities have to be generated from within. A time has come when economic development must get a priority in scheme of things to match with the high level of social advancement already achieved.

## Eastern Coastal Plains

### *Orissa Coast*

This region broadly coincides with the Mahanadi delta. Of all parts of the Eastern Coastal Plains, it is the least developed. Its per capita income is less by nearly one-third of the national average. This is explained largely by an unfavorable position it had during the colonial rule. It formed a part of the Bengal province during 1803-1912 and that of Bihar during 1912-36. Administrative neglect was persistent. Further, it remained virtually isolated from the immediate hinterland of the erstwhile Garhjat states that were under the rule of tribal princes.

By contrast, the region, as the main component of the powerful Kalinga kingdom more than two thousand years ago, was a hearth of rich civilization. Its later history also had glorious phases. Tenth to thirteenth centuries were marked by grand accomplishments in art and architecture to which the Lord Lingaraj temple of Bhubaneswar, Jagannatha temple of Puri, and Sun temple of Konark bear a witness.

The region's present population of 15 million is concentrated on 40,166 km$^2$. More than 95 per cent of the population is Hindu, and over 90 per cent Oriya. Such overwhelming religious and linguistic homogeneity meant a virtual absence of any rival ethnic group.

Floods have been a bane of the region. Before the construction of

**Figure 17.2**
*Eastern Coastal Plains*

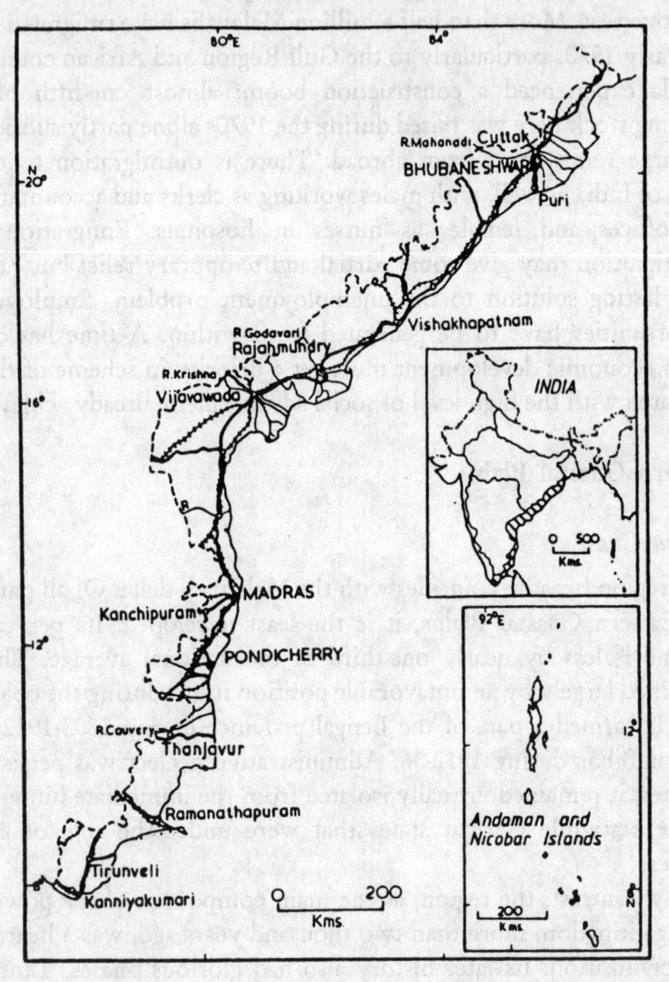


the Hirakud dam across the Mahanadi during the 1950s, at least one destructive flood was an annual feature. July 21 to August 7 is the most vulnerable period when the monsoons are at their climax. Equally devastating are the furious cyclones emanating from the neighboring Bay of Bengal. Such natural hazards eat into the economic vitals of the region.

Marshes and dunes on the coastal fringe, wide alluvium in the middle, and lateritic shelves on the upland margins describe the physiography of the region. Rainfall, on the average, is 150 cm, but is highly variable causing crop failures at times. Famines during the colonial days were not uncommon. The Mahanadi canal system was laid in the later half of the nineteenth century to mitigate the situation.

Agriculture is more intensive on the alluvium; lateritic zone is much less productive. Around half of the total area is under cultivation. Of this, one-half is irrigated, mainly by canals. Rice is the dominant crop, sharing three-fourths of the cropped area. In localities, there is virtual monoculture with 95 per cent of the cropped area devoted to this crop. Jute is the main cash crop. Pulses and oilseeds are also grown. Fishing and salt-making gain prominence in the coastal strip.

Balasore is noted for rice mills and Cuttack for textiles, saw mills and *beedee* (an indigenous cigarette) industry. Paradip has been developed as a port. It is designed for export of iron ore mined in the hinterland.

Bhubaneswar (411,542), the capital, is a new, planned city of Orissa. Cuttack (439,273), some 20 km to the north on the head of the Mahanadi, is ancient and somewhat larger. Earlier this was the state capital. It is still commercially and industrially important. The port of Puri (124,835) is a pilgrimage center.

Hardly one-seventh of the total population is urban; a reflection on the region's low level of economic development. On the other hand, density of population is high, nearly 400. Almost three-fourths of the workforce is directly dependent on agriculture. Population-resource imbalance is evident. Outmigration, particularly to West Bengal, Madhya Pradesh and Assam, has been a regular feature since the colonial days.

### *Andhra Coast*

This is the Telugu-speaking part of the Eastern Coastal Plains. Telugu is compared with Italian for the sweetness of its phonetics. Many of its words have vowel endings, such as *du, mu, vu, and lu.*

During the colonial times, the region was a part of the British province of Madras. It was then known as the Northern Circars. After Independence, in 1953, Andhra was the first state to be constituted on linguistic basis.

The main body of the region is formed by the two deltas of the Godavari and the Krishna. In the north, topography tends to be hilly in close proximity of the sea, and in the south is located the Nellore peneplain carved out by the Penner. Climate is tropical maritime. Rainfall is around 100 cm. It is variable, making irrigation imperative.

About a half of the total area is under cultivation. Irrigation covers a half of it. Canals serve three-fourths of the irrigated area, and tanks most of the remaining one-fourth. In contrast to the Orissa coast, which is food-deficit, the Andhra coast is a surplus producer of rice. It is described as the "granary of south India", one of few areas in the country to have experienced the Green Revolution in rice. Its other important crops include sugarcane, tobacco, oilseeds and vegetables.

Agricultural development in the area got a boost from the extension of canal irrigation by the British in the later half of the nineteenth century. The irrigation base was further strengthened through the various river valley projects undertaken after Independence. Agriculture is now fairly protected against the whims of the monsoons. Cyclones from the Bay of Bengal cause damage to the sugarcane crop at times.

The dominance of water landscape created by braided channels of streams and dense network of canal distributaries is vivid. The flooding of land for rice cultivation adds to this effect. In the coastal strip, salt making also calls for impounding of sea water beds.

Considerable agro-based industries, including rice milling, edible oil processing and fruit canning are important. The port city of Vishakhapatnam (1,051,918) has concentration of industries, including ship building, oil refining, fertilizer and heavy machinery. Vijayawada (845,305), located on the head of the Krishna river, has industries based on tobacco, horticultural products and rice. Rajamundhry (403,781), on the head of Godavari, has engineering goods, tobacco and paper industries. On the whole, about one-fifth of the total population is urban.

With 29 million people on 92,906 km$^2$, the region is densely populated. On an area equal to that of Portugal, it has three times its population. About two-thirds of the workforce is directly dependent upon agriculture causing pressure on the agricultural land. The enterprising farmers of this area responded by migrating to newly colonized area in other parts of Andhra Pradesh and in Karnataka. The landless laborers migrated generally to cities, notably Hyderabad, Mumbai, Chennai (earlier Madras) and Bangalore.

### *Tamil Nadu Coast*

As the hearth of Tamil culture, this coastal region is unique. Over 90 per cent of its population speaks Tamil. Tamil was the first Dravidian language to develop its own script, some two thousand years ago. Among the regional languages of India, it is the oldest. The Tamils produced an early urban civilization, much before the beginning of the Christian era. Their accomplishments in temple architecture, dance and music are well known. A feeling of strong cultural sub-nationalism was natural for such an area. This found an expression in opposition to Hindi as the national language. Prejudice against even the Tamil Brahmins, who are seen as of north Indian origin, also prevails.

Tamil Nadu coast has a population of 34 million on 71,069 km$^2$. Thus, the population density approaches 500. Coupled with the factor of external exposure associated with a coastal location and high literacy rates, this high density caused not only outmigration to other parts of India, especially the cities like Mumbai, Hyderabad, Bangalore and Delhi, but also emigration to other countries, particularly the former British colonies of Sri Lanka, Malaysia and Singapore.

The region is a wide coastal plain with the Kaveri delta in the middle. To its north, the plain owes its origin to the alluvium deposited by the Palar and Pormaiyar rivers, and to its south, the plain is the work of the Vaigai and other rivers. Rainfall regime is rather exceptional. Both advancing (June to September) and retreating (October to December) monsoons bring rainfall, almost in equal measure. A rainfall of 100 to 150 cm is received in the northern part, and of 50 to 100 cm in the southern. The rainfall of only 50 cm, as in the southern part of the region, is atypical of an Indian coastal area. This is explained by the obstruction posed by the Western Ghats to the advancing monsoons, and by the seaward protruding Kaveri delta to the retreating monsoons in respect of this tract.

About one-half to two-thirds of the total area is under cultivation and the irrigated area covers one-half to two-thirds of the cultivated area in the Kaveri delta and in the tract north of it, and one-third to one-half in the south. Canal irrigation dominates in the Kaveri delta; tank and tubewell irrigation takes precedence in other areas.

There is intensive cultivation of rice. The Kaveri delta is one of the few surplus food-producing areas of India. If Punjab is the leader in wheat-based Green Revolution, Kaveri delta deserves the same credit in respect of rice. Bulrush millet is also grown. Among the cash crops, groundnut is important in the northern part, plantain in the delta, and cotton in the southern part. Casuarina plantations provide fuelwood. The coastal economy is a combination of crop cultivation, pomiculture, salt making and sea fishing.

Agro-based industries, such as rice processing, oil milling and cotton textiles, are dispersed throughout the region. Post-Independence era has seen the emergence of several modern industrial units manufacturing diesel engines, railway wagons, coaches and automobiles. Many of these units are in public sector, located in or near Chennai.

Over one-third of the population is urban, living in closely-spaced towns and cities. Chennai (5,361,468) is the capital, port, industrial and educational city. It is the focus of Tamil culture and politics. Though the oldest European city in India, it retains much of traditional way of life. Kanchipuram (169,813), close to Chennai, and Thanjavur (200,216), on the bank of Kaveri river, were the capitals of the historic Pallava and Chola dynasties respectively. Both are famous for their ancient temples and silk products. Tuticorin (284,193) is a major port and a center of pearl diving. Kanniya Kumari (17,206) is located at the southern tip of India where the Arabian Sea and the Bay of Bengal meet.

Pondicherry and Karaikal of the union territory of Pondicherry are located within this region. Its two exclaves—Yanam and Mahe—are situated in Andhra Pradesh and Kerala, respectively. The territory was under the rule of the French till 1954 and bears a strong influence of the French culture. It is scattered over an area of 492 km$^2$ and has a population of 807,045, of which nearly two-thirds is urban. Pondicherry (401,337) is the capital city, with textiles, sugar, paper, liquor and tourism as the main industries. Auroville, an international village, is being developed within Pondicherry since 1968. Its design symbolizes the harmony of all religions and the unity of mankind.

## The Islands

The Andaman and Nicobar Islands, a group of 223 islands scattered in the form of a gentle crescent in the midst of the Bay of Bengal at a distance of 1,255 to 1,980 kms from Calcutta, and the Lakshadweep, a group of 36 islands, located in the Arabian Sea at a distance of 280 to 480 km from the Kerala coast, constitute the oceanic India (see Figure 17.2). Both enjoy the status of union territories, administered directly by the union government. Port Blair (74,810) is the capital of the Andaman and Nicobar Islands and Kavaratti (29,089) of Lakshadweep.

Barring the similarity of their oceanic location and tropical maritime climate, the two island groups differ from each other in several ways. The Andaman and Nicobar Islands are the elevated portions of submerged submarine mountains, a continuation of the Arakan Yoma of Myanmar. These have a hilly topography; the Saddle peak at 732 m is the highest point. On the other hand, the Lakshadweep islands are coral-built on the submerged Aravalli strike, and nowhere rise above five meters in altitude.

Andaman and Nicobar Islands have an area of 8,249 km$^2$ and a population of 279,111. This makes the population density of only 34. By contrast, Lakshadweep islands have an area of about 32 km$^2$ and a population of 51,681, making their density of 1,615 exceptionally high.

Lakshadweep islands are preponderantly Muslim (94 per cent of the population). The native population is Malayalam-speaking. By comparison, Andaman and Nicobar Islands have a diversified religious composition with the Hindus forming 60 per cent of the population, Christians 26 per cent, Muslims 10 per cent, and the rest 4 per cent. The linguistic diversity is even more striking: Bengali-speaking population is 25 per cent, Tamil, Hindi and Malayalam 12 per cent each, Telugu 8 per cent, and the rest are 31 per cent. From 1858 to 1945, these islands were used by the British as a penal colony for the political prisoners and life-convicts from the mainland. This brought in people from every part of India. Marriages took place across religious and linguistic backgrounds, representing a rare biological integration for any part of India. Post-Independence era is experiencing regular immigration associated with initiation of new development projects. Displaced persons from the former East Pakistan have also been settled here. The native aborigines, such as the

Andamese, Onges, Jarawas, Sentinelese and Shompens, are scattered in small numbers over the islands.

While some agriculture in rice, sugarcane and spices is carried on in the Andaman and Nicobar Islands, the economy of Lakshadweep is almost exclusively dependent on coconut plantation and fishing. Coconut covers 85 per cent of Lakshadweep's total area; land under crops (finger millets, sweet potato and plantain) takes less than one per cent of the land. On the other hand, Andaman and Nicobar Islands have 80 per cent of their area under forest. Significant forest-based industries, such as sawn timber, commercial plywood and match splints, have been recently started. In Lakshadweep, coir processing and copra making for export are the major household industries. Some fish-canning units have been recently set up.

Tourism is picking up, particularly since the late 1970s when the Government of India extended special travel facilities to its employees. The rich plant, bird and sea life, the presence of aboriginal tribes, and a reasonable distance from the mainland are major attractions for the tourists. Port Blair is the entry point to Andaman and Nicobar Islands and Kavaratti to Lakshadweep. The two capital towns account for over one-fourth and one-half of the total population of their territories respectively. The former is famous for the Cellular Jail (now a national shrine), Marine Park and Elephant Nursery, while the latter for its beaches.

Above all, the two groups of islands are of strategic importance. These are India's naval outposts in the Bay of Bengal and the Arabian Sea. Their location extends India's territorial jurisdiction into the Indian Ocean bestowing on India the status of a regional seapower.

**Conclusion**

Discussion in the last four chapters (14 to 17) has emphasized the regional diversity of India (as well as a certain amount of functional unity of its diverse regions) produced by considerations of ethnicity, development levels, and administrative controls. Although geographers have traditionally tried to comprehend the regional scene of India based mostly on physiographic considerations but empirical findings based on economic, cultural and political criteria suggest other useful avenues of regional classification. Some areas with similar physiographic development levels differ in their historical experience and development levels, while other areas, though physiographically dissimilar, possess a certain unity of language, administrative control and culture.

However, the paradigm of spatial differentiation based on physico-cultural considerations has also been changing with time. Before Independence, the geographic personality of an area was subject to the dictates of its relief, religious composition of the society, and above all, the system of administrative control (of the British colonial or the native princely rules). After Independence, an expression in regional sentiment based on ethnicity in terms of language, local culture and historical background became more vocal. Regional identities got sharply defined and sought formal recognition. Regionalism was seen as a road to political autonomy and a necessary measure for protecting and promoting economic interests of the "sons of the soil".

This led to a linguistic reorganization of the Indian states in 1956. A periodic reform of the administrative map continued thereafter. Clamors for new states conforming to newly discovered regional identities began to emerge. The two ideologically opposed political streams—the leftists and the rightists—became united in their demand for further regional-political divisions. Although their underlying considerations differed, the leftists view the smaller units as more cohesive for any collective action at the local level; the rightists judge these as favorable to national unity for reason of their greater interdependence as well as dependence on the central government.

Under such a scenario of aggressive regionalism on the one hand and ingrained nationalism on the other, it is not surprising that an average Indian has multiple spatial loyalties: to the ethnic group he or she belongs, to the area of domicile, and to the nation at large. Most of the Indians reside in their ethnic territories and most of the ethnic territories have been organized into states and union territories. It is not without reason that the regional classification adopted in this text conforms mostly to the administrative boundaries.

# Epilogue

## *Emerging India: An Assessment*

India enters the new millennium in a period of transition. The democratic structures, so carefully designed by the founding fathers, are under heavy stresses and strains, threatened by polity fragmented along religious, ethnic and caste lines. Though the Indian economy has taken major strides after independence to achieve food self-sufficiency and to develop a strong industrial base, poverty and inflation have restricted the required growth in the Indian economy. With 320 million persons belows poverty line in 1993-94 (*Manpower Profile, India*, 1998: 343), it is high time for the government to aggressively reduce the fiscal deficit and the burgeoning debt burden on the country.

Despite all this, India at present is the fifteenth largest economy. The World Bank has projected that India will be among the first five developing nations to move up with the world league of major economics by 2020. Unsustainable fiscal deficits of the eighties and excessive external borrowing led to the 1991 crisis when the foreign exchange reserves plunged to barely one billion dollars. The external debt rose four-fold during the decade from 20.5 billion dollars to 83.7 billion dollars in 1990. The 1990s saw India move to a higher growth trajectory with the Indian economy adopting to a highly liberalized market-oriented system. The growth rate of GDP, which was 3 per cent in seventies and 4-5 per cent in eighties, had fallen sharply to below 1 per cent in 1991-92. In 1994-95 it rose to over 7 per cent and in 1998-99 it was 5.8 per cent (*India Today*, March 1999: 43). India is one of the very few countries which has been successful in bringing down inflation from as high as 17 per cent to a low of 1.7 per cent in the first week of August 1999.

Poverty reduction has been the prime focus in the Indian economy. Though there has been a decline in the ratio of poor to the total population but it has been very show at the rate of 2 per cent a year from 1974 to 1994. Owing to the ever-increasing population, however, there has been no reduction in the absolute number of poor which stood at 320 million in 1993-94.

However, the Eighth Plan period saw a growth in per capita income from 3.4 per cent in eighties to 4.9 per cent in the plan period (1992-97). Per capita income has been estimated at Rs. 9,660 in 1997-98 at constant prices while it is Rs. 13,193 at current prices (based on revised series of National Accounts with 1993-94 as the base year).

Responding to the trade policy reforms of 1991-92, the export growth rate increased from 18 to 21 per cent between 1993-94 and 1995-96. But, in the subsequent three years, owing to industrial recession, inadequate corrective measures, Asian crisis and global trade slowdown, it plunged to 12.1 per cent in 1996-97 and 4.4 per cent in 1997-98. Industrial recession, which has been shadowing the Indian economy from 1996-97, slowed down the industrial growth from 12.8 per cent in 1995-96 to 5.6 per cent, 6.6 per cent and 3.8 per cent in the subsequent three years ending March 1999.

Planned development in India did not produce the desired results in eradicating poverty and generating employment but in spite of that literacy in now up to 52 per cent, life expectancy has risen to 62 years, infant mortality has been brought down substantially to 70 per thousand and safe drinking water is now accessible to 85 per cent of population in rural and urban areas.

The economic data available now show that India did fairly well in the last decade or so and the economic reforms did help us to some extent to achieve a sustained growth. The growth of GDP is on increase, growth in agriculture and industry is showing positive trend, inflation has been controlled substantially and there has been a little increase in the exports as well.

## Present Perspective

India today ranks tenth in industrial production, up from its eighteenth position in 1983, but it still accounts for less than one per cent of the world's industrial production. However, as pointed out earlier, its industrial growth rate has been one-half that of several developing nations, although in technology, it ranks among the world leaders (U.S.A., Russia, United Kingdom, Canada and France). India can and does manufacture complex equipment, such as nuclear power stations, supersonic jet fighters, and space satellites. Today, in the field of information technology, particularly in the export of softwares and computer professionals, it has marked its presence in the global marketplace.

Indian economy is still predominantly agricultural with nearly one-third of national income being derived from agriculture and allied activities employing two-third of the total workforce. Country's gross cropped area increased by 20 per cent between 1960 and 1990. India is a major or leading producer of several crops in the world (rice, wheat, jute, sugarcane, cotton and tea). Then, again, its agricultural production in the past increased barely commensurate with population growth. Generally speaking, the living conditions of the population improved moderately, but the agricultural income and per capita income of the farmers, in some areas like Punjab, Haryana, Gujarat and Tamil Nadu, gained substantially. But, the extension of such gains were not shared universally. Punjab and Haryana were the biggest beneficiaries. Since 1976 the nation had been spared the food shortages so common previously. With the foodgrain production touching 200 million tons by the end of 1990s, the attainment of foodgrain self-sufficiency was the most remarkable of India's accomplishments made possible, in a large measure undoubtedly, to nation's inherent political stability. As regards the incidence of illiteracy, poverty and unemployment, the situation has modestly improved. According to the latest 1999 figures given by the National Sample Survey (NSS), the overall literacy percentage in India has gone up to 61 per cent as compared to 52.2 in 1991.

## The Next Decade

The Indian government has a modest but long wish-list for the coming years in the 21st century. India's population which was 844 million in 1991, crossed the billion mark in May 2000. In its struggle to contain the population explosion, birth rates and death rates need to be reduced to 23.7 and 8.4 respectively and the rate of population growth to be reduced to 1.5 per cent a year (down from the current 2.1 per cent). Urban population in the year 2000 is estimated at nearly 315 million, indicating a share of one-third in the total population. Taking into account the estimated backlog of unemployment, the magnitude of additional employment to be generated would be around 130 million. Agriculture will represent one-fourth of the income of gross national product, a reduction from the current nearly one-third; services and mining-manufacturing sectors to increase their share to one-fourth, an increase from the current nearly one-fifth of the total income of gross national product. The foodgrain requirement has been estimated around 250 million tons. An important strategy for agriculture would be to accelerate the use of new technologies on the farms, extension of area in high-yielding varieties, to pump more credit to the farmers, and to develop marketing facilities. The desire for the growth of industry is 8 to 9 per cent a year.

What are the prospects for the achievement of the wish-list? The next decade will be a crucial one, for the decisions today will affect the stability, unity and prospects for democracy. The year 1991 was a watershed for India, when the nation made a radical change in its economic policies. The government decided to discard the protectionist policies of the past and move the country away from bureaucratic control to free market, private enterprise and foreign investments. A sharp devaluation of about 40 per cent of the rupee, as the foreign debt had soared to unacceptable levels in 1990, was designed to make Indian exports more competitive internationally. With the collapse of the Soviet Union, India is re-evaluating its policies, and aligning itself closer to the United States and the European Economic Community. It is also reaching out to China. The old policies are no longer valid. Liberalization and deregulation of trade and agricultural subsidies, pruning of a public debt and opening the doors to foreign investment that were recently undertaken.

Prospects for the realization of administration's wish-list depend undoubtedly on the decisions India takes and implements in the years to come, the durability of its democratic structures, and its strength to withstand its internal stresses. Christopher Thomas evaluates the integrity of India as:

> India's durability is vastly underestimated. Predictions of the breakup of the union have been made since 1947, when the old British Raj was partitioned and India became independent. ...
> ... India has survived war with Pakistan, countless language and religious riots, ... severe poverty and 12 years' conflict in Punjab, where Sikh militants have demanded an independent state. It is certain to outlast the uprising in Kashmir, now approaching its third anniversary ... The assassination last May [1991] of Rajiv Gandhi, the former prime minister ... which many thought would bring political chaos, did not crack the foundations. Indeed, seven months later a ... reasonably stable ... government [of minority party] has embarked on far-reaching economic reforms without as much as murmur of protest from its rivals ... There has been a radical change in the language of Indian politics, however. Everybody knows that reforms are inevitable (*The Times*, London, January 25, 1990, p. 16)

### *References*

Government of India, *Eighth Five Year Plan, 1990-95*. New Delhi, 1985.
Government of India, *Economic Survey, 1989-90*. New Delhi, 1990.
Government of India, *India-1997: A Reference Annual*, New Delhi, 1991.
Heston, Alan, "Poverty in India: Some Recent Policies", in *India Briefing*, 1989, Boulder, Colo. 1990.
Mohammed, Ayoob, "Dateline India: The Deepening Crisis", *Foreign Policy*, Washington, DC, 1992, pp. 166-184.
Muthiah, S. et al. (eds.), *A Social and Economic Atlas of India*. New Delhi, 1987.
Thomas, C. "Focus India", *The Times*, London, January 25, 1991, p. 16.
Wolpert, S. *India*. Berkeley, 1991.
World Bank, *India: Poverty, Employment and Social Services*. Washington, DC, 1989.
*World Population Data Sheet 1997*. Population Reference Bureau, Washington, DC, 1997.

# Index

M



N



O

## P



## R